CLINICAL GUIDE TO CARDIAC AUTONOMIC TESTS

CLINICAL GUIDE TO CARDIAC AUTONOMIC TESTS

Edited by

MAREK MALIK

Department of Cardiological Sciences,
St. George's Hospital Medical School,
London

Kluwer Academic Publishers

Dordrecht / Boston / London

A C.I.P. Catalogue record for this book is available from the Library of Congress.

ISBN 0-7923-5178-9

Published by Kluwer Academic Publishers,
P.O. Box 17, 3300 AA Dordrecht, The Netherlands.

Sold and distributed in North, Central and South America
by Kluwer Academic Publishers,
101 Philip Drive, Norwell, MA 02061, U.S.A.

In all other countries, sold and distributed
by Kluwer Academic Publishers,
P.O. Box 322, 3300 AH Dordrecht, The Netherlands.

Printed on acid-free paper

Printed in the Netherlands.

to

Daniel, Natalie, Michal and David

CONTENTS

PART THREE: CLINICAL PRACTICE

List of Contributors

Solange Akselrod
Center for Medical Physics, School of Physics, Tel Aviv University, Tel Aviv, Israel

A John Camm
Department of Cardiological Sciences, St. George's Hospital Medical School, London, United Kingdom

Xavier Copie
Département de Cardiologie, Hôpital Broussais, Paris, France

Bohumil Fišer
Department of Physiology, Masaryk University, Brno, Czech Republic

Roy Freeman
Department of Neurology, Beth Israel Deaconess Medical Center, Harvard Medical School, Boston, USA

Louis Guize
Département de Cardiologie, Hôpital Broussais, Paris, France

Azad V Ghuran
Department of Cardiology, North Staffordshire City General Hospital, Stoke on Trent, United Kingdom

Roger Hainsworth
Institute for Cardiovascular Research, University of Leeds, Leeds, United Kingdom

Juha EK Hartikainen
Department of Medicine, Kuopio University Hospital, Kuopio, Finland

Antti E Hedman
Department of Medicine, Kuopio University Hospital, Kuopio, Finland

Jean-Yves Le Heuzey
Département de Cardiologie, Hôpital Broussais, Paris, France

Stefan H Hohnloser
JW Goethe University, Department of Internal Medicine, Division of Cardiology, Frankfurt am Main, Germany

Nataša Honzíková
Department of Physiology, Masaryk University, Brno, Czech Republic

Rose Anne Kenny
Cardiovascular Investigation Unit, Royal Victoria Infirmary, University of Newcastle upon Tyne & Institute for the Health of the Elderly, Newcastle upon Tyne, United Kingdom

Laurence Keselbrener
Center for Medical Physics, School of Physics, Tel Aviv University, Tel Aviv, Israel

Thomas Klingenheben
JW Goethe University, Department of Internal Medicine, Division of Cardiology, Frankfurt am Main, Germany

Tom A Kuusela
Department of Applied Mathematics and Department of Physics, Turku University, Turku, Finland

Marek Malik
Department of Cardiological Sciences, St. George's Hospital Medical School, London, United Kingdom

Alberto Malliani
Centro Ricerche Cardiovascolari, CNR, Centro LITA-Vialba, Medicina Interna II, Ospedale "L. Sacco", Università degli Studi, Milano, Italy

Christopher J Mathias
Neurovascular Medicine Unit, Division of Neuroscience & Psychological Medicine, Imperial College School of Medicine at St Mary's & Autonomic Unit, University Department of Clinical Neurology National Hospital for Neurology & Neurosurgery, Queen Square & Institute of Neurology, University College London, London, United Kingdom

Nicola Montano
Centro Ricerche Cardiovascolari, CNR, Centro LITA-Vialba, Ospedale "L. Sacco", Università degli Studi, Milano, Italy

Andrea Mortara
Divisione di Cardiologia, Fondazione "S. Maugeri", IRCCS,
Centro Medico Montescano, Pavia, Italy

Massimo Pagani
Centro Ricerche Cardiovascolari, CNR, Centro LITA-Vialba,
Ospedale "L. Sacco", Università degli Studi, Milano, Italy

Steve W Parry
Cardiovascular Investigation Unit, Royal Victoria Infirmary,
University of Newcastle upon Tyne & Institute for the Health of the Elderly, Newcastle upon Tyne, United Kingdom
(presently supported by the British Heart Foundation, London)

Gian Domenico Pinna
Divisione di Bioingegneria, Fondazione "S. Maugeri", IRCCS,
Centro Medico Montescano, Pavia, Italy

Alberto Porta
Centro Ricerche Cardiovascolari, CNR, Centro LITA-Vialba,
Ospedale "L. Sacco", Università degli Studi, Milano, Italy

Maria Teresa La Rovere
Divisione di Cardiologia, Fondazione "S. Maugeri", IRCCS,
Centro Medico Montescano, Pavia, Italy

Bořivoj Semrád
First Department of Medicine, University Hospital, Masaryk University,
Brno, Czech Republic

Virend K Somers
Cardiovascular Division, Department of Internal Medicine,
University of Iowa, Iowa City, USA

Kari UO Tahvanainen
Department of Clinical Physiology, Tampere University Hospital,
Tampere, Finland

Yee Guan Yap
Department of Cardiological Sciences,
St. George's Hospital Medical School, London, United Kingdom

INTRODUCTION

The existence of physiological mechanisms controlling cardiac function and periodicity has been appreciated for centuries. Rather recently, we started to realise that disturbances of the cardiac autonomic regulations may have profound clinical consequences. The seminal studies showing that the physiological autonomic regulations of the heart are an important part of the natural antitachycardia defence have only been conducted some decades ago. Together with clinical observations of the possible prophylaxis guided by the assessment of cardiac autonomic status, these physiological and pathophysiological investigations demonstrated the potential of cardiac autonomic testing to both research and clinical communities. The research and clinical interest in the assessment of cardiac autonomic status was further promoted by the recent development and advances of technologies that enable detailed dynamic recording of electrocardiograms, continuos blood pressure tracings, neural activity, etc. Indeed, many aspects of cardiac autonomic regulations are so subtle that without proper precise investigation techniques, they would be completely out of our reach.

All these advances in investigation capabilities, physiological understanding and clinical potential led to the development of a whole spectrum of simple as well as rather complex tests that are able to quantify and classify cardiac autonomic reflexes and regulations. Numerous scientific reports have emerged showing the value of the individual tests in various physiological circumstances and clinical conditions. From these reports and from consensus debates organised by scientific societies, technical standards and critically reviewed experience emerged offering a guide to proper conduct, physiological interpretation and clinical appraisal of the individual tests. Although some of these standards, especially those related to direct clinical use of selected autonomic tests, will need to be broaden and deepened in the future, it is now proper to disseminate the present state-of-the-art of cardiac autonomic tests among a wider clinical community.

For all these reasons, I was pleased when I was approached by Kluwer Academic Publishers with the request to edit a book aimed at providing a practical guide to researchers and clinical physicians interested in autonomic investigations of the heart. I was even more pleased when joined in my enthusiasm for the book by the specialists in the field whom I invited to write the individual chapters. To offer the reader a comprehensive coverage of the subject, the book is structured in three parts. The first part deals with the physiology of cardiac autonomic system that creates the background of the particular tests. The second part contains chapters describing specific autonomic tests and investigations. The last part summarises the value of autonomic testing in clinical practice and describes conditions which might alter the results of autonomic investigations.

As with every other multi-authored book, I faced the usual editorial dilemma of finding the proper balance between having the book compact with cross-references between individual chapters, and having the chapters suited for separate reading. Eventually, I felt that with a book aimed at providing a source of standards and references, each chapter should contain a standalone coverage of its subject. Consequently, I am happy to recommend the reader to select chapters corresponding to his or her particular needs and interest. Needless to say, reading the book in its entirety offers much more comprehensive learning of the whole field.

Finally, my deep thanks go all the contributors to the book. Without their kind help, it would have never been published. My apologies go to my family whom editing the book deprived of every moment of my free time.

April 1998
Marek Malik

PART ONE

PHYSIOLOGY

CHAPTER ONE

PHYSIOLOGY OF THE CARDIAC AUTONOMIC SYSTEM

Roger Hainsworth

The term 'autonomic nervous system' is attributed to J N Langley in the early part of this century to describe those nerves that are concerned predominantly with the regulation of bodily functions. These nerves generally function without consciousness or volition, although this distinction from the somatic nervous system is not absolute, for example, some pain sensation is transmitted in autonomic nerves.

Autonomic nerves comprise sympathetic nerves (thoraco-lumbar outflow) and parasympathetic nerves (cranial and sacral outflows). Both divisions contain both afferent and efferent nerves and both myelinated and non-myelinated fibres. In general the effects of the two divisions are complementary, with activity in sympathetic nerves exciting the heart, constricting blood vessels, decreasing gastrointestinal motility and constricting sphincters, and parasympathetic nerves inducing the opposite responses.

The autonomic system supplies both afferent and efferent nerves to the heart. These nerves not only control heart rate and force, but both sympathetic and parasympathetic nerves supply important reflexogenic areas in various parts of the heart which, when excited by either mechanical or chemical stimuli, give rise to reflexes which influence both the heart itself and the state of constriction of blood vessels. Autonomic nerves thus have a pivotal role in the regulation of the cardiovascular system both in ensuring optimal function during

Malik M (Ed) Clinical guide to cardiac autonomic tests, 3–28.

various activities in health, and also in mediating several of the manifestations of cardiac diseases.

RESPONSES TO EFFERENT STIMULATION OF CARDIAC NERVES

Although the heart can function in absence of autonomic influences, both its rate and the force of contraction are altered by changes in activity in the autonomic nerves.

VAGAL EFFECTS

The parasympathetic supply to the heart runs in the vagus (10th cranial) nerves. The cell bodies lie in the dorsal motor nucleus of the vagus and in the nucleus ambiguus, situated in the brain stem. The vagus nerves provide a rich innervation to the sino-atrial node, atrio-ventricular conducting pathways and the atrial myocardium. The question as to whether there is a substantial efferent supply to the mammalian ventricular muscle remains controversial. Anatomical studies do not adequately discriminate between afferent and efferent nerves, and the reported physiological responses in the ventricle to vagal stimulation may largely be secondary to effects on the atrium.

Activity in the efferent vagal nerves slows the heart rate. This is achieved by hyperpolarising the pacemaker cells and slowing their rate of spontaneous depolarisation. At rest there is a degree of vagal 'tone' which reduces heart rate from its intrinsic value of 110 – 120 beats per minute [bpm] to 60 – 80 bpm. The actual heart rate at any time is dependent on the balance between vagal and sympathetic activities. Rates below the intrinsic value indicate that vagal effects predominate. For heart rates to increase above the intrinsic value implies that there must be activity in sympathetic nerves.

High levels of efferent vagal activity produce profound cardiac slowing and possibly even complete atrial asystole. Usually, however, after a few seconds vagal escape occurs and ventricular activity starts, although at a much slower rate. The chronotropic effect of vagal stimulation has a very short latency with the maximum response to a single stimulus occurring in only 400 ms.[1] This implies that a reflex involving vagal efferent responses can influence heart rate on a beat-to-beat basis.

In addition to the obvious chronotropic effects, vagal activity slows atrio-ventricular conduction, possibly causing complete block, and induces a negative inotropic response of atrial muscle. This would decrease ventricular filling and possibly explain the apparent reduction in ventricular contractile force.

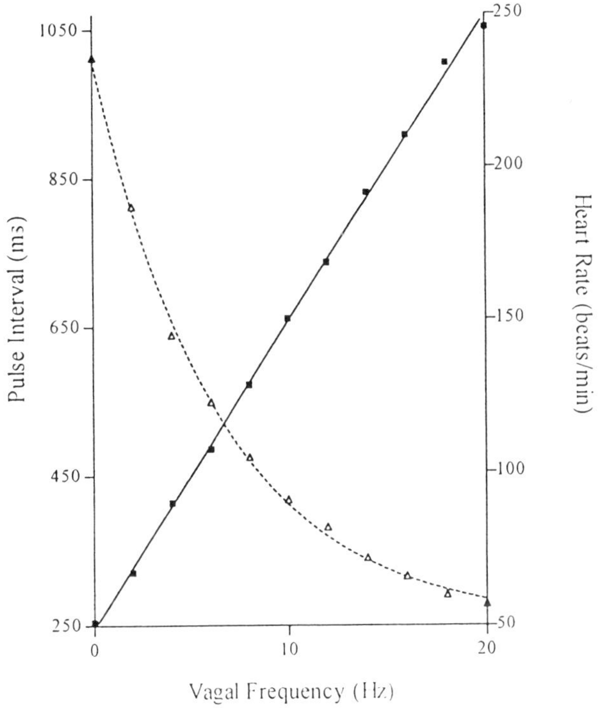

FIGURE 1. Effects of stimulation of vagus nerves at various frequencies in an anaesthetized dog. •—• shows effects on pulse interval, Δ—Δ responses shown as heart rate. Pulse interval is linearly related to stimulus frequency, whereas heart rate shows a hyperbolic relationship (Drawn from data published by Parker et al[3]).

The chronotropic effects of vagal stimulation are graded according to the frequency of efferent nerve impulses. However, the relationship between stimulus frequency and change in heart rate is hyperbolic rather than linear, with the largest changes occurring at low stimulus frequencies.[2] This suggests that it may not actually be the heart rate, which is the appropriate measure to consider. The effect of vagal impulses is to delay the subsequent pacemaker depolarisation and this implies that the interval between depolarisations may be the appropriate variable and, indeed, if cardiac interval instead of heart rate is plotted against stimulus frequency this relationship then does become linear[3,4] (Figure 1).

An important implication of the linearity of the relationship between cardiac interval and vagal activity is that even though a given change in vagal activity at different heart rates may cause similar prolongations of pulse interval, the effects on heart rate would be very different. For example, a prolongation of pulse interval by one-third of a second (333 ms) corresponds to a decrease in heart rate of 15 bpm when the rate is 60 bpm, but at 180 bpm the same change in pulse interval decreases the rate by six times as much (90 bpm). Conversely a given change in heart rate at a high initial rate corresponds to a smaller response of the pulse interval than when the heart is slower. Care must therefore be taken when interpreting the significance of different heart rate responses when the initial rate is different.

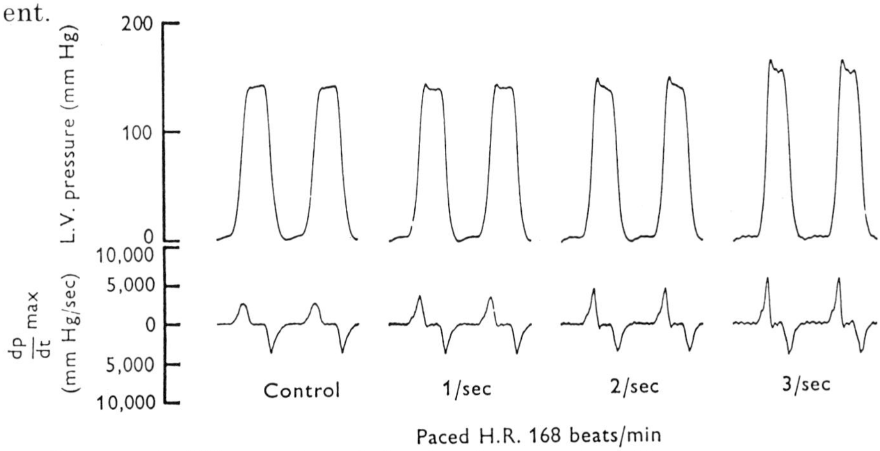

FIGURE 2. Effects of graded stimulation of cardiac sympathetic nerves on left ventricular pressure and left ventricular dP/dt. Heart paced and aortic pressure controlled. Sympathetic stimulation increases rate of rise of ventricular pressure and shortens duration of systole (From Furnival et al[5]).

SYMPATHETIC EFFECTS

The sympathetic supply to the heart originates in the cells of the intermediolateral column of the spinal cord in the upper thoracic region. Sympathetic innervation supplies all regions of the heart:

pacemaker and conducting tissue, and atrial and ventricular myocardium.

Increased sympathetic activity increases heart rate by increasing the rate of depolarisation of pacemaker cells. As for the effects of vagal stimulation, it is the cardiac interval that is more closely related to stimulus frequency. High levels of sympathetic drive are responsible for the high heart rates seen in maximal exercise. In the dog, at least, the right sympathetic nerves are more effective at increasing heart rate than the left.[5] The responses to sympathetic stimulation, however, differ from those to vagal stimulation in respect of the latency of the response. Whereas vagal activity can delay the very next heart beat, sympathetic responses develop and decay more slowly and there may be little change for 5 seconds and maximal responses may not occur for as long as 20 – 30 seconds. This is not due to the latency of the efferent neural discharge but rather to the slow response of the heart.

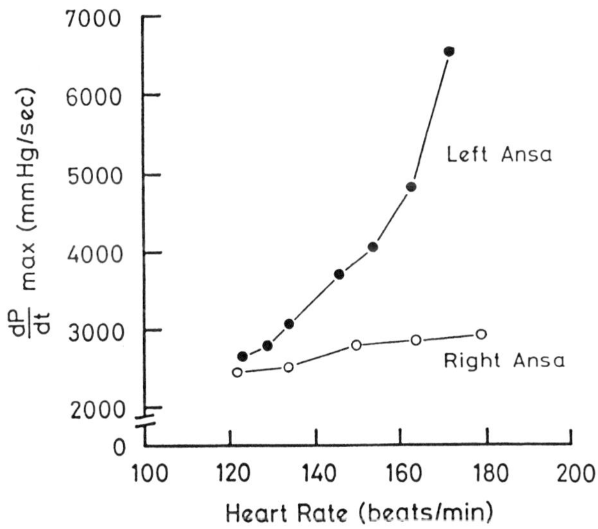

FIGURE 3. Comparison of the responses of heart rate and dP/dt max to graded stimulation of left and right sympathetic nerves in a dog. Note that the left nerves have a greater effect on the inotropic state whereas the right nerves are more effective in changing heart rate (From Furnival et al[5]).

Sympathetic stimulation also increases the rate of stimulus spread throughout the heart, increases the force of contraction at any given myocardial fibre length, and reduces the duration of systole. One way of assessing cardiac inotropic state is to measure the maximal rate of increase of ventricular pressure (dP/dt max) under controlled conditions, and this increases in relation to stimulus frequency (Figure 2). Dog experiments have shown that stimulation of the left sympathetic nerves has a greater effect on inotropic state in contrast to right nerve stimulation that mainly increases heart rate (Figure 3).

Physiological Significance of Cardiac Autonomic Control

The relevance of heart rate lies in its relationship to cardiac output:

$$\text{Cardiac output} = \text{stroke volume} \times \text{heart rate}$$

This equation is mathematically unarguable, but it may be physiologically misleading because it assumes that stroke volume and heart rate are independent of each other. This is frequently not so and changes in heart rate may induce reciprocal changes in stroke volume, thereby limiting any effect on cardiac output.

Several studies, both in animals and humans,[6–8] have investigated the effects on cardiac output of changes in heart rate induced by atrial pacing and these have shown that changes in rate between about 80 and 150 bpm are accompanied by proportionate decreases in stroke volume so that cardiac output remains little changed. Above 150 bpm, diastolic filling time is drastically reduced and cardiac output falls sharply. At normal or high cardiac filling pressures changes in heart rate between about 80 and 50 bpm do have some effect on cardiac output, and below 50 bpm, stroke volume is relatively fixed and cardiac output is then directly related to heart rate. The reason for the apparently anomalous effect of heart rate can be appreciated from Figure 4. As heart rate slows, diastolic filling time increases so that the ventricular end-diastolic volume increases and, from Starling's law, stroke volume also increases. Rapid heart rates reduce filling time and consequently reduce stroke volume.

The influence of heart rate on cardiac output is critically dependent on venous filling pressures. When this is low, for example following blood loss or during orthostatic stress, cardiac filling time is unlikely to be a limiting factor even at very low heart rates, and this explains why attacks of vasovagal syncope are usually not prevented by pacing.

The situation, however, is quite different when venous filling pressure is high as seen particularly during physical exercise. High heart rates during exercise are achieved by activity in sympathetic nerves and one important consequence of an increased cardiac inotropic state is the shortening of ventricular systole. Cardiac acceleration is still mainly at the expense of shortened diastole but there is a sufficient reduction in systolic time for some preservation of diastole too even at heart rates in excess of 200 bpm. This is an important difference from the tachycardias induced by abnormal pacemaker activity and is of importance for achieving maximal cardiac output during strenuous exercise.

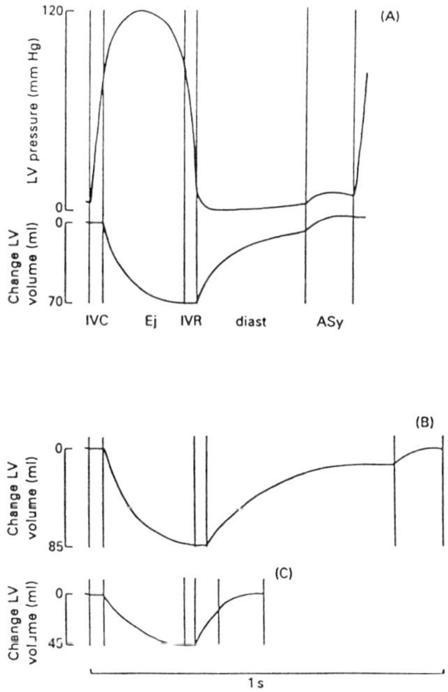

FIGURE 4. Diagram to show changes in left ventricular pressure and volume during a cardiac cycle and the effect of changing heart rate. In B the slower heart rate compared to A results in longer diastolic time, more ventricular filling, and consequently an increased stroke volume. Conversely in C when heart rate is increased stroke volume is smaller. The effects of changes in heart rate on stroke volume mean that cardiac output remains relatively constant (From Hainsworth[69]).

CARDIAC AFFERENT NERVES AND REFLEXES

The cardiac chambers are richly innervated with afferent fibres from both parasympathetic and sympathetic divisions The heart is thus able to sense both the extent of its filling and the pressure that it generates and can initiate appropriate regulatory responses. It is important to appreciate, however, that stimulation of each of the various regions within the heart and the lungs gives rise to a specific pattern of responses. It is, therefore, not valid to group all receptors sensitive to changes in blood pressures and volumes within the thorax as if they were a homogeneous population.

The experimental preparations, which are necessary to apply discrete stimuli to the various reflexogenic areas, are inevitably very complex. For this reason almost all our knowledge concerning the function of cardiac receptors and reflexes has had to come from animal experiments.

ATRIAL REFLEXES

Innervation and Electrophysiology

Afferent nerves, with both myelinated and non-myelinated fibres, running in both autonomic divisions abundantly innervate both atria. However, the only specific nerve endings to have been positively described are the complex unencapsulated endings, which are attached to myelinated vagal fibres and are situated mainly at the junctions of the great veins and atria but also in the atrial appendages[10] (Figure 5).

The discharge characteristics of atrial receptors were categorised by Paintal.[11] They were denoted as type A if their discharge occurred only during atrial systolic (corresponding to the 'a' wave), as type B when their discharge occurred during the filling phase ('v' wave) or intermediate when they exhibited both characteristics. Pure type A fibres are relatively uncommon and the discharge characteristics of most fibres can be altered by changes in cardiac filling and contraction.[12] Despite the existence of different discharge characteristics, the overall net afferent input shows a linear correlation with the atrial pressure. Atrial receptors are also very slowly adapting so they are able to sense sustained changes in the level of atrial filling. Because one of the major factors which influences atrial filling is the level of blood volume, atrial receptors are often referred to as 'volume receptors'.

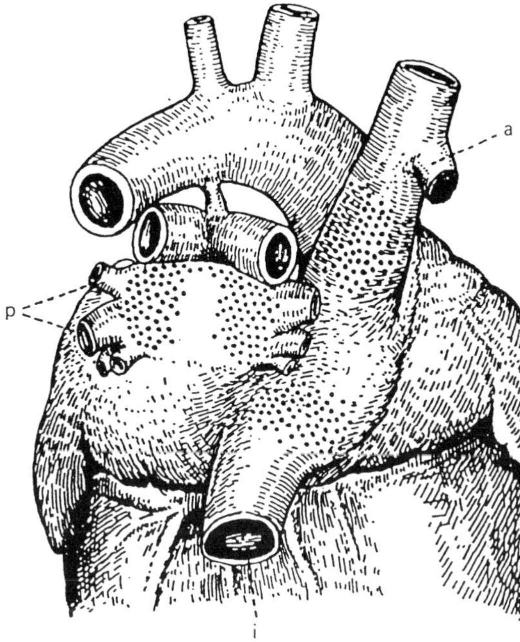

FIGURE 5. View from posterior aspect of heart showing location of atrial receptors. These receptors, shown as dots, are concentrated at the junctions of the veins with the atria. *p* – pulmonary veins; *i* – inferior vena cava; *a* – azygos vein (From Nonidez[10]).

In addition to the myelinated vagal afferents, the atria are also supplied with non-myelinated vagal fibres and with fibres running in the sympathetic system. Specific nerve endings have not been described. Non-myelinated nerves discharge with a lower frequency than myelinated nerves and the threshold for stimulation is generally higher.

Reflex Responses

Bainbridge[13] in 1915 reported that rapid intravenous infusions to anaesthetised dogs resulted in an increase in heart rate. The basis for this response was established by Ledsome and Linden[14] who directly stimulated atrial receptors by distending small balloons at the vein-atrial junctions and obtained a marked and consistent reflex tachycardia. This response was subsequently shown to be due to stimulation of myelinated vagal afferents and to involve an efferent pathway in the sympathetic nerves to the sino-atrial node.[15] The atrial receptor reflex is unusual in that it does not appear to involve efferent vagal pathways, and is thus slower in onset than, for example, baroreceptor reflexes. It does not appear to involve sympathetic fibres to the myocardium and therefore does not control cardiac inotropic state.

Atrial receptors also do not appear to affect resistance vessels in most regions of the body[16] (Figure 6). However the exception is that they inhibit renal nerve activity causing an increase in renal blood flow and increasing salt and water excretion.[17] The other way in which atrial receptors control the kidney is by inhibition of vasopressin release from the posterior pituitary gland, and this results in an increase in the water loss by the kidney.[18]

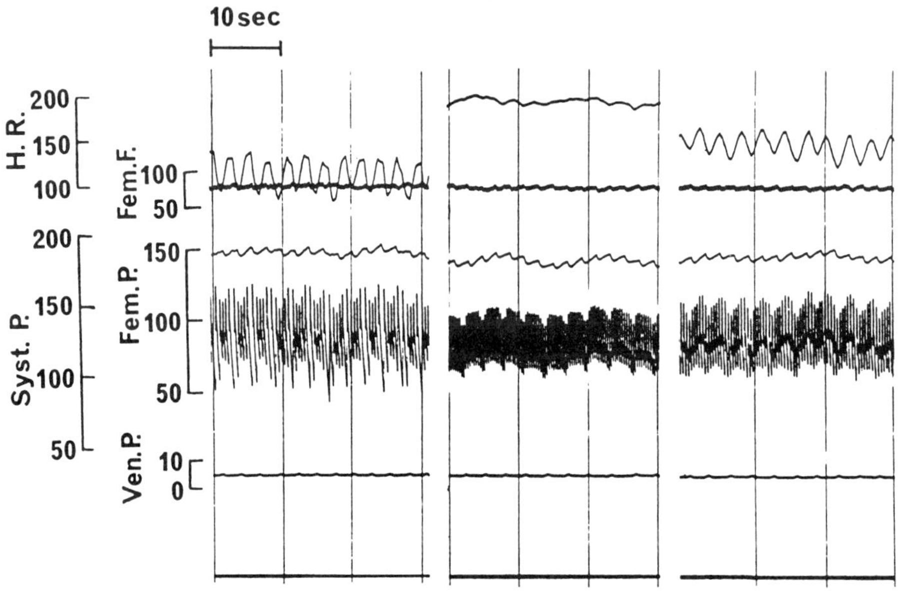

FIGURE 6. Response in a dog to stimulation of left atrial receptors by distension of small balloons at the pulmonary vein-atrial junctions. Traces show heart rate (bpm); femoral perfusion flow (held constant, ml/min) and pressures in perfused femoral artery, systemic artery and femoral vein. Atrial receptor stimulation (centre panel) causes a large increase in heart rate, but little effect on vascular resistance (perfusion pressure nearly constant at constant flow) (From Carswell et al[16]).

It should be noted that the effects of stimulation of atrial receptors are reflex responses involving neural pathways. Distension of the atria causes release into the circulation of atrial natriuretic peptide. This may also contribute to the natriuresis when the atria are distended but this is not a reflex response. The physiological significance of atrial natriuretic peptide is still unclear.

Physiological and Pathophysiological Significance of Atrial Reflexes

Stimulation of atrial receptors, by causing a reflex increase in heart rate would reduce cardiac volume and conceivably could work to cause the heart to function at an optimal level of filling. The effect on the kidney is likely to contribute to the control of extracellular fluid and plasma volume.

Atrial receptors may be implicated in some of the effects in cardiac diseases. One example is in supraventricular tachycardias where the atria become distended and the resulting stimulation of atrial receptors could explain the diuresis that is sometimes observed. In chronic congestive heart failure the atria remain distended. The expected diuresis does not occur because atrial receptors have become desensitised[19].

VENTRICULAR REFLEXES

Innervation and Electrophysiology

The afferent innervation of the ventricles is almost entirely by non-myelinated nerves and most of these supply the left ventricle.[20,21] Some afferent innervation runs in the sympathetic nerves and these fibres may have multiple endings that may even terminate on different cardiac chambers.[22] Some ventricular efferents may be spontaneously active with a pulse-related discharge, although under normal conditions most are believed to be silent and those that do discharge have a very low frequency.[23] Ventricular receptors may be classified as having mechanosensitivity, chemosensitivity or both.

Mechanosensitive ventricular nerves may be excited by large increases in ventricular pressures, with diastolic distension being more effective than high systolic pressures.[23–25] Increases in systolic pressure have little effect until it increases to 150 – 200 mm Hg or more and even then the increase in discharge is often only transient. Discharge from ventricular afferents may also be influenced by the cardiac inotropic state, as the result of sympathetic activity or inotropic agents.[26,27] Although most ventricular afferents are thought to decrease their activity as ventricular volume decreases, Oberg and Thoren[28,29] suggested that some may actually increase their discharge at small volumes, particularly if this is accompanied by an increased inotropic state. It should be pointed out that relatively few nerves respond in this way and that overall the discharge frequency is likely to be reduced.

Chemosensitive afferents can be strongly stimulated by injection into the coronary arteries of various chemicals including veratridine and capsaicin. They may also be excited by some substances that may be released during ischaemia or hypoxia, such as bradykinin and prostaglandins.[30,31]

Reflex Responses

There have been numerous reports that mechanical stimulation of ventricular receptors results in reflex depressor responses.[32] However, most can be criticised because of the poor localisation or unphysiological nature of the stimuli applied.

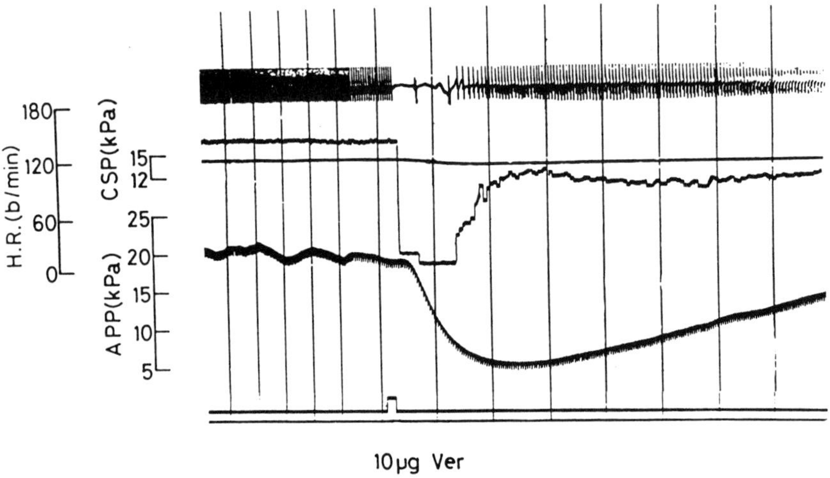

FIGURE 7. The coronary chemoreflex (Bezold-Jarisch reflex). Veratridine injected into the coronary circulation results in profound bradycardia and vasodilatation shown as a decrease in perfusion pressure to a hind limb perfused at constant flow. CSP and APP, carotid sinus pressure and limb arterial perfusion pressure (1 kPa = 7.5 mm Hg) (From McGregor et al[34]).

In contrast to the uncertainty concerning effects of mechanosensitive receptors, stimulation of chemosensitive afferents results in profound bradycardia and vasodilatation (Figure 7). This is the Bezold-Jarisch reflex and may be elicited when several noxious chemicals enter the coronary circulation. Localised injections of veratridine into various branches of the coronary circulation revealed that responses were only obtained to injection into coronary branches supplying the left ventricle and none from the atria or right ventricle.[33]

Much less is known about the effects of sympathetic afferents from the heart. They have been shown to be excited by similar chemical and mechanical interventions as stimulate the vagal afferents. They are believed to mediate cardiac pain and the reflex responses probably include increases in heart rate and blood pressure.[32]

Physiological and Pathophysiological Significance of Ventricular Reflexes

The physiological role of ventricular reflexes remains uncertain. Stimulation by mechanical events does not seem to cause large responses, at least not when the stimuli change within physiological limits. It is possible that gross myocardial distension or very large increases in systolic pressure or inotropic state may cause reflex bradycardia and vasodepression although whether this is of importance in regulation in healthy individuals is still uncertain.

Ventricular receptors have been implicated in causing vasovagal syncope.[34] It was claimed that the increased cardiac inotropic state, resulting from sympathetic activity, accompanied by a small cardiac volume sufficiently excited ventricular receptors to lead to a profound reflex vasodilatation and bradycardia. This now seems unlikely for several reasons. Firstly, relatively few ventricular afferents are excited in this way.[29] Secondly, experiments in which dog hearts were bypassed and the sympathetic nerves stimulated did not cause these responses.[35] Furthermore, similar syncopal reactions have been shown to occur in people who have had heart transplants and who therefore had denervated ventricles.[36]

Patients with aortic stenosis sometimes experience attacks of syncope. When these patients exercise there may be an abnormal vasodilatation and this has been attributed to an abnormal excitation of ventricular mechanoreceptors.[37]

Ventricular chemosensitive nerves may be stimulated in myocardial ischaemia and infarction. However, the reported effects of this, in both experimental animals and humans, have been variable. This is probably explained by the likelihood that several opposing mechanisms are simultaneously involved. The experience of cardiac pain as well as the buffering effects of arterial baroreceptors may modify the direct responses from stimulation of reflexes arising in the heart. A further complication is that obstruction to blood flow in any major coronary vessel is likely to impair myocardial function and this in itself may result in hypotension. The responses also may depend on which region of the heart is affected. Occlusion of the circumflex branch frequently results in bradycardia and vasodilatation, responses that do not usually occur in response to occlusion of the anterior descending branch.[38,39] It has also been shown that ischaemia involving the posterior and inferior walls of the left ventricle are more likely to result in bradycardia.[40,41] Occlusion of the anterior descending branch, on the other hand, has been reported to predispose to arrhythmias.[42] This may also excite sympathetic afferent nerves and this causes reflex increases in efferent sympathetic discharge.[43,44] It has been suggested that sympathetic reflexes may also be excited at the end of cardiopulmonary bypass and may sometimes give rise to marked hypertensive responses.[45]

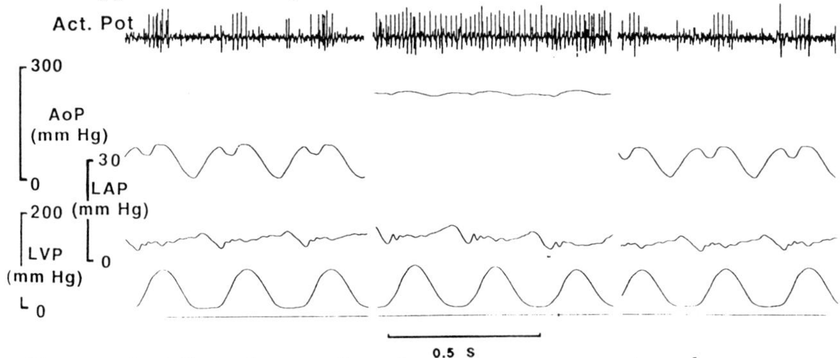

FIGURE 8. Afferent recording from a coronary artery baroreceptor. Traces show action potentials, aortic root pressure (= coronary arterial pressure), left atrial and ventricular pressures. Afferent discharge is related to coronary arterial rather than ventricular pressure and the threshold for discharge is less than 7kPa (52 mm Hg).

CORONARY MECHANORECEPTORS

In the dog some myelinated vagal afferents have been shown to discharge in relation to coronary arterial pressure.[25,46] Careful probing of the coronary arteries revealed that most coronary receptors were situated on the left coronary artery and the proximal ends of its main branches. The phasic discharge of these receptors was related to coronary arterial rather than to left ventricular pressure (Figure 8) and they behaved very much like other arterial baroreceptors, except that their operating range was much lower.[25]

Stimulation of coronary afferents causes reflex vasodilatation.[47] They seem to be of similar potency to carotid baroreceptors. However, a possibly important difference is the lower pressures required to induce reflex response.[48]

Physiological and Pathophysiological Significance of Coronary Reflexes

Coronary baroreceptors are likely to function in the same way as other arterial baroreceptors in controlling blood pressure mainly by inducing reflex vasodilatation. If in humans, like in the dog, the coronary baroreceptors operate mainly in the hypotensive region this could have importance for their function. Carotid and aortic baroreceptors rapidly reset so that they function over the existing level of blood pressure, and buffer against both hypotension and hypertension. The low range of pressures which excite the coronary receptors suggest that they do not reset to the same degree. If this is the case, they may have a more important role in controlling the actual level of an individual's blood pressure rather than in damping down pressure fluctuations.

The significance of coronary baroreceptors in disease states is unknown. However, the left coronary artery and the proximal part of its branches are commonly affected by atheroma and atherosclerosis and these could reduce their function, possibly contributing to the development of hypertension. These vessels also may be distended during coronary angioplasty and this could stimulate and/or destroy the coronary baroreceptors. This might induce hypotension during the procedure and lead to hypertension afterwards.

OTHER REFLEXES CONTROLLING CARDIAC AUTONOMIC NERVOUS ACTIVITY

BARORECEPTORS

There have been many studies, in both animals and humans, of the reflex responses to stimulation of baroreceptors (for reviews see references 49–52). Baroreceptors are stretch receptors and are situated in the walls of some arteries including the carotid sinuses, aortic arch and coronary arteries. They respond by increasing the discharge in their afferent nerves when the vessel is stretched by an increase in blood pressure. It is important to note that they are very sensitive to changing pressures and show a rapid discharge during the rising phase of the pressure pulse and are usually silent when the pulse is falling. This implies that they can detect the magnitude of the pulse pressure and, therefore, indirectly respond to changes in cardiac stroke volume. This provides a mechanism for inducing reflex response to changes in cardiac output as would occur, for example, during postural changes.

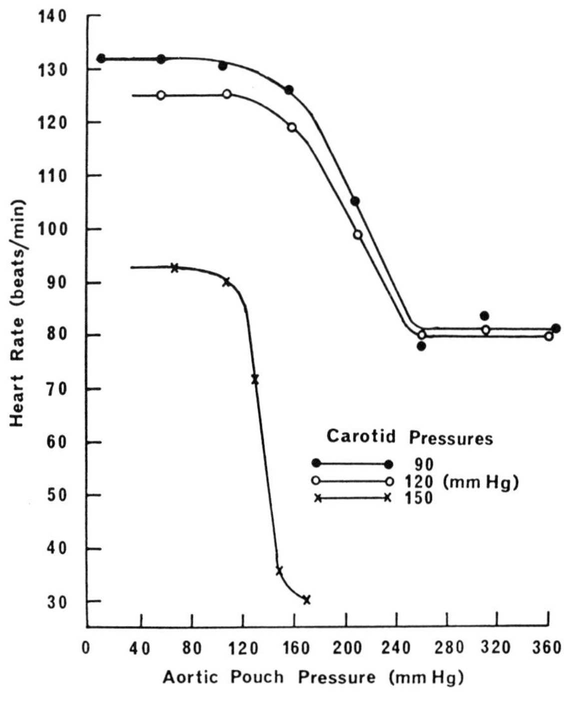

FIGURE 9. Summation of the heart rate response to stimulation of aortic arch and carotid sinus baroreceptors. An increase in carotid pressure, particularly to 150 mm Hg, decreases heart rate and increases the sensitivity of the changes in heart rate to changes in aortic pressure. These results can be explained by simple addition of the two reflex inputs on a non-linear total stimulus-response relationship (From Hainsworth et al[55]).

If the level of blood pressure changes, baroreceptors reset so that their operating range encompasses the prevailing level of blood pressure. Early resetting occurs in minutes after a pressure change. Chronic resetting takes place over months and this results also in a decrease in the sensitivity of the reflex.[53] The phenomenon of resetting implies that baroreceptors probably are more concerned with stabilising fluctuations in blood pressure rather than with determining the absolute level.

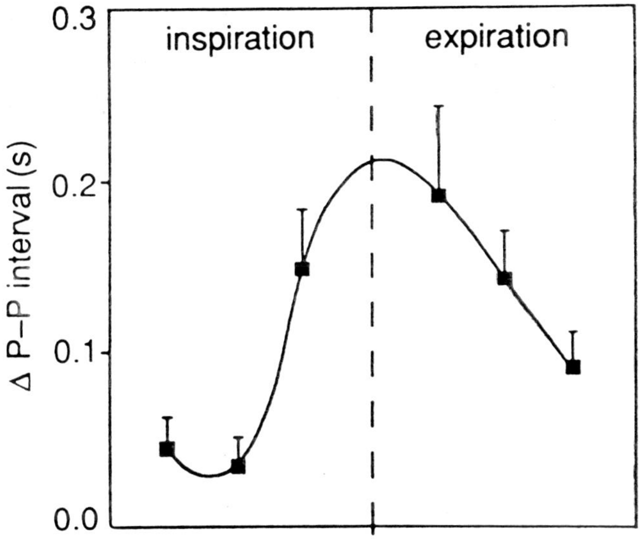

FIGURE 10. Changes in cardiac interval in response to a brief carotid baroreceptor stimulus applied at various phases of the respiratory cycle. The stimulus was applied by suction applied to an air-tight chamber over the anterior and lateral aspects of the subject's neck. Note the virtual absence of a response during early and mid-inspiration. Results of means ± SE from six human subjects (From Eckberg et al[56]).

Stimulation of baroreceptors increases vagal activity and inhibits sympathetic activity. The latency of the response of pulse interval is very short. Eckberg[54] reported that the maximal cardiac prolongation following a brief stimulus to human carotid receptors, effected by applying suction to an airtight chamber fitted to the neck, occurred within 0.75 seconds. This short latency implies that the response is vagally mediated. Sympathetic effects occur more slowly although this is not due to a delay in the response of the sympathetic efferent nerves, but rather to the time for the heart (and vascular smooth muscle) to respond. The very short latency of the baroreceptor-cardiac interval reflex means that baroreceptors control heart rate on a beat-to-beat basis.

Baroreceptor reflexes show a threshold pressure, below which no response occurs, an operating range and a saturation level. Inputs from the various baroreceptor areas summate to induce larger reflex responses than would be obtained individually[55] (Figure 9). There is interaction between the baroreceptor reflex and respiratory activity. The mechanism for this is complex and involves activity in central respiratory neurones as well as reflex inputs from lung receptors. The effect is that the baroreceptor reflex is inhibited during inspiration. Eckberg[56] confirmed this in humans by applying a brief stimulus to carotid baroreceptors at different phases of the respiratory cycle and showed almost complete inhibition during early inspiration (Figure 10).

Baroreceptors also control the activity in sympathetic nerves to the atrio-ventricular conducting mechanism and the myocardium, and thus control cardiac inotropic state and the duration of contraction.

CHEMORECEPTORS

Peripheral chemoreceptors are found in the carotid bodies, situated near the carotid sinuses and in the aortic bodies adjacent to the arch of the aorta. Central chemoreceptors are in the floor of the medulla. Peripheral chemoreceptors initiate a very low rate of afferent nervous activity at normal arterial blood gas levels, but become strongly stimulated during arterial hypoxia and hypercapnia. Central chemoreceptors are sensitive to the hydrogen ion concentration in the cerebrospinal fluid, which in turn depends mainly on arterial carbon dioxide tension. The main function of chemoreceptors is to control respiration but they also influence the cardiovascular system.

Stimulation of central chemoreceptors increases sympathetic activity to the heart and blood vessels and potentiates the baroreceptor reflex[57] (Figure 11). Stimulation of carotid chemoreceptors induces a primary reflex bradycardia and negative inotropic effect, although in spontaneously breathing animals this is usually masked by the opposite reflex responses from the effects of the increased respiration[58]. Stimulation of aortic body chemoreceptors seems to cause increased sympathetic activity generally, including that to the heart[59]. The net effect of generalised hypoxia and/or hypercapnia is to increase sympathetic efferent activity.

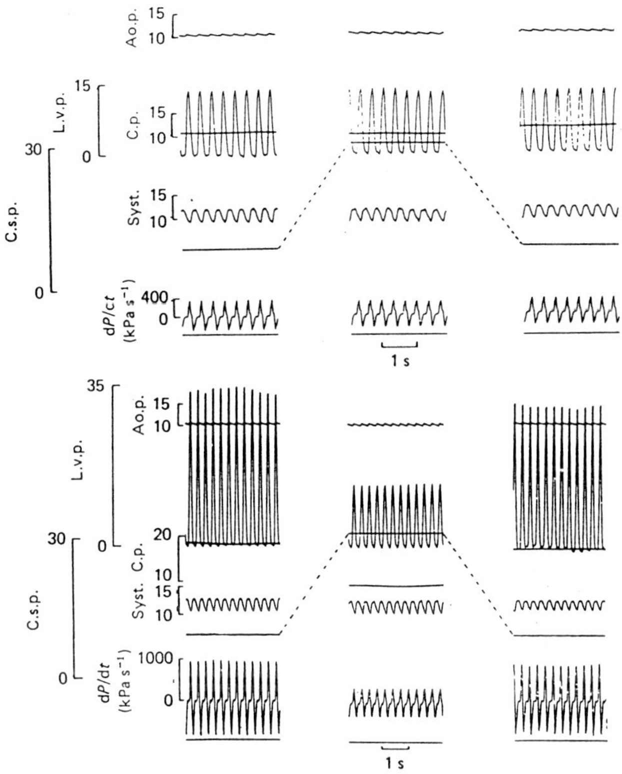

FIGURE 11. Effects of changes in carotid sinus pressure on cardiac inotropic state at different levels of cerebral arterial P_{CO_2} in an anaesthetized dog. Traces are of aortic pressure (held constant), left ventricular, cephalic perfusion, systemic arterial and carotid sinus pressures and ventricular dP/dt. Heart rate was paced throughout. Under these conditions the maximum value of dP/dt is a reliable inotropic index. The upper and lower traces were with cerebral arterial P_{CO_2} at 4.9 and 8.4 kPa respectively. Hypercapnia increased cardiac inotropic state and the response to baroreceptor stimulation was enhanced. Pressures in kPa; 1kPa = 7.5 mm Hg (From Karim et al[59]).

COMPLEX INTERVENTIONS AFFECTING CARDIAC AUTONOMIC ACTIVITY

EXERCISE

Immediately at the start of exercise there is an increase in heart rate that is attributed to vagal withdrawal and believed to be the result of central command. Metaboreceptors in exercising muscle contribute to the sustained vagal inhibition and sympathetic excitation. They also seem to inhibit the baroreceptor reflex. During progressive exercise vagal activity decreases and sympathetic activity increases resulting in increases in heart rate to maximal levels which decline with increasing age. Physical training induces adaptive response in the cardiovascular system which include slowing of heart rate at rest and submaximal work, although it has little or no effect on maximal heart rate. Training increases heart size and stroke volume. Inactivity and weightlessness cause opposite changes.

For details describing the cardiovascular effects of exercise and exercise training more comprehensive reviews are available.[60–64]

VALSALVA

The Valsalva manoeuvre involves making a forced expiratory effort against a closed glottis or obstruction and this causes a complex series of changes in the cardiovascular system.[65] Typically intrathoracic and intra-abdominal pressure may increase to 40 – 100 mm Hg. The increased pressures generated by this manoeuvre compress the heart and great vessels and impede venous return from outside the thorax and abdomen. After an initial increase in arterial pressure caused by the direct compression of the heart and arteries, arterial pressure, and particularly pulse pressure, decreases as venous return declines. This reduces baroreceptor stimulation causing tachycardia and vasoconstriction. At the end of the Valsalva the sudden increase in blood returning to the heart is pumped into a vasoconstricted circulation causing a marked increase in blood pressure particularly systolic. This strongly stimulates baroreceptors inducing reflex bradycardia. The cardiac slowing is a vagal response and persists for several seconds until the blood pressure decreases as the sympathetic vasoconstriction more slowly declines.

The Valsalva manoeuvre is used as a test of the integrity of cardiovascular autonomic nerves.

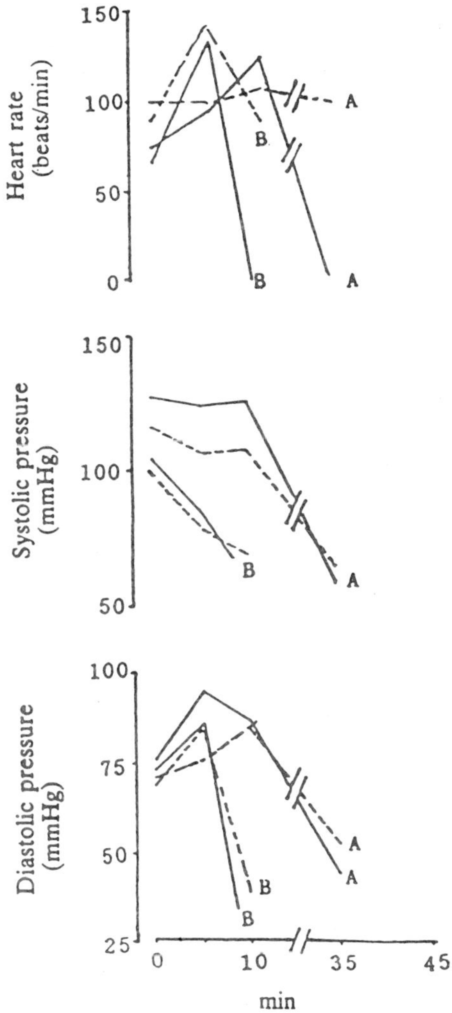

FIGURE 12. Responses of two patients with orthostatic intolerance to head-up tilting before and after implanting cardiac pacemakers. Before implantation tilting resulted in syncope accompanied by asystole. After implantation, although bradycardia was prevented, syncope still occurred, and at the same time after tilting (From El Bedawi et al[9]).

DECREASES IN VENOUS RETURN

The rate at which blood fills the heart is dependent mainly on the venous filling pressure. This is relatively high in supine individuals and during exercise. It is low when blood volume is low due to haemorrhage or when blood and plasma fluid accumulate in dependent regions of the body during orthostatic stress. The effects of decreasing venous return may be studied by use of a tilt-table, subatmospheric pressure applied to the lower part of the body, or the combination of both.[66]

The effect of the decreased cardiac filling is to decrease cardiac output. Initially there is little effect on blood pressure, although heart rate and vascular resistance increase, effected through the reduced pulse pressure providing smaller stimulus to baroreceptors. The sensitivity of the baroreceptor reflex may be potentiated under these conditions, possibly by reflexes initiated by venous congestion.[67] If the stress continues the vasoconstriction and tachycardia may suddenly switch to vasodilatation and bradycardia: the vasovagal reaction.[68] Rarely the vagal bradycardia may be so pronounced as to cause a period of asystole, and usually the vasodilatation is by far the major component and prevention of heart slowing by artificially pacing usually has no effect on the susceptibility to an attack[9] (Figure 12). Despite much speculation the mechanism precipitating vasovagal syncope remains a mystery.[69]

REFERENCES

1. Levy MN, Martin PJ, Iano T. Effects of single vagal stimuli on heart rate and atrioventricular conduction. *Am J Physiol* 1970; **218:**1256–1262.
2. Rosenblueth A, Simeone FA. The interrelations of vagal and accelerator effects on the cardiac rate. *Am J Physiol* 1934; **110:**42–55.
3. Parker P, Celler BG, Potter EK, McClosky DI. Vagal stimulation and cardiac slowing. *J Auton Nerv Syst* 1984; **11:**226–231.
4. Mary DASG, Hainsworth R. Methods for the study of cardiovascular reflexes. In: Hainsworth R, Mark AL (Eds) Cardiovascular Control in Health and Disease. London, W B Saunders Co, 1993.
5. Furnival CM, Linden RJ, Snow HM. Chronotropic and inotropic effects on the dog heart of stimulating the efferent cardiac sympathetic nerves. *J Physiol* 1973; **230:**137–153.
6. Rushmer RF. Constancy of stroke volume in ventricular responses to exertion. *Am J Physiol* 1959; **196:**745–750.
7. Miller DE, Gleason WL, Whalen RE, Morris JJ, McIntosh HD. Effect of ventricular rate in the cardiac output in the dog with chronic heart block. *Circ Res* 1962; **10:**658–663.

8. Bevegard S, Jonsson B, Karlof I, Lagergren H, Sowton E. Effects of changes in ventricular rate on cardiac output and central pressures at rest and during exercise in patients with artificial pacemakers. *Cardiovasc Res* 1967; **1**:21–33.
9. El Bedawi KM, Wahbha MMAE, Hainsworth R. Cardiac pacing does not improve orthostatic tolerance in patients with vasovagal syncope. *Clin Auton Res* 1994; **4**:233–237.
10. Nonidez JF. Identification of the receptor areas in the venae cavae and the pulmonary veins which initiate cardiac acceleration (Bainbridge's reflex). *Am J Anat* 1937; **61**:203–231.
11. Paintal AS. A study of right and left atrial receptors. *J Physiol* 1953; **120**:596–610.
12. Kappagoda CT, Linden RJ, Mary DASG. Atrial receptors in the cat. *J Physiol* 1976; **262**:431–446.
13. Bainbridge FA. The influence of venous filling upon the rate of the heart. *J Physiol* 1915; **50**:65–84.
14. Ledsome JR, Linden RJ. A reflex increase in heart rate from distension of the pulmonary vein-atrial junctions. *J Physiol* 1964; **170**:456–473.
15. Linden RJ, Kappagoda CT. Atrial receptors. Cambridge University Press, Cambridge 1982.
16. Carswell F, Hainsworth R, Ledsome JR. The effects of distension of the pulmonary vein-atrial junctions upon peripheral vascular resistance. *J Physiol* 1970; **207**:1–14.
17. Sreeharan N, Kappagoda CT, Linden RJ. The role of renal nerves in the diuresis and natriuresis caused by stimulation of atrial receptors. *Q J Exp Physiol* 1981; **66**:431–438.
18. Bennett KL, Linden RJ, Mary DASG. The effect of stimulation of atrial receptors on the plasma concentration of vasopressin. *Q J Exp Physiol* 1983; **68**:579–589.
19. Zucker IH, Earle AM, Gilmore JP. The mechanism of adaptation of left atrial stretch receptors in dogs with chronic congestive heart failure. *J Clin Invest* 1977; **60**:323–31.
20. Paintal AS. The study of ventricular pressure receptors and their role in the Bezold reflex. *Q J Exp Physiol* 1955; **40**:348–363.
21. Coleridge HM, Coleridge JCG, Kidd C. Cardiac receptors in the dog, with particular reference to two types of afferent ending in the ventricular wall. *J Physiol* 1964; **174**:323–339.
22. Maksymowicz, W, Szulczyk P. Properties of mechanoreceptor afferent fibers in left inferior cardiac nerve. *Brain Res* 1983; **262**:209–215.
23. Thames MD, Donald DE, Shepherd JT. Behavior of cardiac receptors with non-myelinated vagal afferents during spontaneous respiration in cats. *Circ Res* 1977; **41**:694–701.
24. Thoren P. Characteristics of left ventricular receptors with nonmedullated vagal afferents in cats. *Circ Res* 1977; **40**:231–237.
25. Drinkhill MJ, Moore J, Hainsworth R. Afferent discharges from coronary arterial and ventricular receptors in anaesthetized dogs. *J Physiol* 1993; **472**:785–800.

26. Gupta BN, Thames MD. Behavior of left ventricular mechanoreceptors with myelinated and non-myelinated afferent vagal fibers in cat. *Circ Res* 1983; **52**:291–301.
27. Sleight P, Widdicombe JG. Action potentials in afferents fibres from pericardial mechanoreceptors in the dog. *J Physiol* 1965; **181**:259–269.
28. Oberg B, Thoren P. Studies on left ventricular receptors signalling in non-medullated vagal afferents. *Acta Physiol Scand* 1972; **85**:145–163.
29. Oberg B, Thoren P. Increased activity in left ventricular receptors during hemorrhage or occlusion of caval veins in the cat. A possible cause of the vaso-vagal reaction. *Acta Physiol Scand* 1972; **85**:164–173.
30. Kaufman MP, Baker DG, Coleridge HM. Stimulation by bradykinin of afferent vagal C-fibers with chemosensitive endings in the heart and aorta of the dog. *Circ Res* 1980; **46**:476–484.
31. Baker DG, Coleridge HM, Coleridge JCG. Vagal afferent C fibres from the ventricle. In: Hainsworth R, Kidd C, Linden RJ (Eds) Cardiac Receptors. Cambridge Univ Press, Cambridge 1979, 117–137.
32. Hainsworth R. Reflexes from the heart. *Physiol Rev* 1991; **71**:617–658.
33. Dawes GS. Studies on veratrum alkaloids. VII. Receptor areas in the coronary arteries and elsewhere as revealed by the use of veratridine. *J Pharmacol Exp Ther* 1947; **89**:325–342.
34. McGregor KH, Hainsworth R, Ford R. Hind limb vascular responses in anaesthetized dogs to aortic root injections of veratridine. *Q J Exp Physiol* 1986; **71**:577–587.
35. Al-Timman JKA, Drinkhill MJ, Hainsworth R. Reflex vascular responses to changes in left ventricular pressure, heart rate and inotropic state in dogs. *Exp Physiol* 1992; **77**:455–469.
36. Fitzpatrick AP, Banner N, Cheng A, Yacoub M, Sutton R. Vasovagal reactions may occur after othotopic heart transplantation. *J Am Coll Cardiol* 1993; **27**:203–214.
37. Mark AL, Kioschos JM, Abboud FM, Heistad DD, Schmid PG, Burr JW. Abnormal vascular responses to exercise in patients with aortic stenosis. *J Clin Invest* 1973; **52**:1138–1146.
38. Rutlen DL, Underwood SR. Reflex influence of selective coronary artery occlusion on the total capacitance vasculature in the dog. *J Clin Invest* 1984; **73**:241–250.
39. Thames MD, Jarecki M, Donald DE. Neural control of renin secretion in anesthetized dogs. Interaction of cardiopulmonary and carotid baroreceptors. *Circ Res* 1978; **43**:237–245.
40. George M, Greenwood TW. Relation between bradycardia and the site of myocardial infarction. *Lancet* 1967; **2**:739–740.
41. Webb SW, Adgey AA, Pantridge JF. Autonomic disturbance at onset of acute myocardial infarction. *Br Med J* 1972; **3**:89–92.
42. Constantin L. Extra cardiac factors contributing to hypotension during coronary occlusion. *Am J Cardiol* 1963; **11**:205–217.
43. Brown AM, Malliani A. Spinal sympathetic reflexes initiated by coronary receptors. *J Physiol* 1967; **12**:685–705.

44. Pagani M, Pizzinelli P, Bergmaschi M, Malliani A. A positive feedback sympathetic pressor reflex during stretch of the thoracic aorta in conscious dogs. *Circ Res* 1982; **50**:125–132.
45. Fouad FM, Estafanous FG, Gravo EL. Possible role of cardioaortic reflexes in post coronary bypass hypertension *Am J Cardiol* 1979; **44**:866–872.
46. Brown AM. Mechanoreceptors in or near the coronary arteries. *J Physiol* 1965; **177**:203–214.
47. Al-Timman JKA, Drinkhill MJ, Hainsworth R. Reflex responses to stimulation of mechanoreceptors in the left ventricle and coronary arteries in anaesthetized dogs. *J Physiol* 1993; **472**:769–784.
48. McMahon NC, Drinkhill MJ, Hainsworth R. Vascular responses to stimulation of carotid aortic and coronary artery baroreceptors with pulsatile and non-pulsatile pressures in dogs. *Exp Physiol* 1996; **81**:969–982.
49. Persson PB, Kirchheim HR. Baroreceptor reflexes. Springer, Berlin 1991.
50. Zucker IH, Gilmore JP. Reflex control of the circulation. CRC Press, Boca Raton 1991.
51. Hainsworth R, Mark AL. Cardiovascular reflex control in health and disease. Saunders, London 1993.
52. Eckberg DL, Sleight P. Human baroreflexes in health and disease. Oxford University Press, Oxford 1992.
53. Chapleau MW, Abboud FM. Mechanisms of adaption and resetting of the baroreceptor reflex. In: Hainsworth R, Mark AL (Eds) Cardiovascular Reflex Control in Health and Disease. Saunders, London 1993, 165–194.
54. Eckberg DL. Temporal response patterns of the human sinus node to brief carotid baroreceptor stimuli. *J Physiol* 1978; **258**:769–782.
55. Hainsworth R, Ledsome JR, Carswell F. Reflex response from aortic baroreceptors. Am *J Physiol* 1970; **218**:423–429.
56. Eckberg DL, Kifle YT, Roberts VL. Phase relationship between normal human respiration and baroreflex responsiveness. *J Physiol* 1980; **304**:489–502.
57. Hainsworth R, McGregor KH, Rankin AJ, Soladoye AO. Cardiac inotropic responses from changes in carbon dioxide tension in the cephalic circulation of anaesthetized dogs. *J Physiol* 1984; **357**:23–35.
58. Hainsworth R, Jacobs L, Comroe JH Jr. Afferent lung denervation by brief inhalation of steam. *J Appl Physiol* 1973; **34**:708–714.
59. Karim F, Hainsworth R, Sofola OA, Wood LM.. Responses of the heart to stimulation of aortic body chemoreceptors in dogs. *Circ Res* 1980; **46**:77–83.
60. Mitchell JH, Schmidt RF. Cardiovascular reflex control of afferent fibres from skeletal muscle receptors. In: Shepherd JT, Abboud FN (Eds) Handbook of Physiology. The Cardiovascular System, Volume 3. American Physiological Society, Bethesda 1983, 623–658.
61. Astrand PO. Quantification of exercise capability and evaluation of physical capacity in man. *Prog Cardiovasc Dis* 1976; **19**:51–67.
62. Vatner SF, Pagani M. Cardiovascular adjustments to exercise: hemodynamics and mechanisms. *Prog Cardiovasc Dis* 1976; **29**:91–188.

63. Ray CA, Mark AL. Sympathetic adjustments to exercise: insights from microneurographic recordings. In: Hainsworth R, Mark AL (Eds) Cardiovascular Reflex Control in Health and Disease. Saunders, London 1993, 137–164.
64. Saltin B, Blomqvist G, Mitchell JH, Johnson RL, Willenthal K, Chapman CB.. Response to exercise after bedrest and after training. *Circulation* 1968; **37**(supp 7):1–55.
65. Sharpey-Shafer EP. Effects of respiratory acts on the circulation. In: Hamilton WF, Dow P (Eds) Handbook of Physiology, Section 2, Volume 3. American Physiological Society, Washington 1964, 1875–1886.
66. El-Bedawi KM, Hainsworth R. Combined head-up tilt and lower body suction: a test of orthostatic tolerance. *Clin Auton Res* 1994; **4**:41–47.
67. Doe CPA, Drinkhill MJ, Myers DS, Self DA, Hainsworth R. Reflex vascular responses to abdominal venous distension in the anesthetized dog. *Am J Physiol* 1996; **271**:H1049–H1056.
68. Lewis T. Vasovagal syncope and the carotid sinus mechanism. *Br Med J* 1932; **1**:873–876.
69. Hainsworth R. Syncope and fainting. In: Mathias CJ, Bannister R (Eds) Autonomic Failure. Oxford University Press, Oxford 1998.

CHAPTER TWO

PATHOPHYSIOLOGY OF CARDIAC AUTONOMIC DISORDERS

Christopher J Mathias

The heart has a dual autonomic supply, with efferent pathways that include the parasympathetic (vagus) nerves and sympathetic nerves from the thoracic paravertebral ganglia.[1] These extrinsic autonomic neural pathways are closely linked to baroreceptor reflex activity, with changes in blood pressure playing a key role in either increasing or diminishing activity of one or other pathway. There also are intrinsic pathways that are closely involved with function of the heart. A variety of afferent information, a number of cerebral centres, and factors within the heart, ranging from the state of receptors to local modulatory factors, can influence how autonomic nerve activation effects cardiac function. This chapter will begin by providing an outline classification of both localised and generalised disorders that affect cardiac autonomic function. This will be followed by a brief description of the key autonomic disorders, the pathophysiological basis of the ensuing problems and how understanding of the abnormalities aids clinical management.

CLASSIFICATION OF AUTONOMIC DISORDERS

Disorders affecting cardiac autonomic function (Table 1) may be

Malik M (Ed) Clinical guide to cardiac autonomic tests, 29–47.

TABLE 1.
Outline classification and examples of disorders that cause cardiac autonomic dysfunction

PRIMARY
- ACUTE/SUBACUTE DYSAUTONOMIAS
  - Pure cholinergic dysautonomia
  - Pure pandysautonomia
  - Pandysautonomia with neurological features
- CHRONIC AUTONOMIC FAILURE SYNDROMES
  - Pure autonomic failure
  - Multiple system atrophy (Shy-Drager syndrome)
  - Autonomic failure with Parkinson's disease

SECONDARY
- CONGENITAL
  - Nerve growth factor deficiency
- HEREDITARY
  - *Autosomal dominant trait*
    - Familial amyloid neuropathy
    - Porphyria
  - *Autosomal recessive trait*
    - Familial dysautonomia – Riley-Day syndrome
    - Dopamine beta-hydroxylase deficiency
    - Friedrich's ataxia
- METABOLIC DISEASES
  - Diabetes mellitus
  - Chronic renal failure
  - Chronic liver disease
  - Thyroid disease (thyrotoxicosis & myxoedema)
  - Vitamin B_{12} deficiency
  - Alcohol-induced
- INFLAMMATORY
  - Guillain-Barre syndrome
  - Transverse myelitis
- INFECTIONS
  - Bacterial – tetanus
  - Viral– human immunodeficiency virus infection
  - Parasitic – Trypanosomiasis Cruzi; Chagas' disease
  - Prion – fatal familial insomnia

TABLE 1.
Continued

SECONDARY
NEOPLASIA
Brain tumours – esp of third ventricle or posterior fossa
Paraneoplastic – adenocarcinomas:
lung, pancreas, and Lambert-Eaton syndrome
SURGERY
Organ transplantation – heart, kidney
Vagotomy and drainage procedures – 'dumping syndrome'
Regional sympathectomy – splanchnic
TRAUMA
Spinal cord transection
MISCELLANEOUS NEUROLOGICAL DISORDERS
Subarachnoid haemorrhage
Epilepsy
Narcolepsy
POSTURAL TACHYCARDIA SYNDROME
NEURALLY MEDIATED SYNCOPE
Vasovagal syncope
Carotid sinus hypersensitivity
Micturition syncope
Cough syncope
Swallow syncope
Associated with glossopharyngeal neuralgia

Adapted from Mathias.[2,3]

localised or generalised.[2,3] A prime example of the former is cardiac transplantation, where extrinsic nerves are severed but intrinsic networks are preserved. In Chagas disease that complicates infection with Trypanosomiasis Cruzi, the cholinergic intrinsic plexuses in the heart and gut are targeted. There are a number of generalised disorders where cardiac autonomic pathways are affected either directly or indirectly, along with other parts of the autonomic nervous system. These include the primary autonomic failure syndromes, of

Table 2.
Examples of drugs, chemicals, poisons and toxins that may cause cardiac autonomic dysfunction

DECREASING SYMPATHETIC ACTIVITY
- CENTRALLY ACTING
  - Clonidine
  - Methyldopa
  - Reserpine
  - Barbiturates
  - Anaesthetics
- PERIPHERALLY ACTING
  - Sympathetic nerve ending
    - guanethidine, bethanadine
  - Alpha-adrenoceptor blockade
    - phenoxybenzamine
  - Beta-adrenoceptor blockade
    - propranolol

INCREASING SYMPATHETIC ACTIVITY
- Amphetamines
- Releasing noradrenaline
  - tyramine
- Uptake blockers
  - imipramine
- Monoamine oxidase inhibitors
  - tranylcypromine
- Beta-adrenoceptor stimulants
  - isoprenaline

DECREASING PARASYMPATHETIC ACTIVITY
- Antidepressants
  - imipramine
- Tranquillisers
  - phenothiazines
- Antidysrhythmics
  - disopyramide
- Anticholinergics
  - atropine, probanthine, benzotropine
- Toxins
  - botulinum

Table 2.
Continued

INCREASING PARASYMPATHETIC ACTIVITY	
Cholinomimetics	
	carbachol, bethanechol, pilocarpine, mushroom poisoning
Anticholinesterases	
Reversible carbonate inhibitors	
	pyridostigmine, neostigmine
Organophosphorus inhibitors	
	parathion
MISCELLANEOUS	
Alcohol, thiamine (vitamin B_1 deficiency)	
Mercury poisoning ('Pink' disease)	
Ciguatera toxicity	
Jellyfish and marine animal venoms	
First dose of certain drugs	

Adapted from Mathias.[3]

unknown aetiology, that are grouped under the rarer subacute/acute dysautonomias and the more frequently encountered chronic autonomic failure syndromes, the most common of which is multiple system atrophy (Shy-Drager syndrome). A variety of secondary autonomic disorders results from a defined neurological lesion (spinal cord transection), may complicate a disease (diabetes mellitus) or are caused by a genetic or enzymatic defect (such as familial amyloid polyneuropathy and dopamine beta-hydroxylase deficiency). There is a further group of patients with an intermittent disorder, neurally mediated syncope, where there is increased activity of the cardiac vagi causing bradycardia and often withdrawal of sympathetic nerve activity resulting in vasodepression and syncope. A further group that is increasingly being recognised because of postural (orthostatic) intolerance is the postural tachycardia syndrome. Various drugs can result in impairment of cardiac autonomic function. These include drugs that act upon parasympathetic or sympathetic pathways, and also those that result in a neuropathy, as may occur with alcohol (Table 2). Abnormalities of cardiovascular autonomic function also occur in a variety of disorders, such as hypertension and congestive cardiac failure; these diseases will not be discussed further.

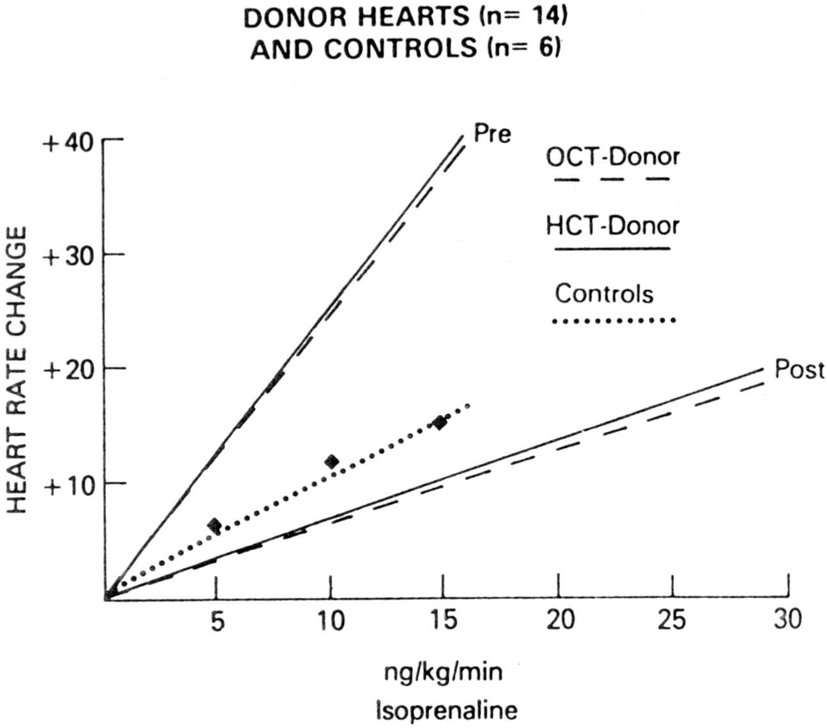

FIGURE 1. Changes in heart rate during infusion of isoprenaline in the donor hearts of patients with heterotopic (HCT) and orthotopic (OCT) cardiac transplants before (pre) and after (post) β-adrenoceptor blockade. The response in 10 normal subjects is also plotted. There is a greater response in both HCT and OCT cardiac transplantees before β-blockade (From Yusuf et al[4]).

LOCALISED CARDIAC AUTONOMIC DISORDERS

The cardiac parasympathetic and sympathetic pathways are severed in cardiac transplantees. Thus, the heart is unable to respond to a variety of stimuli that depend upon these extrinsic autonomic pathways. Examples include the abnormal cardiac rate changes which occur during postural change and the Valsalva manoeuvre, during various stimuli which rapidly increase sympathetic nervous activity, and when there is a change in respiratory rate as with deep breathing. There may be upregulation of cardiac receptors, especially adrenoreceptors, that result in enhanced responses to agonists such as isoprenaline[4] (Figure 1). Endogenous stimuli such as exercise (that elevates plasma noradrenaline and adrenaline levels), can raise heart rate in cardiac transplantees. Studies in patients who have had both heterotopic and orthotopic cardiac transplantation, indicate that mild exercise results in a similar rise in heart rate in both denervated donor hearts and innervated recipient hearts, although the heart rate rise in the denervated heart is slower as it is dependent upon an elevation in circulating catecholamines; the reverse, a slower return to baseline levels following cessation of exercise occurs in the denervated heart.[5]

In Chagas disease, the intrinsic plexuses to the heart are affected. The reasons for this are not entirely clear and include the effects of neurotoxins released from parasites or, more likely, an immunological process specifically targeting intrinsic cholinergic plexuses in the heart and gut.[6] There are abnormalities of conduction with bradycardia and a tendency to various dysrhythmias.

PRIMARY AUTONOMIC FAILURE SYNDROMES

These include the acute/subacute and the chronic autonomic failure syndromes.

Acute/Subacute Autonomic Dysautonomias

These are relatively rare disorders, with involvement of one or both branches of the autonomic nervous system. When both pathways are involved, without additional neurological involvement, the term 'pure pandysautonomia' has been used. When only the cholinergic system is targeted, the term 'pure cholinergic dysautonomia' is favoured, as it involves not only the parasympathetic nervous system but also sympathetic cholinergic fibres to sweat glands. The lesion in such cases is often presynaptic, as cholinergic receptors are responsive when cholinomimetics are administered. In the third variety there is additional neurological involvement, usually of the peripheral nervous system.

In pure panydysautonomia, in addition to postural hypotension, there is often involvement of the cardiac vagi. The latter is a characteristic of pure cholinergic dysautonomia with the basal heart rate often elevated, despite lack of hypotension. Testing indicates inactivity of the cardiac vagi with preservation of the cardiac sympathetic adrenergic pathways. Thus, during postural change heart rate may

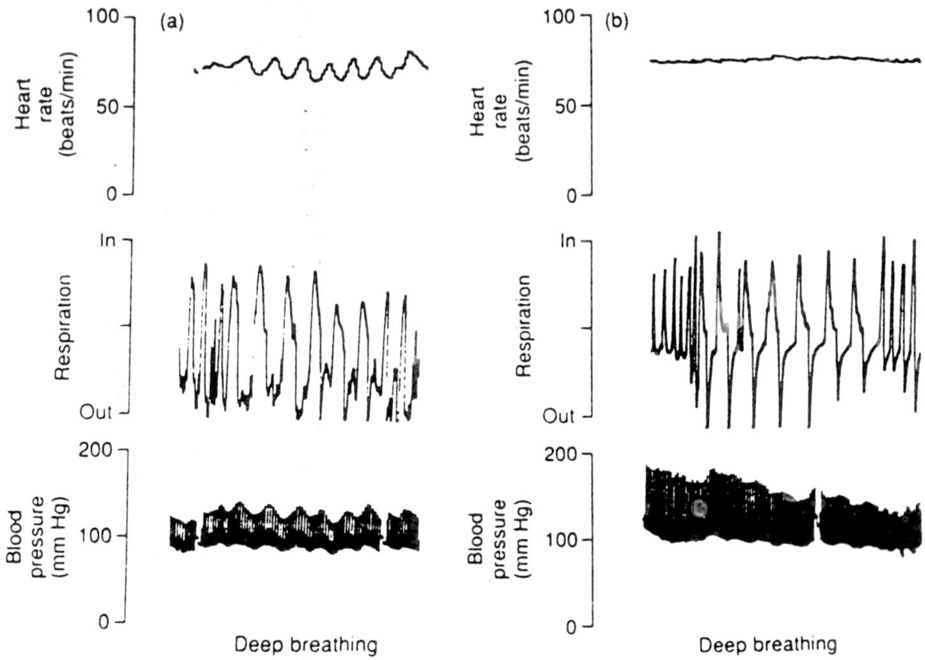

FIGURE 2. The effect of deep breathing on heart rate and blood pressure in a normal subject (a) and in a patient with autonomic failure (b). Sinus arrhythmia is not present in the patient. Respiratory changes are indicated in the middle panel (From Mathias et al[9]).

rise modestly, as it may during various pressor tests activating sympathetic efferent pathways; the heart rate changes during the Valsalva manoeuvre however usually are impaired, and sinus arrhythmia is not present.

The cause of these disorders is not known. An immunological basis has been suggested, in part based on the apparent ability of intravenous administration of immunoglobulin to reverse the autonomic features.[7, 8]

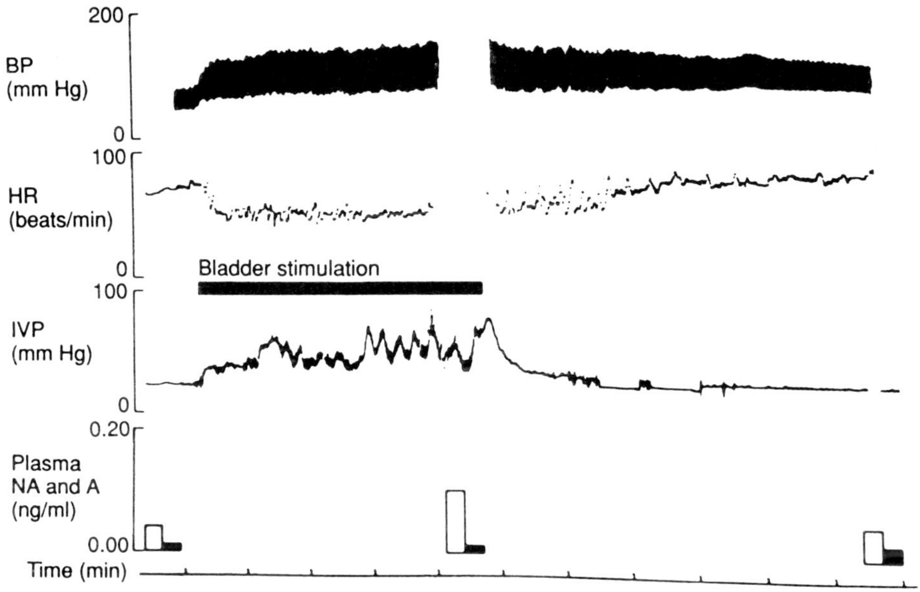

FIGURE 3. Blood pressure (BP), heart rate (HR), intravesical pressure (IVP) and levels of plasma noradrenaline (NA – open histograms) and adrenaline (A – filled histograms) in a tetraplegic patient before, during and after bladder stimulation induced by suprapubic percussion of the anterior abdominal wall. The rise in blood pressure is accompanied by a fall in heart rate as a result of increased vagal activity in response to the rise in blood pressure. Plasma NA but not A levels rise, suggesting an increase in sympathetic neural activity independently of adrenal medullary activation. (From Mathias et al[15]).

CHRONIC AUTONOMIC FAILURE SYNDROMES

There are three varieties in each of which the aetiology is unknown. In pure autonomic failure the lesions are thought to be peripheral, there is no additional neurological involvement and in addition to sympathetic vasoconstrictor failure there is usually cardiac parasympathetic involvement (Figure 2), with appropriate abnormalities on testing.[9] The more common condition is multiple system atrophy (synonymous with the Shy-Drager syndrome), which is a sporadic and unpredictably progressive disorder, with various components of autonomic function impaired at different stages of the disease.[3,10] Cardiac autonomic function may not be affected as severely as in pure autonomic failure. There is substantial data indicating that the autonomic lesion in multiple system atrophy is predominantly central.[11,12] There is a small group of patients with idiopathic Parkinson's disease and autonomic failure, in whom peripheral cardiac autonomic pathways also may be involved.

SECONDARY AUTONOMIC FAILURE/DYSFUNCTION

These include a wide range of disorders in which cardiac autonomic function may be affected in varying degrees. A brief description of some of the major disorders is provided below.

Diabetes Mellitus

This is a common metabolic disorder that can result in a peripheral neuropathy, that often affects cardiac autonomic function.[14] This often involves the cardiac vagi resulting in a resting tachycardia and lack of response to various stimuli that should normally modify heart rate. Cardiac parasympathetic impairment may not necessarily be associated with cardiac sympathetic failure. Diabetic patients are prone to coronary artery disease and myocardial ischaemia, and the autonomic imbalance may predispose them to cardiac dysrhythmias that contribute to sudden death.

Spinal Cord Lesions

A variety of autonomic abnormalities affecting the heart and blood vessels may occur in patients with cervical and high thoracic spinal cord lesions.[15] In such patients the autonomic pathways to the heart and blood vessels are severed from cerebral control; however the baroreceptor afferents are preserved, while the parasympathetic efferents to the heart remain intact and functional. Thus, in such patients a fall in blood pressure during postural change, because of inability of the brain to activate peripheral sympathetic efferent pathways, results in a rise in heart rate although this is mainly due to vagal withdrawal and the elevation is usually not higher than 120 beats per minute, despite a substantial fall in blood pressure. The reverse, autonomic dysreflexia, may occur when isolated spinal cord sympathetic pathways are activated by stimulation of skin, skeletal muscle or viscera below the level of the lesion. This can result in a paroxysmal and severe rise in blood pressure, with a compensatory bradycardia (Figure 3).

Patients with high cervical lesions, especially involving the 4th and 5th cervical segments and who need respirator support, are prone to severe bradycardia and even cardiac arrest, especially in the early phases following injury when they are in a state of 'spinal shock'. In this phase, even isolated spinal cord sympathetic reflexes, such as those that cause autonomic dysreflexia, do not appear to be operative. In these patients disconnecting the respirator, or tracheal suction as for intermittent tracheal toilette, can result in activation of vagal afferents that then increase vagal efferent activity.[16,17] This is not opposed by the sympathetic nerves as would occur normally. The vagi may be relatively overactive because of hypoxia, and their effects on heart rate cannot be reversed by the 'inflation' reflex because these patients have respiratory paralysis. Thus, activation of afferents can rapidly result in severe bradycardia and hypotension (Figure 4), with potentially disastrous sequelae. The importance of the final pathway is emphasised by the effectiveness of the cholinergic (muscarinic) blocker, atropine, in preventing bradycardia and cardiac arrest. The management involves prevention of these episodes, initially by ensuring that there is adequate oxygenation. If bradycardia still occurs, intravenous atropine may be needed. Parasympathomimetic drugs such as neostigmine and carbachol, which may be used to reverse bladder and bowel paralysis in spinal shock, should be avoided or used with caution. Isoprenaline raises heart rate but may reduce blood pressure markedly because of enhanced vasodilatation following vascular β_2 adrenoceptor stimulation.[9,15] Temporary demand pacemakers may be needed. Similar problems may occur in chronic tetraplegics undergoing general anaesthesia, when vagal afferents are stimulated during intubation.[18] It is important that such patients are given an adequate amount of atropine prior to intubation.

In patients with incomplete spinal lesions at different levels, various forms of cardiac autonomic dysfunction may occur depending upon the degree of damage to sympathetic pathways.

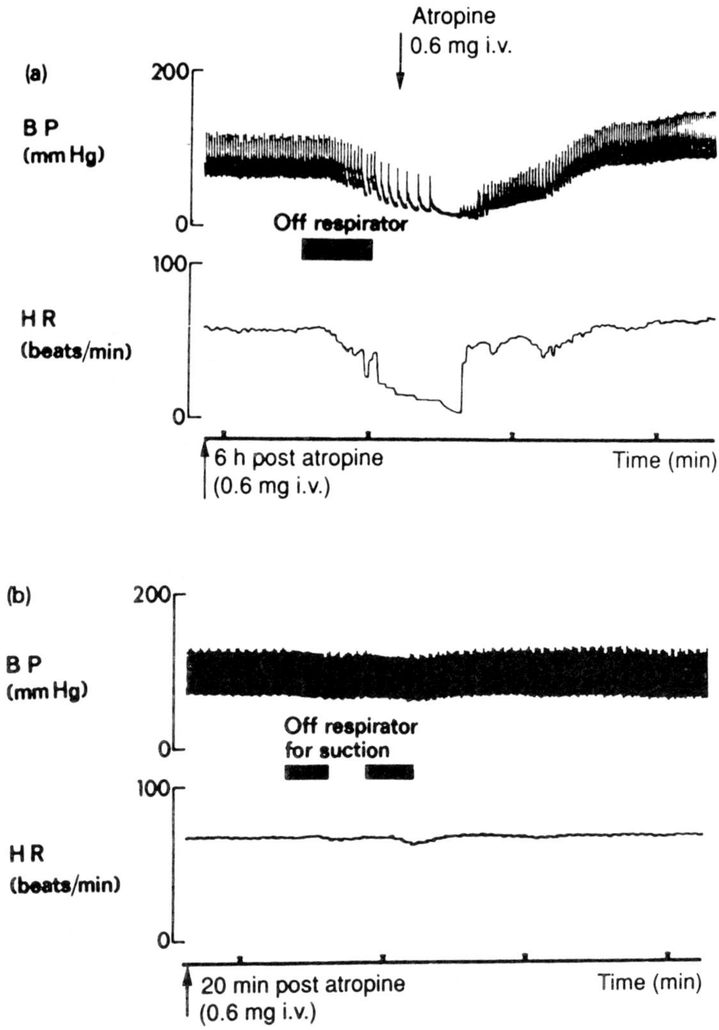

Figure 4. **(a)** The effect of disconnecting the respirator (as is required for aspirating the airways), on the blood pressure (BP) and heart rate (HR) of a recently injured tetraplegic patient (with a C4/5 lesion) in spinal shock, 6 hours after the last dose of intravenous atropine. Sinus bradycardia and cardiac arrest (also observed on the electrocardiograph) were reversed by reconnection, intravenous atropine and external cardiac massage (From Frankel et al[16]). **(b)** The effect of tracheal suction 20 minutes after atropine in the same patient. Disconnection from the respirator and tracheal suction did not lower either heart rate or blood pressure (From Mathias[17]).

THE GUILLAIN-BARRE SYNDROME, TETANUS AND SUBARACHNOID HAEMORRHAGE

These are examples of disorders where there may be rapidly fluctuating changes in cardiac autonomic activity. In post-inflammatory polyneuritis (the Guillain-Barre syndrome), there may be either severe bradycardia or marked tachycardia due to altered activity of parasympathetic or sympathetic nerves,[19,20] along with extreme lability of blood pressure. Autonomic dysfunction appears to be associated with increased morbidity and mortality. In patients with tetanus, sympathetic overactivity may result in tachycardia and hypertension[21], especially in those who are on respiratory support. Various neurological disorders due to intracranial pathology may result in abnormal cardiac autonomic dysfunction. An example is subarachnoid haemorrhage where electrocardiographic abnormalities appear to be related to increased sympathetic activity.[22]

AMYLOIDOSIS

This includes primary and secondary amyloidosis. In familial amyloid polyneuropathy, which is an autosomal dominant condition, an abnormal protein, transthyretin, is deposited in both somatic and autonomic peripheral nerves.[23] In addition, amyloid may be deposited in the heart and blood vessels. The cardiac vagus is often impaired at an early stage, resulting in a high resting heart rate and lack of response to various stimuli. Cardiac conduction defects may occur. Hepatic transplantation currently is the only method known to halt the progression of familial amyloid polyneuropathy. Detection of the autonomic abnormalities in these patients is of importance as they may contribute to peri-operative morbidity and mortality.

NEURALLY MEDIATED SYNCOPE

This is an intermittent disorder in which there often is a rapid fall in heart rate and blood pressure; the former is due to increased vagal activity and the latter to withdrawal of sympathetic tone. Between episodes, standard testing of autonomic function usually indicates no abnormality. There are a variety of causes of neurally-mediated syncope. In vasovagal syncope, that often occurs in the young (Figure 5), there may be a recognised emotional component with stimuli such as venepuncture; even the exposure to a needle or discussion of venepuncture may result in a typical episode. Insertion of a cardiac pacemaker alone may not prevent such syncopal episodes, indicating the importance of withdrawal of sympathetic nerve activity in causing hypotension. Similar changes may occur in carotid sinus

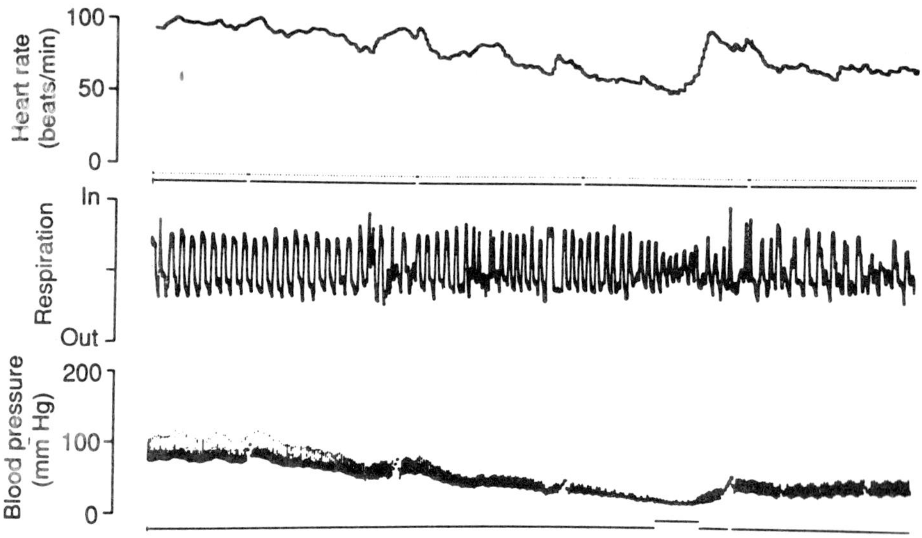

FIGURE 5. Blood pressure changes towards the end of a period of head-up tilt in a patient with recurrent episodes of vasovagal syncope. Blood pressure which was previously maintained begins to fall; there also is a fall in heart rate. Initially there are relatively minor changes in respiratory rate (middle panel) which can be derived from the signal above it, each minor dot indicating a second and the bolder mark indicating a minute. The patient was about to faint and was put back to the horizontal (indicated by elevated time signal below) and then to 5° head-up tilt. Blood pressure and heart rate recover but still remain lower than previously. This patient had no other autonomic abnormalities on detailed testing. Blood pressure was measured non-invasively by the Finapres (From Mathias et al[9]).

hypersensitivity, which increasingly is being recognised in the elderly,[24] especially when there is no other cause for recurrent falls. Such patients may have a classical history of syncope induced or associated with neck movements or tightening of the collar. There are

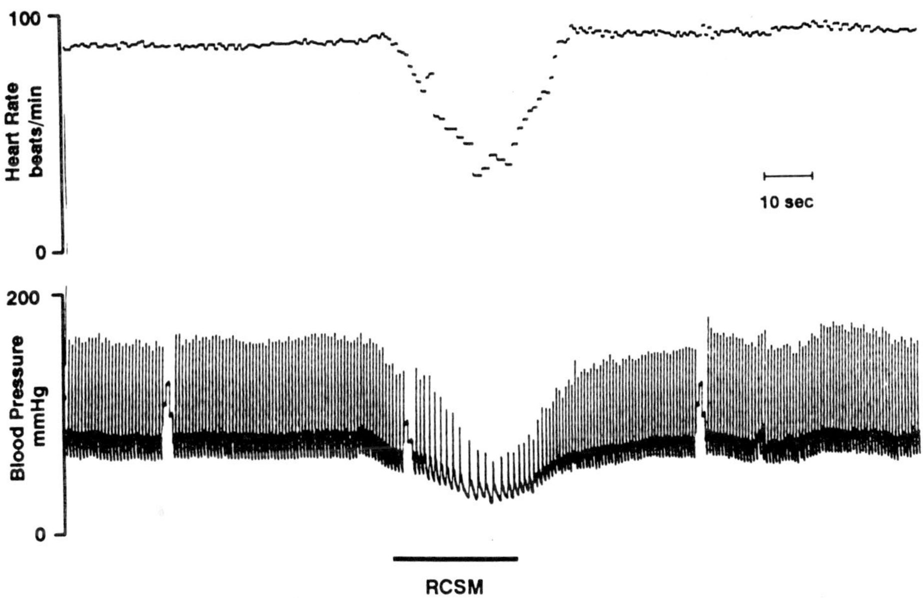

FIGURE 6. Heart rate and blood pressure before, during and after right carotid sinus massage (RCSM) in a patient with syncopal episodes. There is a fall in both heart rate and blood pressure during carotid sinus massage, typical of the mixed (cardio-inhibitory and vasodepressor) form of this disorder. The breaks in the record indicate in-built calibration by the Finapres machine (From Mathias[25]).

three major forms of carotid sinus hypersensitivity. In the cardio-inhibitory form there is marked bradycardia and in the vasodepressor form, which is rarer, there is substantial hypotension without bradycardia. The majority fall into the mixed form with a combination of bradycardia and hypotension[25] (Figure 6). In the cardio-inhibitory form a cardiac pacemaker is needed; in the mixed form it is often used but may not be successful in preventing hypotension. When hypotension occurs (as in the vasodepressor form in particular), vasopressor agents or even denervation of the carotid sinus may be needed.

Neurally mediated syncope may be induced by stimuli ranging from coughing and laughing, to swallowing and the onset of glossopharyngeal neuralgia. It may be difficult to distinguish from other causes of syncope, including epilepsy.[26] A precise diagnosis, with delineation of the mechanisms responsible, is of importance in management.

POSTURAL TACHYCARDIA SYNDROME (ORTHOSTATIC INTOLERANCE)

In this disorder, usually seen in women between the ages of 15 and 50, there are symptoms of postural (orthostatic) intolerance usually without postural hypotension, but associated with a postural tachycardia of 30 beats per minute or greater.[27,28] There are associations with mitral valve prolapse, the chronic fatigue syndrome and panic attacks. Investigations suggest that the pathophysiological basis may include hypovolaemia, incomplete autonomic denervation, adrenoreceptor changes and hyperventilation. The various descriptions suggest that it is a heterogenous disorder.[29]

DRUGS, POISONS AND TOXINS

These may impair cardiac autonomic function either by a direct action upon the neural pathways or receptors involved, or though a neuropathy (Table 2). Centrally-acting drugs, such as clonidine and methyldopa, may depress heart rate. Others acting peripherally, such as the β-adrenoreceptor blockers, may cause a marked bradycardia even when administered locally, as occasionally occurs in the treatment of glaucoma. Drugs may raise heart rate either through increasing sympathetic activity, or by a reduction in cardiac cholinergic activity. Botulinum toxin inhibits release of acetylcholine and results in tachycardia. Cholinomimetics and anticholinesterases increase parasympathetic activity and can lower heart rate. Alcohol may affect cardiac autonomic function by causing a neuropathy. Consumption of reef fish containing ciguatera toxin may cause bradycardia through increased vagal tone; this can be reversed by atropine.

Certain drugs, that include anti-hypertensive agents such as the ACE-inhibitors (Captopril) and α-adrenoceptor blockers (Prazosin), may result in syncope especially after the first dose. It has been postulated that C-fibre afferents in the heart are activated, causing increased cholinergic activity and withdrawal of sympathetic activity, resulting in bradycardia and vasodilatation. This is thought to be similar to the Bezold-Jarisch reflex and thus could be an example of drug-induced neurally mediated syncope.

CONCLUDING REMARKS

Understanding the pathophysiological processes is of importance in evaluation of the various disorders in which cardiovascular autonomic function is impaired. This applies to currently recognised autonomic diseases[10] and also to a range of cardiovascular disorders that include hypertension and congestive cardiac failure. Investigation should be directed towards evaluation of cardiac autonomic dysfunction and also evaluation of the primary cause, if present, as treatment of the underlying disorder additionally, may be of particular importance in overall management.

REFERENCES

1. Armour JA. Anatomy and function of peripheral autonomic neurones involved in cardiac regulation. In: Shepherd JT, Vatner SF (Eds) Nervous control of the heart. Horwood, Amsterdam 1996, 29–48.
2. Mathias CJ. Autonomic neuropathy – aspects of diagnosis and management. In: Asbury AK, Thomas PK (Eds) Peripheral Nerve Disorders II, Butterworth-Heinemann, Oxford 1995, 95–117.
3. Mathias CJ. Disorders of the autonomic nervous system. In: Bradley WG, Daroff RB, Fenichel GM, Marsden CD (Eds) Neurology in Clinical Practice. Butterworth-Heinemann, Boston 1982, 1953–1981.
4. Yusuf S, Theodropoulos S, Mathias CJ, Dhalla N, Wittes J, Mitchell A, Yacoub M. Increased sensitivity of the denervated transplanted human heart to isoprenaline both before and after beta-adrenergic blockade. *Circulation* 1987; **75**:696–704.

5. Yusuf S, Theodoropoulos S, Dhalla N, Mathias CJ, Teo KK, Wittes J, Yacoub M. Influence of beta blockade on exercise capacity and heart rate response after human orthotopic and heterotopic cardiac transplantation. *Am J Cardiol* 1989; **64**:636–641.
6. Fernandez A, Hontebeyrie M, Said G. Autonomic neuropathy and immunological abnormalities in Chagas disease. *Clin Autonom Res* 1992; **2**: 409–412.
7. Heafield MT, Gammage MD, Nightingale S, Williams AC. Idiopathic dysautonomia treated with intravenous gammaglobulin. *Lancet* 1996; **347**:28–29.
8. Smith AAJ, Vermeulen M, Koelman JHTM, Wieling W. Unusual recovery from acute panautonomic neuropathy after immunoglobulin therapy. *Mayo Clinic Proc* 1997; **72**: 333–335.
9. Mathias CJ, Bannister R. Investigation of autonomic disorders. In: Mathias CJ, Bannister R (Eds) Autonomic Failure: A Textbook of Clinical Disorders of the Autonomic Nervous System. 4th Edition. Oxford University Press, Oxford. (In press).
10. Mathias CJ. Autonomic disorders and their recognition. *New Eng J Med* 1997; **336**: 721–724.
11. Kimber JR, Watson L, Mathias CJ. Distinction of idiopathic Parkinson's disease from multiple system atrophy by stimulation of growth hormone release with clonidine. *Lancet* 1997; **349**: 1877–1881.
12. Mathias CJ, Polinsky RJ. Separating the primary autonomic failure syndromes, multiple system atrophy and pure autonomic failure from Parkinson's disease. In: G Stern (Ed) Advances in Neurology. Lippincott-Raven, Philadelphia. (In press)
13. Goldstein DS, Courtney Holmes CMT, Cannon RO, Eisenhofer G, Kopin IJ. Sympathetic cardioneuropathy in dysautonomias. *New Eng J Med* 1997; **336**: 696–702.
14. Watkins PJ, Edmonds M. Diabetic autonomic failure. In: Mathias CJ, Bannister R (Eds) Autonomic Failure: A Textbook of Clinical Disorders of the Autonomic Nervous System. 4th Edition. Oxford University Press, Oxford. (In press).
15. Mathias CJ, Frankel HL. Autonomic disturbances in spinal cord lesions. In: Mathias CJ, Bannister R (Eds) Autonomic Failure: A Textbook of Clinical Disorders of the Autonomic Nervous System. 4th Edition. Oxford University Press, Oxford. (In press).
16. Frankel HL, Mathias CJ, Spalding JMK. Mechanisms of reflex cardiac arrest in tetraplegic patients. *Lancet* 1975; **2**:1183–1185.
17. Mathias CJ. Bradycardia and cardiac arrest during tracheal suction – mechanisms in tetraplegic patients. *Eur J Intens Care Med* 1976; **2**:147–156.
18. Welply N, Mathias CJ, Frankel HL. Circulatory reflexes in tetraplegics during artificial ventilation and during general anaesthesia. *Paraplegia* 1975; **13**:172–182.
19. Palterman TG, Wright I, Doyle DV, Amiel S. Electrocardiographic abnormalities an dautonomic dysfunction in Guillain-Barre syndrome. *Br Med J* 1982; **284**: 1231–1232.

20. Oakley CM. The heart in the Guillain-Barre syndrome. *Br Med J* 1984; **288**:94.
21. Corbett JL, Kerry JH, Prys-Roberts C, Smith AC, Spalding JMK. Cardiovascular disturbances in severe tetanus due to overactivity of the sympathetic nervous system. *Anaesthesia* 1969; **24**:198–212.
22. Cruickshank JM, Neil-Dwyer G, Lane J. The effect of oral propranolol upon the ECG changes occurring in subarachnoid haemorrhage. *J Neurolog Neurosurg Psychiat* 1974; **37**:755–759.
23. Reilly MM, Thomas PK. Amyloid neuropathy. In: Mathias CJ, Bannister R (Eds) Autonomic Failure: A Textbook of Clinical Disorders of the Autonomic Nervous System. 4th Edition. Oxford University Press, Oxford. (In press).
24. McIntosh SJ, Lawson J, Kenny RA. Clinical characteristics of vasodrepressor, cardio-inhibitory and mixed carotid sinus syndrome in the elderly. *Am J Med* 1993; **95**:203–208.
25. Mathias CJ. Autonomic dysfunction and the elderly. In: Grimley-Evans J (Ed) Oxford Textbook of Geriatric Medicine. Oxford University Press, Oxford. (In press).
26. Lempert T. Recognising syncope: pitfalls and surprises. *J Royal Soc Med* 1996; **89**:372–375.
27. Biaggioni I. Orthostatic intolerance syndrome, vasoregulatory aesthenia and other hyperadrenergic states. In: Robertson D, Biaggioni I (Eds) Disorders of the autonomic nervous system. Horwood, Amsterdam 1995, 271–286.
28. Low PA, Schondorf R, Novak V, Sandroni P, Opfer-Gehrking TL, Novak P. Postural tachycardia syndrome. In: Low PA (Ed) Clinical autonomic disorders. Lippincott-Raven, Philadelphia 1997, 681–697.
29. Khurana RK. Orthostatic intolerance and orthostatic tachycardia: a heterogenous disorder. *Clin Autonom Res* 1995; **5**:12–18.
30. Burnett JW, Weinrich D, Williamson JA, Fenner PJ, Lutz L, Bloom DA. Autonomic neurotoxicity of jellyfish and marine animal venoms. *Clin Autonom Res* 1998; **8**:125–130.

PART TWO

AUTONOMIC TESTS

CHAPTER THREE

BASIC AUTONOMIC TESTS

Stefan H Hohnloser and Thomas Klingenheben

Sudden cardiac death remains a major public health problem in the Western countries. This devastating event is responsible for approximately 300000 deaths each year in the US. Most patients who die have underlying structural heart disease with coronary disease being the most frequently encountered diagnosis. Recently, compelling evidence has accumulated that links the autonomic nervous system and sudden cardiac death. In the clinical setting, therefore, two technologies have been developed over the last 10 years, which help to non-invasively assess cardiac autonomic tone in patients prone to sudden cardiac death. The first technique is the assessment of heart rate variability from the surface ECG or from 24 hour Holter recordings.[1] The second method is the determination of baroreflex sensitivity,[2] which is most commonly achieved using the method devised by Smith and co-workers in 1966.[3] Both methods are dealt with in detail in subsequent chapters of this book.

Another way of assessing autonomic nervous system function is to perform provocative manoeuvres known as cardiovascular reflex tests. In Table 1, a variety of tests are listed, which have been proposed to assess autonomic nervous system dysfunction. Whereas assessment of heart rate variability or baroreflex sensitivity are being considered laboratory tests requiring the respective technical equipment, cardiovascular reflex tests can be performed at the bedside in every day clinical routine. The purpose of this chapter is to describe the pathophysiological background, the clinical performance, and po-

Malik M (Ed) Clinical guide to cardiac autonomic tests, 51–65.

tential clinical relevance of these tests. Particular emphasis will be given to those tests for which scientifically valid data combined with clinical experience is available.

TABLE 1
Procedures used for evaluating autonomic reflexes

Valsalva manoeuvre
Slow deep breathing
Cold face test
Heart rate response to standing
Blood pressure response to sustained handgrip
Mental arithmetic stress

PHYSIOLOGY OF REFLEXES USED FOR EVALUATION OF AUTONOMIC NERVOUS SYSTEM

Several excellent reviews on the physiology of reflex arcs involved in cardiovascular autonomic control have recently been published[4–8]. In essence, the autonomic nervous system is usually explored by evaluating reflex arcs. A reflex arc involves a stimulus, a receptor, an afferent nerve, central processing, an efferent nerve and an end-organ response. In addition to the reflex arc, there are several synapses involved throughout the pathway and different neurotransmitters at each synaptic cleft. It is important to mention that where possible, the confounding variables, standardisation of stimulus, and normal end-organ function is established before one elicits the reflex arc to test the autonomic nervous system.

The arterial and cardiopulmonary baroreceptor reflexes are the autonomic cardiovascular reflex arcs most often evaluated in humans.[4] The basic components of the arterial and cardiopulmonary baroreceptor reflexes that modulate cardiovascular function are illustrated in Figure 1. These two reflex arcs derive their names from the anatomical location of their afferent fibres. The afferent limb of each reflex arc can be activated by stimulation of mechanoreceptors sensitive to changes in stretch or transmural pressure. Mechanoreceptors can be found in the carotid and aortic arch (arterial baroreceptors)

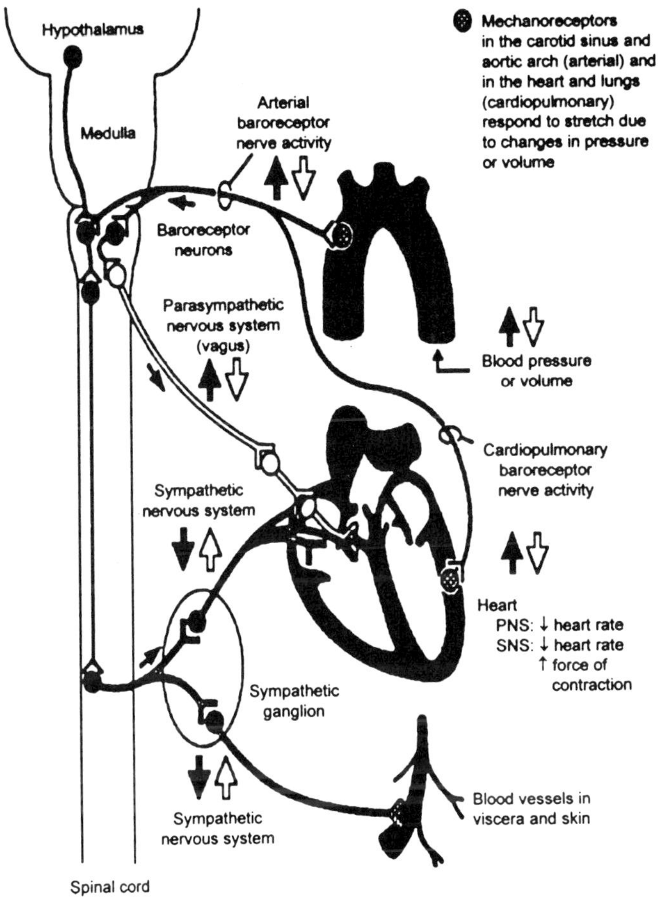

FIGURE 1. Basic components of the arterial and cardiopulmonary baroreceptor reflexes. Increases in blood pressure or in circulating volume result in enhanced activity in the arterial and cardiopulmonary afferent baroreceptor nerves terminating in the medulla (brain stem). Internuncial neurons in the brain stem cause increased firing of preganglionic and postganglionic sympathetic neurons to the heart and vessels. This results in heart rate decrease, diminished cardiac contractility, and decline in tonic constriction of vessel resistance, tending to return the blood pressure or apparent blood volume toward normal. A decline in blood pressure or in apparent blood volume produces the reverse effects on parasympathetic and sympathetic nerve activity in an effort to increase blood pressure toward normal (Modified from Taylor[4]).

and also in the great veins and right atrium and possibly the lungs and pleura (cardiopulmonary) as well as in the cardiac ventricles (cardiac baroreceptors).[4] The current understanding of these highly complex functions of reflex arcs derives mainly from sophisticated studies in experimental models with strictly controlled conditions. However, there are also increasing data available from studies in humans which now include direct recordings of sympathetic neuronal firing rates after baroreceptor activation or deactivation.[5,7] These findings support the notion that autonomic regulation of cardiovascular function in humans and in experimental animals bear a great deal of similarity.[4]

EVALUATING AUTONOMIC FUNCTION: STANDARDISATION OF TESTS

It is of paramount importance to standardise test procedures of autonomic function as far as possible in order to reduce the great variability in results of various procedures. It is known that eating, drinking coffee, smoking, volume status, upright posture, medicines, and exercise can affect the cardiovascular autonomic system. For instance, in both normotensive and hypertensive subjects it has been demonstrated that moderate levels of exercise produce effects on cardiovascular reflexes for 8 hours and more. In an ideal situation, therefore, studies on autonomic function should be performed with the patient being in stable clinical conditions for at least 48 hours after a fasting period of 12 hours. Moreover, the studies should be performed in the morning in a quiet relaxed atmosphere after careful instruction of the patient regarding the performance of the test. Only if those requirements are fulfilled, the retest variation of cardiovascular parameters in autonomic testing may be sufficiently high for clinical purposes.[9]

VALSALVA MANOEUVRE

One of the most widely used tests of human autonomic function is the Valsalva manoeuvre a test which has been studied extensively in normal subjects and in patients with cardiovascular diseases. The procedure holds great promise since it is safe and can be performed without sophisticated equipment. Moreover, it yields reproducible

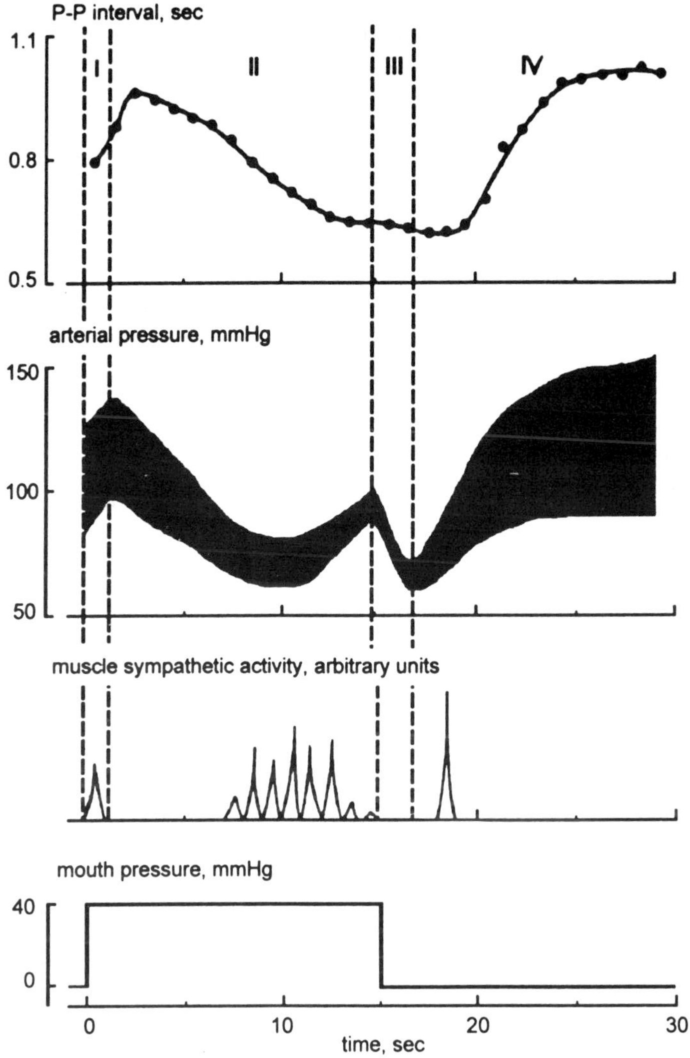

FIGURE 2. Illustration of normal responses to the Valsalva manoeuvre. According to the proposal of Hamilton et al,[16] four distinct phases of response can be separated (Adopted from Eckberg[15]).

quantitative results.[10] This manoeuvre provides an excellent test of the integrity of both arterial and cardiopulmonary baroreflex arcs.[11–15]

Methodology

The Valsalva manoeuvre comprises an abrupt transient voluntary elevation of intrathoracic and intra-abdominal pressures provoked by straining. For this purpose, the proband/patient blows into a mouthpiece of a manometer to 40 mm Hg for 15 sec with continuous ECG monitoring before, during, and after the procedure. There are numerous modifications of this protocol regarding for instance mouth pressure or duration of straining.[15] Despite these methodological variations, there appears to be a nearly universal acceptance of the proposal of Hamilton, Woodbury, and Harper[16] that responses to the manoeuvre be divided in 4 phases. During the Valsalva manoeuvre, intrathoracic pressure rises, venous return to the heart diminishes, stroke volume falls, and venous pressure rises. This results in the occurrence of the following 4 phases (Figure 2): (1) an evanescent rise of arterial pressure and reduction of heart rate immediately after the onset of straining; (2) a fall, and later partial recovery of arterial pressure, and a speeding of heart rate during the period of straining; (3) on release of the strain, a sudden drop of arterial pressure equivalent to the fall in intrathoracic pressure and elevation of heart rate immediately following the release of straining; and (4) a terminal elevation of arterial pressure above control levels and slowing of heart rate. The characteristic sequence of these changes is shown in Figure 3. Although interpretation of the test is most accurately performed by considering changes in heart rate and in pulse pressure, often the so-called Valsalva ratio (VR) is used to estimate the result of the Valsalva manoeuvre. This ratio is usually calculated from the relation of the longest RR interval obtained during continuous ECG monitoring after strain, to the shortest RR interval during strain. A ratio of ≥ 1.2 is accepted as the normal value (Table 2).

Interpretation of test results

Characteristic abnormalities in the hemodynamic response to the Valsalva manoeuvre have been described in patients with autonomic failure, regardless of the underlying cause.[17,18] These include failure of the heart rate to increase appropriately during phase 2 and diminished or even absent hypertensive overshoot during phase 4. Abnormal hemodynamic responses to Valsalva are also observed in patients with severe congestive heart failure.[18–20] Here, phase 1 and 3 are normal indicating normal transmission of the elevated intrathoracic pressure into the arterial tree during phase 1 and sudden loss of this with the release of strain during phase 3. However, due to the fact that the heart operates on the flat portion of its Starling curve, the impedance of the venous return during phase 2 does not affect stroke volume. Therefore, the baroreceptor reflex is not activated and there is no overshoot upon release of the strain.

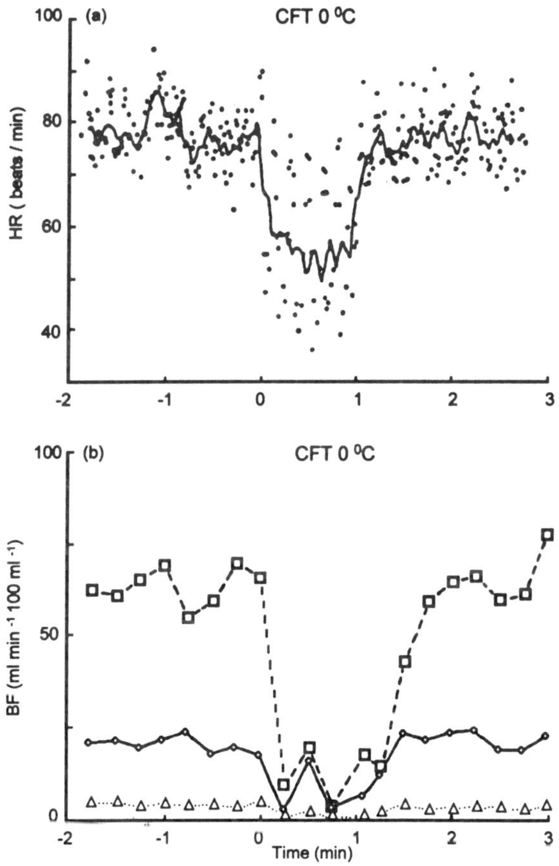

Figure 3. Responses in heart rate (HR – upper panel) and blood flow (BF – lower panel) during standardised cold face test. Blood flow is measured in the finger (circles), toe (squares), and calf (triangels) (Adopted from Health et al[32]).

There are, however, limitations for the widespread use of the Valsalva manoeuvre. A major drawback is the individual and inter-individual reproducibility particularly when normal subjects and patients with cardiovascular diseases are compared.[21–25] This has some investigators led to describe the simplicity of the Valsalva manoeuvre as 'illusory'[15] since it represents a complex test that involves several types of receptors (baroreceptors, cardiopulmonary receptors, and chemoreceptors). The Valsalva manoeuvre has been used to test autonomic reflexes in two studies involving survivors of myocardial infarction.[21,25] In the first study,[21] patients with a recent myocardial infarction were evaluated. The main finding was that the blood pressure rise during phase 4 of the Valsalva manoeuvre was significantly blunted compared to normal subjects. The authors took this finding as evidence for autonomic imbalance after myocardial infarction. The second study evaluated patients one year after their index infarction[25]. These patients were found to have less reflex tachycardia than did control subjects during the blood pressure decrease in phase 2 of the manoeuvre. None of these studies, however, allowed evaluation of the prognostic value of the Valsalva manoeuvre for future death or arrhythmic events due to the small patient samples.

TABLE 2
Normal and abnormal values of simple bedside cardiovascular reflex tests

Test	Normal	Borderline	Abnormal
Valsalva manoeuvre (Valsalva ratio)	≥ 1.2	N/A	≤ 1.20
Heart rate variation (Δ heart rate)	≥ 15 bpm	11–14 bpm	≤ 10 bpm
Heart rate response to standing (30 : 15 ratio)	≥ 1.04	1.01 – 1.03	≤ 1.0
BP response to standing (postural fall in systolic BP)	≤ 10 mmHg	11–29 mmHg	≥ 30 mmHg
Sustained handgrip (increase in diastolic BP)	≥ 16 mmHg	11–15 mmHg	≤ 10 mmHg

BP – blood pressure, bpm – beats per minute.

SLOW DEEP BREATHING

Inspiration and expiration performed at a constant rate (one full cycle every 10 seconds) accentuates the normal respiratory sinus arrhythmia observed in most individuals.[13] The routine examination of this manoeuvre is performed at a standardised rate of 6 breaths per minute. The heart rate is continuously monitored and the difference between maximum and minimum heart rate or the ratio of these two values is recorded. To quantify the test score the difference between maximal and minimal heart rate for each of the six cycles is determined and averaged to obtain the inspiratory – expiratory (I – E) difference in beats/min. This reflex is predominantly mediated by changes in parasympathetic (vagal) activity in humans.[4] Both cardiac vagal inhibition and activation are tested. Abnormal test results have been documented in a variety of dysautonomias. Loss of normal respiratory sinus arrhythmia is often the earliest sign of autonomic neuropathy in diabetes mellitus.[26,27]

Other investigators have shown that this test also yields prognostic value in patients prone to sudden cardiac death.[28] In a study of Andresen and co-workers, 97 survivors of acute myocardial infarction underwent a deep breath test 5–8 days after the acute event.[28] An heart rate difference between inspiration and expiration of < 9 beats was defined as a pathological response which was observed in 38/97 patients (39%). Compared to individuals with a normal response on deep breath test, these patients were older and had a higher incidence of arterial hypertension and diabetes mellitus. Over a follow-up period of 2 years, total mortality in the group of patients with an abnormal deep breath test was 26 % compared to only 11 % (p=0.04) in individuals with normal reflexes. These results emphasise the importance of reduced vagal activity as a marker of increased risk after myocardial infarction which has been derived using methods such as heart rate variability or baroreflex sensitivity testing in several other large trials.[1,2,29]

THE COLD FACE TEST (DIVING REFLEX)

The cold face test (CFT) was proposed by Khurana et al[30] as a simple clinical test to assess parasympathetic-cardiac and sympathetic-vascular smooth muscle pathways. The cold stimulus is applied to the face using cold compresses usually in form of commercially available gel packs. In contrast to facial immersion, the CFT is simple to administer, does not require breath-holding, and can be performed in a supine position that allows for plethysmographic blood flow measurements for assessment of peripheral sympathetic pathways. Activation of temperature-sensitive afferent fibres in the face leads to increased activity of sympathetic efferent fibres to the heart and vasculature; this causes marked increases in peripheral vascular resistance with a resultant rise in both systolic and diastolic blood pressure, and activation of cardiac vagal efferents resulting in profound bradycardia.[31–35] In Figure 4, a typical example of changes in heart rate and blood flow during CFT is shown. According to a recent study of Heath and Downey,[32] bilateral application of compresses at 0°C provided an effective and unobnoxious standard for assessment of trigeminal – vagal – cardiac and trigeminal – sympathetic – vascular smooth muscle pathways. In their study, there was a very good repeatability in responses to cold face tests performed on the same subject on different days, at different times of day and even at different times of the year.[32] Accordingly, the authors report several advantages of this test. First, the test can easily been performed and is not stressful for the patient. Second, it represents an additional test of the vagal-cardiac pathway that differs from the Valsalva manoeuvre in that its transducers and afferent pathways are independent of pressure and stretch receptors in the thoracic cavity and the afferent vagal fibres. These attributes make the CFT useful for differentiating between abnormal vagal-cardiac function and abnormal transducer-afferent nerve function. Furthermore, it simultaneously tests the sympathetic-vascular smooth muscle pathway. However, as a potential limitation, the CFT may underestimate total vagal-cardiac response in trained individuals or probands with otherwise high resting vagal tone[32]. At present, however, there are no data on the utility of the cold face test in patient populations with cardiac diseases, particularly with respect to its potential usefulness as a prognostic marker after myocardial infarction.

HEART RATE RESPONSE TO STANDING

Assumption of the standing position after supine rest for 30 minutes results in venous pooling and a transient decrease in cardiac output eliciting reflex activation of the sympathetic nervous system and a withdrawal of cardiac parasympathetic tone.[36] This is reflected by characteristic changes in heart rate which increases abruptly towards a primary peak at around 3 sec, increases further to a secondary peak at about 12 sec, declines to a relative badycardia at about 20 sec, and then gradually rises again. Generally, in quantifying the initial heart rate response to standing, the secondary heart rate peak is used. The ratio between the highest and lowest heart rate in the first 30 sec from the onset of standing is generally used to quantify the relative bradycardia. The characteristic heart rate response can also be expressed by the 30:15 ratio which is the ratio of the longest RR interval around the thirtieth beat after standing up to the shortest RR interval around the fifteenth beat. In the clinical setting, interpretation of this test imposes difficulties, however, since variability of heart rate is determined in a very complex situation, which is also influenced by sympathetic tone.

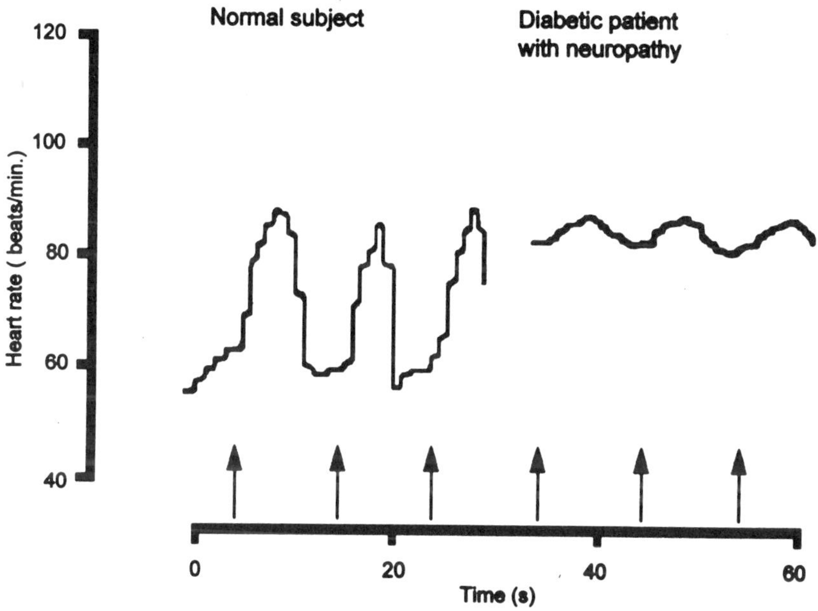

Figure 4. Original tracings of variations in heart rate recorded with regular deep breathing at 6 breaths/min in a normal subject and a patient with diabetic neuropathy. The arrows indicate the onset of inspiration (Adopted from Mackay[27])

BLOOD PRESSURE RESPONSE TO SUSTAINED HANDGRIP

This test is performed by maintaining handgrip at 30% of the maximum voluntary contraction for up to a maximum of 5 minutes utilising a handgrip dynamometer. Blood pressure is measured every 5 minutes. The difference between the diastolic blood pressure just before release of the handgrip and before starting is taken as the measure of response.[37] Normal and abnormal values for this and other simple bedside cardiovascular reflex tests are summarised in Table 2.

MENTAL ARITHMETIC STRESS

During this test, the subject is asked to perform a series of subtractions with measurements of skin temperature and heart rate performed continuously. The largest experience with this test exists for patients with diabetes mellitus. In a recent study, diabetics with and without autonomic neuropathy were compared by means of mental arithmetic stress.[38] While mental arithmetic rapidly reduced skin temperature in normals and subjects without evidence for neuropathy, no effects was found in patients with neuropathy. There were also characteristic differences in terms of heart rate changes between the 3 groups. The authors therefore concluded that this simple test could be considered as an additional tool for evaluating the presence of orthosympathetic autonomic neuropathy in diabetes. There are, however, no systematic data on the prognostic value of this test for instance in patients with cardiovascular diseases at increased risk of sudden death.

CONCLUSIONS

The role of non-invasive cardiovascular reflex testing appears to be established in various clinical conditions. The greatest experience has been accumulated in the assessment of diabetic autonomic neuropathy. Five year mortality in this condition is approximately 50 % with a substantial proportion of deaths being sudden and unexpected, i.e. presumably due to ventricular tachyarrhythmias. Given the well established role of the autonomic nervous system in the genesis of sudden death, autonomic testing is of particular relevance in this context. Some preliminary results concerning the prognostic value of basic autonomic tests in patients with cardiac diseases, particularly in those with a recent myocardial infarction, are promising.[28] However, future research is necessary to establish their definite role within the framework of risk stratification in this setting. It remains to be determined whether these simple bedside tests yield equivalent prognostic information as tests like heart rate variability derived from 24 hour ambulatory ECG recordings[1] or baroreflex sensitivity measurements[2] both of which have been proven as important risk stratifiers for future death and arrhythmic events in survivors of acute myocardial infarction.[29]

REFERENCES

1. Hohnloser SH, Klingenheben Th, Zabel M, Li Y-G. Heart rate variability used as an arrhythmia risk stratifier after myocardial infarction. *PACE* 1997; **20**:2594–2601.
2. La Rovere MT, Schwartz PJ. Baroreflex sensitivity as a cardiac and arrhythmia mortality risk stratifier. *PACE* 1997; 20:2602–2613.
3. Smyth HS, Sleight P, Pickering GW. Reflex regulation of arterial pressure during sleep in man: a quantitative method of assessing baroreflex sensitivity. *Circ Res* 1969; **24**:109–121.
4. Taylor A. Autonomic control of cardiovascular function: clinical evaluation in health and disease. *J Clin Pharmacol* 1994; **34**:363–374.
5. Victor RG, Mark AL. Interaction of cadiopulmonary and carotid baroreflex control of vascular resistance in humans. *J Clin Invest* 1985; **76**:1592–1598.
6. Mark AL. The Bezold-Jarish reflex revisited: Clinical implications of inhibitory reflexes originating in the heart. *J Am Coll Cardiol* 1983; **1**:90–102.
7. Somers VK, Mark AL, Abboud FM. Interaction of baroreceptor and chemoreceptor reflex control of sympathetic nerve activity in normal humans. *J Clin Invest* 1991; **87**:1953–1957.
8. Abboud FM. Intergration of reflex response in the control of blood pressure and vascular resistance. *Am J Cardiol* 1979; **44**:903–911.
9. Braune S, Schulte-Mönting J, Schwerbrock S, Lücking CH. Retest variation of cardiovascular parameters in autonomic testing. *J Autonom Nerv Syst* 1996; **60**:103–107.
10. Levin AB. A simple test of cardiac function based upon the heart rate changes induced by the Valsalva manoeuvre. *Am J Cardiol* 1966; **18**:90–99.
11. Shimada K, Kitazumi T, Ogura H, Sadakane N, Ozawa T. Effects of age and blood pressure on the cardiovascular responses to the Valsalva manoeuvre. *J Am Geriatr Soc* 1986; **34**:431–434.
12. Pickering TG, Sleight P. Quantitative index of baroreflex activity in normal and hypertensive subjects using the Valsalva manoeuvre. *Br Heart J* 1969; **31**:392 (abstract).
13. Piha SJ. Cardiovascular autonomic reflex tests: Normal responses and age-related reference age-values. *Clin Physiol* 1991; **11**:277–290.
14. Ryder REJ, Hardisty CA. Which battery of cardiovascular autonomic function tests. *Diabetologia* 1990; **33**:177–179.
15. Eckberg DL. Parasympathetic cardiovascular control in human disease: a critical review of methods an results. *Am J Physiol*, 1980; **239**:H581–H591.
16. Hamilton WF, Woodbury RA, Harper HT, Jr. Physiologic relationships between intrathoracic, intraspinal and arterial pressures. *J Am Med Assoc* 1959; **107**:853–856.

17. Bannister R. Autonomic Failure: A textbook of clinical disorders of the autonomic nervous system. Oxford University Press, New York 1983.
18. Nishimura RA, Tajik AJ. The Valsalva manoeuvre and response revisited. *Mayo Clin Proc* 1984; **61**:211–217.
19. Taylor AA, Davies AO, Mares A, Raschko J, Pool J, Nelson EB, Mitchell JR. Spectrum of dysautonomia in mitral valvular prolapse. *Am J Med* 1988; **86**:267–274.
20. Zema MJ, Caccavano M, Kligfield P. Detection of left ventricular dysfunction in ambulatory subjects with the bedside Valsalva manoeuvre. *Am J Med* 1983; **75**:241–248.
21. Bennett T, Wilcox RG, Hampton JR. Cardiovascular reflexes in patients after myocardial infarction. Effect of long-term treatment with beta-adrenoceptor antagonists. *Br Heart J* 1980; **44**:265–270.
22. Korner PI, Tonkin AM, Uther JB. Valsalva constrictor and heart rate reflexes in subjects with essential hypertension and with normal blood pressure. *Clin Exp Pharmacol Physiol* 1979; **6**:97–110.
23. Parisi AF, Harrington JJ, Askenazi J, Pratt RC, McIntyre KM. Echocardiographic evaluation of the Valsalva manoeuvre in healthy subjects and patients with and without heart failure. *Circulation* 1976; **54**:921–927.
24. Robertson D, Stevens RM, Friesinger GC, Oates JA. The effect of the Valsalva manoeuvre on echocardiographic dimension in man. *Circulation* 1977; **55**:596–602.
25. Kirby BJ. Circulatory reflexes in myocardial infarction *Br Heart J* 1977; **39**:168–172.
26. Ewing DJ, Martyn CN, Young RJ, Clarke BF. The value of cardiovascular autonomic function tests: 10 years experience in diabetes. *Diabetes Care* 1985; **8**:491–498.
27. Mackay JD. Respiratory sinus arrhythmia in diabetic neuropathy. *Diabetologia* 1983; **24**:253–256.
28. Andresen D, Brüggemann T, Behrens S, Ehlers C. Heart rate response to provocative manoeuvres. In: Malik M, Camm AJ (Eds) Heart Rate Variability. Futura, Armonk 1995, 267–274.
29. La Rovere MT, Mortara A, Bigger JT, Marcus FI, Malik M, Nohara R, Schwartz PJ, Camm AJ, Hohnloser SH, for the ATRAMI Investigators. Prognostic value of autonomic markers on life-threatening arrhythmic events after myocardial infarction: results from the ATRAMI Study. *Circulation* 1996; **94** (Suppl.): 3658.
30. Khurana R.K., Watabiki S, Hebel JR, Toro R, Nelson E. Cold face test in the assessment of trigeminal-brain-stem-vagal function in humans. *Ann Neurol* 1980; **7**:144–149.
31. Taylor AA, Mitchel JR. Laboratory evaluation of autonomic nervous system function. In: Laragh JR, Brenner BM (Eds) Hypertension: Pathophysiology, Diagnosis and Management. Raven, New York 1990, 1461–1478.
32. Health ME, Downey JA. The cold face test (diving reflex) in clinical autonomic assessment: Methodological considerations and repeatability of responses. *Clin Sci* 1990; **78**:139–147.

33. Strauss MB. Physiological aspects of mammalian breath-hold diving: A review. *Aerospace Med* 1970; **41**:1362–1374.
34. Kawakami Y, Natelson BH, DuBois AB. Cardiovascular effects of face immersion and factors affecting diving reflex in man. *J Appl Physiol* 1967; **23**:964–970.
35. Hestad DD, Abboud FM, Eckstein JW. Vasoconstrictor response to simulated diving in man. *J Appl Physiol* 1968; **25**:542–549.
36. Stevens PM. Cardiovascular dynamics during orthostasis and the influence of intravascular instrumentation. *Am J Cardiol* 1966; **17**:211–218.
37. Ewing DJ. Analysis of heart rate variability and other non-invasive tests with special reference to diabetes mellitus. In: Bannister R, Mathias CJ (Eds) Autonomic Failure. Oxford University Press, Oxford 1992, 312–333.
38. Locatelli A, Franzetti I, Lepore G, Maglio ML, Gaudio E, Caviezel F, Pozza G. Mental arithmetic stress as a test for evaluation of diabetic sympathetic autonomic neuropathy. *Diabetic Medicine* 1989; **6**:490–495.

CHAPTER FOUR

TILT TABLE TESTING

Steve W Parry and Rose Anne Kenny

Pathophysiological studies of tilt table aided orthostatic stress on humans were fifty years old before the evolution of head-upright tilt table testing into the diagnostic test of choice in vasovagal syncope. Variations in orthostatic tolerance demonstrated by tilt table testing were observed in normal subjects[1,2] and in association with pregnancy,[3] anxiety,[4] prolonged inactivity[5,6] and immersion.[7] Tilt table testing was also used in the assessment of physiologic responses to orthostatic stress,[8,9] weightlessness[10] and parachuting[11] by workers in aerospace and aviation medicine. Throughout these investigations, prolonged orthostasis was noted to provoke syncope in subgroups of those studied and subsequently the tilt table was used to investigate several components of the vasovagal response, notably cardiac output,[12,13] changes in forearm blood flow[14] and venous tone.[15] Several authors commented on the occurrence of syncope in varying proportions of the subjects studied,[1,3,8,11,15] while others commented on an increase in the frequency of syncope during tilting in those with a history suggestive of the vasovagal syndrome.[11,16] It was not until 1986, however that the intellectual leap from interesting clinical observation to diagnostic utility was made, with the publication of the Westminster group's landmark study on the usefulness of head-up tilt in the diagnosis of unexplained syncope.[17] In this series,[17] 67% of patients with otherwise unexplained syncope developed syncopal symptoms associated with hypotension and bradycardia during pro-

Malik M (Ed) Clinical guide to cardiac autonomic tests, 67–99.

longed head-up tilt. Only 10% of asymptomatic controls had similar reactions. The head-up tilt table test has since withstood extensive clinical and experimental scrutiny, and is currently the instrument of choice in the diagnosis of vasovagal syncope.[18–21]

PATHOPHYSIOLOGICAL RATIONALE FOR HEAD-UP TILT TABLE TESTING

THE VASOVAGAL RESPONSE

The study of the 'simple faint' began in the early years of this century, when Cotton and Lewis published their descriptions of the haemodynamic phenomena associated with simple fainting attacks.[22] While Gowers[23] was the first to use the term 'vasovagal' in association with such attacks, it was Lewis who later coined the phrase 'vasovagal syncope'[24] to describe the characteristic clinical features of enhanced parasympathetic and diminished sympathetic tone with resultant hypotension, bradycardia and syncope. Lewis also showed that the fall in heart rate was abolished by atropine but with little effect on hypotension,[24] introducing the concepts of distinct vasodepressor and cardioinhibitory components of the response. Although vasovagal syncope has since been induced and studied experimentally with the aid of a variety of agents, including venous occlusion,[25,26] isoproterenol,[27] head-up tilt,[17] lower body negative pressure[28] and haemorrhage,[25,26] the precise pathophysiological mechanism of the vasovagal response has yet to be elucidated. The multiple afferent pathways through which syncope is triggered have yet to be definitively identified,[30,31] though a complex interaction of neurocardiovascular and neurohumoral events mediated by the central and autonomic nervous systems is likely.[18,29–34] The following account will thus describe vasovagal syncope in terms of central and peripheral provocative stimuli, central nervous modulation of the syncopal response and the pathophysiology of the syncopal episode.[30,34]

CENTRAL PROVOCATION OF VASOVAGAL SYNCOPE

Higher cortical stimulation of vasovagal syncope through the experience of strong emotional stress,[35] severe visceral or somatic pain,[31] or noxious odours[30] is well documented. The exact mechanism by which emotional stimuli induce syncope is unknown, though animal experiments strongly implicate the limbic system.[30,31,34] Neurones originating in the amygdala are thought to relay stressor-related information to hypothalamic and brain-stem autonomic nuclei,[36] while stimulation of the limbic sympathoinhibitory area directly from higher centres or via somatic and visceral afferent signals causes bradycardia, loss of sympathetic tone and hypotension.[30,31,37,38] It has been postulated that such central provocation of vasovagal syncope in response to overwhelmingly noxious stimuli is the human equivalent of the "playing dead" reaction seen in some animal species.[37]

PERIPHERAL PROVOCATION OF VASOVAGAL SYNCOPE

Vasovagal syncope of non-central origin almost invariably (but not exclusively[39]) occurs in the upright position, during which 70% of human blood volume is below heart level.[40] On the assumption of the upright posture, 700 ml of blood is displaced downward by gravitational forces[31], the majority being stored in the capacitance vessels.[15] In the short term, compensatory mechanisms to maintain arterial blood pressure in the face of such relative central hypovolaemia depend on the manipulation of cardiac chronotropy and inotropy and peripheral vascular resistance by the arterial baroreceptor reflex arcs.[32,41] Rapidly acting humoral factors, in particular renin-angiotensin and epinephrine-norepinephrine,[41] also have an important role. Sympathetic efferent signals to systemic blood vessels appear to be of most import in maintaining normotension in response to orthostatic stress[32], as evidenced by the impairment of blood pressure recovery on standing following central sympathetic vasomotor blockade with clonidine[42] and the induction of orthostatic hypotension in human subjects following sympathetic ganglia blockade.[43]

Orthostatic stress with venous pooling and distorted responses to the resultant relative central hypovolaemia thus provide the substrate for the clinical and pathophysiological features of the vasovagal response in susceptible individuals, though the exact triggers remain obscure. For almost a century, the concept of the heart as the sensory organ responsible for the initiation of the neurocardiovascular cascade leading to syncope held sway.[44–47] Von Bezold and Hirt[48] noted bradycardia and hypotension in experimental animals following intra-cardiac injection of veratrum alkaloids, findings that were later confirmed and extended by Jarisch,[49] among others,[45,46] to show that mechanical stimulation of the left ventricle, particularly the infero-posterior wall, provoked a similar reaction. In the course of their work on hypovolaemia induced by haemorrhage in cats, Oberg and

Thoren noted that an increase in afferent discharge from unmyelinated vagal C fibres originating in the left ventricle coincided with a paradoxical bradycardia interrupting the physiological tachycardia.[50,51]. Moreover, denervation of the ventricle prevented the bradycardia,[51] leading them to the conclusion that acute hypotension provokes massive sympathetic activation resulting in vigorous ventricular contraction on a relatively empty ventricle, with subsequent mechanical deformation of the ventricular wall and inappropriate activation of mechanoreceptors provoking the vasovagal response.[34,51] Further work from the animal literature supports this contention; the vasodepressor response initiated by acute central hypovolaemia caused by experimental obstruction was found to be modulated by cardiac vagal afferents.[52]

While this line of reasoning appears compelling at first glance, it should be noted that vagal deafferentation in canine models of hypovolaemia has little effect on the induction of paradoxical sympathoinhibition. Extrapolation of the so-called left ventricular mechanoreceptor reflex[41] to humans is thus open to debate, although indirect evidence for this train of events is derived from several quarters. Sharpey-Schafer et al[47] demonstrated a reduction in cardiac output in subjects with syncope secondary to fear and nausea, while Weissler et al[12] showed similar results in subjects with 'vasodepressor syncope' induced by head-up tilt and nitrates. Echocardiographic studies during head-up-tilt induced syncope have demonstrated vigorous left ventricular contractions[53] and diminished left ventricular volume in patients with a history of vasovagal syncope but not in normal controls.[53–55] These observations are contradicted in a recent study showing that myocardial contractility did not change during tilt-induced syncope measured with a microaccelerometer located within the tip of a unipolar ventricular pacing lead.[56]

Further doubt has been cast on this model by the demonstration of vasovagal syncope in orthotopic heart transplant recipients.[57–59] While four of thirteen patients originally reported [57,59] showed evidence of vagal reinnervation, the rest did not, casting grave doubt on the uniqueness of the role of cardiac vagal afferent initiation of the vasovagal response. Nonetheless, mechanoreceptors are present in several other central sites[30,33] that would not necessarily suffer the same degree of denervation as the transplanted myocardial tissue. Others have suggested that sudden collapse of underfilled central venous vessels and atria may provoke syncope through mechanical stimulation of venous baroreceptors.[60] The frequent association of venipuncture with vasovagal syncope may result not only from centrally mediated emotional factors but also from afferent signals derived from as yet undetected peripheral venous baroreceptors.[32] Enhanced susceptibility to tilt-induced syncope following intravascular instrumentation noted by several authors[11,61] adds further intriguing circumstantial evidence to this postulate.

CENTRAL NERVOUS MODULATION OF THE SYNCOPAL EPISODE

Integration of afferent traffic from cardiovascular baro-, chemo- and mechanoreceptors, higher cortical centres, the hypothalamus and spinal cord is achieved predominantly within the nucleus tractus solitarius of the medulla.[62] This region is also within close proximity of the area postrema, which has a less restrictive blood brain barrier to facilitate modulation by circulating humoral factors.[30] Along with its multiple afferent pathways, the nucleus tractus solitarius has a number of vital efferent projections, including those to the hypothalamus, limbic system, reticular formation, and preganglionic sympathetic and parasympathetic nuclei.[41] Hypotension/bradycardia induced by stimulation of the nucleus tractus solitarius and its connections is thought to be mediated through a complex interaction of excitatory and inhibitory stimulation of these structures.[30,41,63]

PATHOPHYSIOLOGY OF THE SYNCOPAL EPISODE

Sympathetic and Parasympathetic Events

The syncopal episode represents the final common pathway for the triggering and modulating factors described above. Vagal excitation results in bradycardia and/or sinus arrest, and promotes an increase in gastrointestinal motility with subsequent nausea. Hypotension often antedates bradycardia, however, and as Lewis demonstrated, parasympatholysis with atropine does not forestall the vasodepressor response.[24] There is currently little doubt that the main contribution to hypotension during vasovagal syncope is derived from sympathetically mediated vasodilatation,[15,18,25,26,29,30–32,34,41,45,47] though debate continues as to whether this is active or passive.[32] Evidence for active vasodilatation is derived from Barcroft et al's classic studies.[25,26] During post-haemorrhagic syncope, forearm blood flow is greater in syncopal versus non-syncopal subjects,[26] while procaine block of the nerves innervating forearm muscles abolishes the vasodilator reaction.[25] Roddie[64] similarly suggested that cholinergic sympathetic vasodilator fibres were responsible for increased skeletal muscle blood flow during syncope as it was diminished by sympathectomy.[64] However, microneurographic recordings of sympathetic impulses to peroneal muscle during vasovagal syncope clearly demonstrated cessation of sympathetic traffic, which reappeared following the termination of an attack.[65]

Neurohumoral Events

Vasovagal syncope provokes measurable changes in the levels of a number of neurohumoral elements. The most striking changes occur in circulating catecholamines: epinephrine levels increase from 60–100% in syncopal subjects,[54,66–68] while norepinephrine levels remain relatively constant.[30,67] The epinephrine levels are high enough to produce β-adrenergic dilatation of resistance vessels in both splanchnic and skeletal vascular beds,[32] while the rise in concentration occurs prior to overt heart rate and blood pressure changes, raising the possibility that this event is part of the trigger mechanism for syncope.[30] The therapeutic success of β-blocker treatment may be in part due to antagonism of the effects of the epinephrine surge.[32] Plasma vasopressin is also increased during vasovagal syncope[69,70] but its usual pressor effect may be off-set by augmentation of baro- and mechanoreceptor inhibitory influences on cardiovascular regulation.[70] Plasma β-endorphins have been found to increase significantly prior to tilt-induced syncope[71] though attempts to ameliorate syncope with naloxone have met with variable success.[32]

Recent work has highlighted the importance of serotonin in the vasovagal response.[18] Intracerebroventricular serotonin induces hypotension, inhibits renal sympathetic nerve activity and enhances adrenal sympathetic activity.[72] Administration of the serotonin receptor blocker methysergide produces a marked pressor effect during hypovolaemia induced by haemorrhage.[73] Serotonin reuptake inhibitors have also proven therapeutically effective in small series of patients with vasovagal syncope.[123]

PHYSIOLOGICAL AND PATHOPHYSIOLOGICAL RESPONSES TO HEAD-UP TILT TABLE TESTING

Head-up tilt table testing provides a controlled, powerful orthostatic stimulus, which simulates the maximal venous pooling required for the provocation of the peripheral type of vasovagal syncope in susceptible individuals. In response to head-up tilt, normal subjects exhibit the venous pooling described above,[15,31] with a concomitant fall in central venous pressure, stroke volume and cardiac output.[74] Arterial normotension is maintained by vasoconstriction of skeletal muscle and splanchnic vascular beds.[75] This sequence of events is initially similar in patients with vasovagal syncope, but eventually the characteristic symptoms and neurocardiovascular features of the syncopal episode supervene as illustrated in Figure 1. Several lines of evidence suggest that such tilt-induced syncope is analogous to spontaneous vasovagal syncope. Firstly, the prodromal symptoms of light-headedness, nausea, pallor, etc are similar in both conditions. Secondly, the sequence of heart rate and blood pressure changes are virtually identical.[18,19] Finally, the neuroendocrine changes described above are similar in both spontaneous and tilt-induced syncope, in particular the epinephrine surge prior to syncope.[54,67] The first two of these factors, namely symptom reproduction and characteristic haemodynamic changes, are thus the cornerstone for the clinical diagnostic use of the head-up tilt table test.

INDICATIONS FOR HEAD-UP TILT TABLE TESTING

Head-up tilt table testing should be considered in patients with recurrent syncope, or in high risk patients (e.g. those suffering serious injury during syncope or those experiencing syncope while driving) suffering a single syncopal episode where no other cause for syncope is suggested by history, examination or cardiovascular and neurological investigations. Tilt table testing is also useful in the assessment of elderly patients with recurrent unexplained falls[18] and in the further evaluation of patients with an established cause of syncope who have symptoms suggestive of vasovagal syncope.[19] Further indications for tilt testing are discussed in depth later in this chapter.

CONTRAINDICATIONS TO HEAD-UP TILT TABLE TESTING

Head-up tilt table testing is remarkably safe with few reported adverse events.[19,109] Relative contraindications include clinically severe left ventricular outflow obstruction, critical mitral stenosis or proximal coronary artery stenoses, and known severe cerebrovascular stenosis. Contraindications to isoproterenol provoked tilt table testing are discussed in the appropriate section.

METHODOLOGICAL ISSUES

Any discussion on the utility of head-up tilt table testing is hampered by the variety of tilt table methodologies used in different laboratories, the often small numbers of subjects studied and the lack of a definitive diagnostic test for vasovagal syncope. This section will attempt to summarise the issues of sensitivity and specificity of the head-up tilt table test, its reproducibility and the effects of pharmacological and mechanical provocative agents.

DISCRIMINATORY POWER OF HEAD-UP TILT TABLE TESTING IN HEALTHY SUBJECTS VERSUS PATIENTS WITH UNEXPLAINED SYNCOPE: SENSITIVITY AND SPECIFICITY OF THE TILT TEST

The observation of syncope during tilt testing in subjects with no prior history of syncope[1,2,11,17] prompted attempts to clarify the diagnostic role of prolonged head-up tilt table testing in both symptomatic and asymptomatic populations. Exact estimates of the sensitivity and specificity of the head-up tilt table test are difficult to calculate because of the absence of a 'gold-standard' diagnostic tool for vasovagal syncope. Nonetheless, the ability of the test to discriminate between healthy subjects and symptomatic patients is well documented. Specificity, in the absence of pharmacological provocation and at tilt angles between 60°–70°, is generally up to 90%, whilst sensitivity ranges between 32% and 85%.[19,27,29,77] For example, Kenny et al's original study reported tilt-induced syncope in only 10% of control subjects but in 67% of those with unexplained syncope.[17] Schvartz et al reported syncope in just 3 (8%) of 36 healthy subjects undergoing head-up tilt,[76] while Fitzpatrick et al observed only 2 syncopal episodes (7%) among 27 normal volunteers and 5 (16%) of 34 patients with syncope secondary to conducting tissue disease.[77] De May and Enterling found only 8 syncopal episodes (20%) among 40 normal subjects,[78] while Raviele et al reported no syncopal episodes in 35 control subjects during studies on the effects of nitrates on the positivity of tilt testing.[79]Natale et al reported specificities of between 80% and 92% in their study of the effects of various angles of tilt and isoproterenol provocation on the results of tilt-table testing in 150 healthy

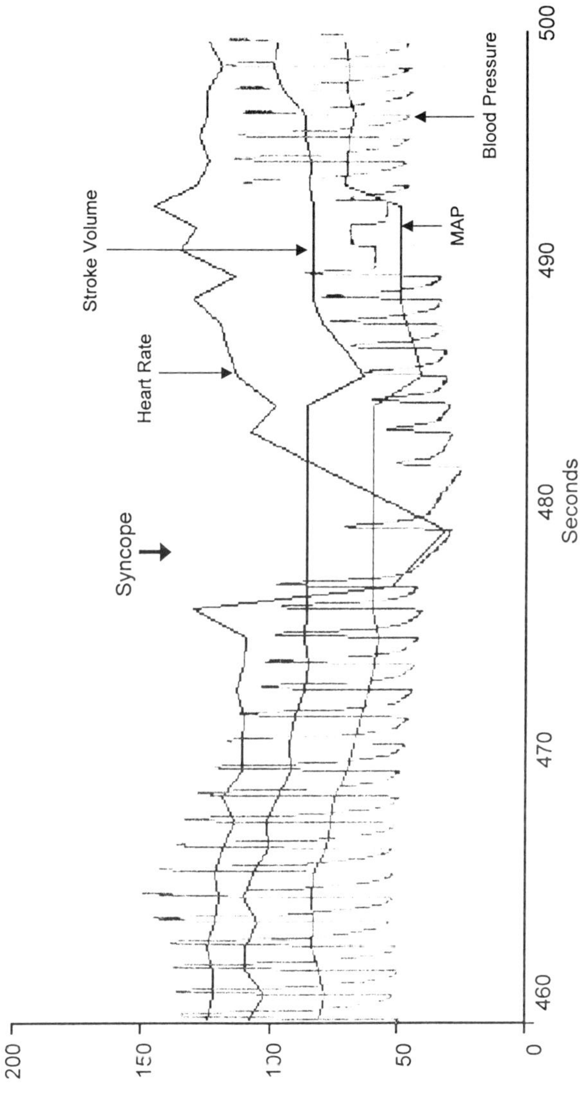

FIGURE 1. Cardiovascular Parameters during Positive Head-Up Tilt table Testing. Blood pressure, heart rate, mean arterial pressure (MAP) and stroke volume (SV) in a 36 year old man with vasovagal syncope during tilt testing. Flat portions of SV and MAP graphs represent calibration adjustments during asystole. *Data derived from Portapres monitoring system and reproduced with kind permission of TNO-TPD Biomedical Instrumentation, Amsterdam.*

volunteers,[80] while Kaufman showed that head-up tilt testing provoked syncope in 37% of a group of 201 subjects with unexplained syncope but in only 6% of 102 control subjects.[34] Grubb et al elicited syncope in none of the 6 healthy volunteers they subjected to tilt with and without isoproterenol provocation, but demonstrated syncope in 60% of previously symptomatic patients.[81]

Thus while head-up tilt table testing exhibits a high level of diagnostic specificity,[19,77–81] sensitivity is more difficult to assess. Sensitivity is influenced by several factors including age and the confounding effects of invasive blood pressure monitoring and intravenous cannulae.[11,61,103] Fouad et al[82] demonstrated sensitivity of 57% in their examination of 44 young patients with unexplained syncope, while others have reported sensitivities ranging from 35% to 85%,[19,29] with the median closer to the upper number. This figure compares well with standard cardiological diagnostic tests including the 12-lead ECG and exercise testing.[29] The effects of pharmacological provocation on sensitivity and specificity will be discussed later.

REPRODUCIBILITY

Adequate reproducibility of the head-up tilt table test is vital in determining both its diagnostic usefulness and its utility in the evaluation of therapy. Short and long term reproducibility studies have been attempted with mixed results. The short-term reproducibility studies performed to date have produced variable results. Fish et al[83] found that 67% of patients with an initially positive tilt test had a second positive test after a thirty minute rest period. Though superficially impressive, the reproduction of symptoms and degrees of cardioinhibition and vasodepression was inconsistent, highlighting the variability of the responses to tilt testing within individual subjects, and providing a note of caution in the use of the test in the evaluation of therapy.[83] De Mey and Enterling[84] and Morillo et al[85] had similar reservations. In contrast, Chen et al[86] found that 87% of 23 patients undergoing tilt testing had a similar symptomatic and haemodynamic response during a second tilt test thirty minutes after the original test. Interestingly, of the 8 patients who were initially tilt negative, none became tilt positive.[86] Longer-term studies over periods of one day[87] to four years[88] have reported reproducibility of between 62% and 85%,[87–90] although one study reported one day reproducibility as low as 35%.[91] A recent report from Folino et al[92] may explain at least some of the variability in reproducibility studies. When the historical and clinical characteristics of a large series of 745 patients with recurrent syncope were examined, the time interval between last spontaneous syncopal episode and the day of the tilt test negatively influenced the proportion of positive tests (>30days = 45.1% vs. <30 days = 77.2%; $p<0.001$).[92] Furthermore, markedly symptomatic patients (i.e. those with more than 3 syncopal episodes in the 2 months prior to testing) were more likely to have a positive tilt test (83.9% vs. 64.5%; $p<0.001$).[92] While this is a retrospective uncontrolled study, these observations are a clinical commonplace and would benefit from further investigation.

PHARMACOLOGICAL AND MECHANICAL PROVOCATIVE AGENTS

The use of provocative agents in an attempt both to shorten the duration and increase the sensitivity of the head-up tilt table test is now well established. While pharmacological provocateurs have been in use since 1989, mechanical provocation with lower body negative pressure has received recent attention.

Isoproterenol

The finding of dramatic elevation of circulating catecholamines immediately prior to syncopal episodes[54,66–68] led to the suggestion that the administration of exogenous catecholamine may precipitate vasovagal syncope during head-up tilt table testing in susceptible individuals.[93] Almquist et al reported on the use of graded isoproterenol infusion (to a maximum of 5μg per minute) during tilt testing in 24 syncopal patients and 18 normal controls.[27] 9 of 11 patients with negative electrophysiological studies and tilt tests had syncope with hypotension/bradycardia during the administration of isoproterenol, while only 2 of 18 control subjects experienced syncope under similar conditions.[27] These results have since been confirmed (though again with a variety of protocols) in many other centres,[94–98] with sensitivity generally being high with the addition of isoproterenol.[27,85,95,96,98] High-dose regimes maintained high sensitivity but at the expense of a fall in the specificity of the test.[95,99,100] Specificity appears to be maintained with low-dose isoproterenol infusion as illustrated by studies examining graded dosage regimes[95,98] and a recent report evaluating this approach in normal subjects.[101] It is interesting to note that exogenous epinephrine does not appear to have the same provocative effect as isoproterenol. A double blind crossover study of the efficacy of epinephrine and isoproterenol in conjunction with head-up tilt table testing failed to show any benefit with epinephrine.[94]

Isoproterenol tilt is remarkably safe with few reported adverse effects, although recent data from our group shows that isoproterenol is poorly tolerated in older subjects with unexplained syncope, possibly secondary to an age-related increase in β-receptor sensitivity. Contraindications to isoproterenol include coronary artery disease, uncontrolled hypertension, left ventricular outflow obstruction and significant aortic stenosis. Isoproterenol should be used with caution in known supraventricular and ventricular arrhythmias.

Nitroglycerine

Weissler et al used oral sodium nitrite with head-up tilt to induce and study syncope in normal volunteers as early as1957,[12] but it was not until the mid 1990's that first intravenous[102] then sublingual[79,103,104] glyceryl trinitrate was used diagnostically with head-up tilt in the evaluation of vasovagal syncope. The rationale for the use of nitrates in this setting is based on their venodilator properties. In vasovagal syncope sufferers the venodilation is thought to lead to small ventricular volumes, vigorous myocardial contractility and activation of the syncope cascade through ventricular mechanoreceptor activation.[103] In the studies published to date, sublingual glyceryl trinitrate was administered during a second tilt period following a negative prolonged passive head-up tilt; case detection was enhanced in all studies[79,102–104] with specificities ranging from 85%–94%, and sensitivities of 51%–81%.[79,103,104] This compares favourably with both passive tilt and isoproterenol tilt, and has the advantages of ease of administration and avoidance of intravascular instrumentation, a factor known to adversely affect the specificity of head-up tilt table testing.[11,61] Our own experience in over a thousand subjects shows this procedure to be well tolerated in all age groups, with no significant adverse events reported to date. Further studies are needed to compare the relative efficacies of isoproterenol and sublingual glyceryl trinitrate.

Edrophonium

Edrophonium is thought to potentiate the vagal efferent response during vasovagal syncope by inhibiting acetylcholine degradation in the heart and peripheral vasculature.[105] Two reports have commented on the utility of bolus doses of edrophonium to provoke syncope during head-up tilt table testing.[100,105] Early results are interesting, but as yet there appears to be little benefit over more established provocateurs.[100]

Lower Body Negative Pressure

Negative suction of the lower part of the body in combination with head-up tilt table testing (Figure 2) has long been used in physiological studies of the effects of microgravity on cardiopulmonary reflexes and responses to orthostatic stress.[106] The controlled fall in central venous pressure achievable with lower body negative pressure makes it an ideal candidate for peripheral stimulation of vasovagal syncope. Indeed, given sufficient lower body negative pressure during tilt, up to 85% of healthy individuals experience syncope via this method.[106] It also has the advantages of avoidance of intravascular instrumentation[11,61] and the potential adverse effects of medication. Hainsworth and El-Bedawi[107] examined the utility of 20mmHg lower body negative pressure in combination with head-up tilt in 79 patients with unexplained syncope and 40 normal controls. At 10 minutes, hypotension/bradycardia was induced in 85% of patients but in only 23% of asymptomatic controls.[107] A further study using supine lower body negative pressure reproduced syncope in 8 (73%) of 11 previously tilt-positive patients with neurocardiogenic syncope but in only 1 of 11 volunteers.[108] Lower body negative pressure thus offers a potentially promising alternative to pharmacological provocative agents, but further work is necessary before this practice is widely adopted to ensure that specificity is not sacrificed.

HEAD-UP TILT TABLE TEST METHODOLOGY

In order to maximise the usefulness of the head-up tilt test, careful attention must be paid to the conditions under which the test is carried out, as these critically influence outcome. Patient preparation, environment, equipment, angle and duration of the tilt test will be discussed, as will the indications and protocols for provocative agents and the criteria for test positivity. The methodology described here has arisen through a mixture of consensus and evidence – where the latter is available it will be discussed in detail.

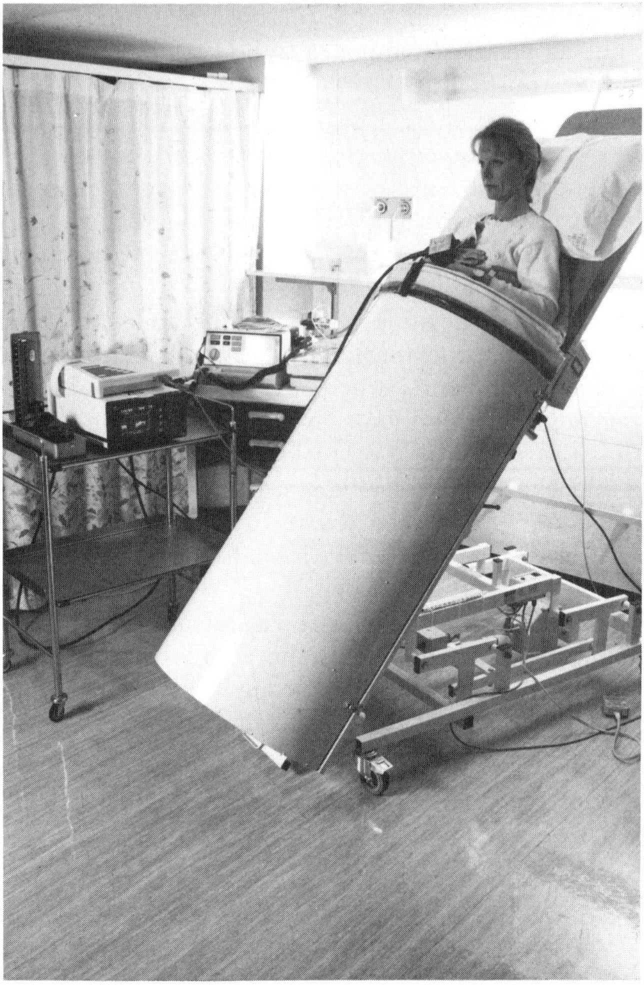

FIGURE 2. Lower body negative pressure application during head-up tilt table testing.

PATIENT PREPARATION

Patients, particularly those over 60 years of age, should be fasted for several hours prior to the procedure, in order to avoid the post-prandial hypotension and heart rate acceleration commonly seen during orthostatic stress.[19,109] The procedure should be explained to the patient who should rest supine for 20 to 45 minutes[19,61,77]. This period is particularly important if intra-vascular instrumentation is required.[11,61] Where practicable, cardioactive drugs and those likely to affect blood volume or the autonomic nervous system should be discontinued for at least five half-lives prior to the test[109] unless these are implicated as an attributable cause of syncope. If prolonged fasting has occurred, intravenous fluid supplementation may be necessary to avoid the confounding effects of relative dehydration and hypotension.[19]

LABORATORY ENVIRONMENT

The room should be quiet, with dim lighting and at a comfortable temperature. The aim is to minimise external stimuli and hence autonomic excitation during the test.[109]

TILT TABLE

The tilt table should be of the foot-plate support type, and should permit calibrated tilt angles of 60° to 90°, electrically or mechanically (Figure 3). The assumption of the upright position should occur smoothly, and take approximately 15 seconds.[19] Similarly, a mechanism should exist for rapid resumption of the supine position. The patient should be gently secured with appropriate strapping to prevent falling and injury during syncope.[109] The patient should be instructed to avoid movement of the lower extremity musculature and joints in order to maximise venous pooling.[19,109] The saddle type tilt table is now be of historical interest only because of the high incidence of false positive results (67% of normal controls).[77]

MONITORING EQUIPMENT

Electrocardiographic monitoring using at least three leads should occur at five minute intervals or continuously during any change from baseline heart rate and blood pressure. Blood pressure recording has been revolutionised by the development of non-invasive beat to beat finger photoplethysmographic monitoring[110] (Figure 4); this type of device should be used in preference to all others if locally available. Intra-arterial monitoring and sphygmomanometric measurement are still widely used, and if a non-continuous method is used, blood pressure should be recorded as frequently as is practicable, particularly if symptoms supervene.

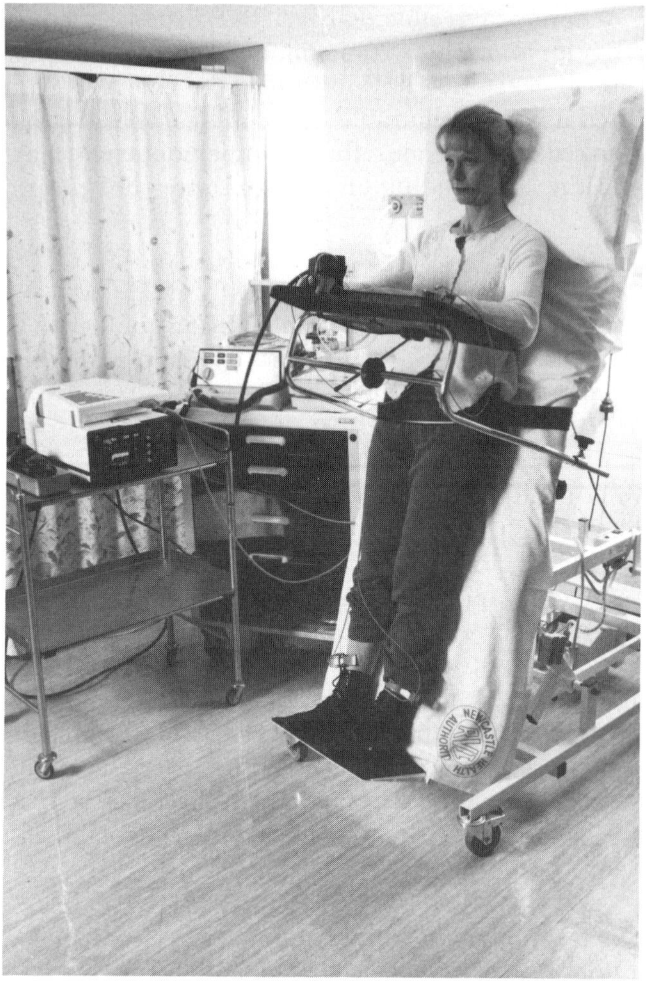

FIGURE 3. Akron footplate-assisted tilt table.

SUPERVISION OF THE HEAD-UP TILT TABLE TEST

Ideally, a qualified nurse, medical technician or physician experienced in the practicalities and interpretation of the test and the management of potential complications should be present at all times. Resuscitation equipment should be available, and be to the standard of that available in cardiovascular exercise testing laboratories.[19]

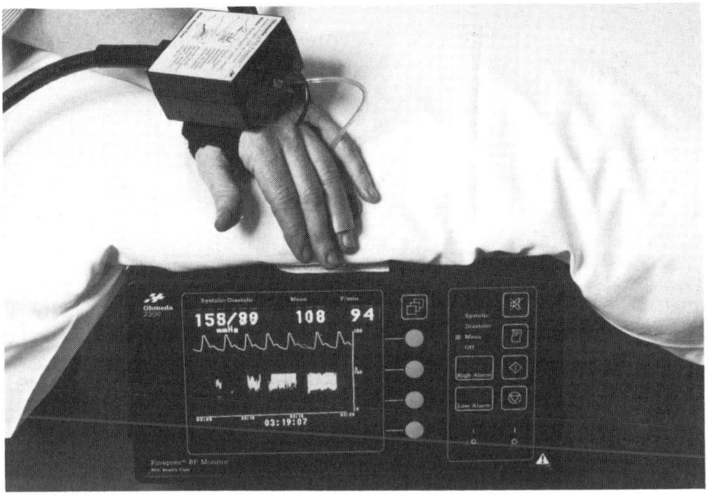

FIGURE 4. Digital photoplethysmographic blood pressure monitoring system (Finapres, Ohmeda, Englewood, NJ, USA).

Tilt Table Angle and Duration of the Head-Up Tilt Table Test

Tilt table angle and the duration of the tilt test are crucial determinants of its positivity, sensitivity and specificity.[19,99] Differences between centres in these variables make comparisons between studies virtually impossible. Nonetheless, several studies have attempted to define the appropriate angle and duration of tilt testing. The available evidence to date suggests that tilt angles of between 60° and 80° are optimal in provoking sufficient orthostatic stress without increasing the incidence of false positive results.[19,77,111] Fitzpatrick et al[77] found that 75% of patients with unexplained syncope had a positive result at 60° head-up tilt, while only 30% were positive at 45°. Tilt angles in excess of 80°considerably reduce the specificity of the test. For example, Davies et al[112] observed a 35% syncope incidence in normal volunteers using a tilt angle of 85°, giving an approximate specificity of 65%[111] against Fitzpatrick and Sutton's reported specificity of 90% using a 60° tilt angle.[113] Most authorities thus recommend tilt angles of between 60° and 80°.[19,77,111] Our own practice utilises a 70° angle for head-up tilt table testing.[61,103]

The ideal duration of head-up tilt table test is similarly contentious, with reported length of tilt varying from 10 to 60 minutes advocated by various centres.[17,18,27,61,77,80,97,98,102,103,111,113] There is little hard data available on the relative merits of different drug free, passive tilt test periods, though most recent reports favour prolonged passive head-up tilt (30–45 minutes).[61,19,77,100] Fitzpatrick et al noted a significant increase in the number of false positive results with shorter tilt periods.[77] The mean time to syncope in their patients was 24 minutes; [77] assuming a normal distribution plus two standard deviations, a duration of approaching 45 minutes would be optimal at present.[19,111] Where the initial drug free tilt test is non-diagnostic, pharmacological provocation or lower body negative pressure should subsequently be considered. Supportive evidence for the ideal protocol for isoproterenol and glyceryl trinitrate provocation is similarly scarce. Most centres employ a graded increase in isoproterenol dosage from 1 to 5 μg per minute with specified supine reequilibration periods between increments.[96–99] Ongoing studies from our group may lead to a decrease in the duration of head-up tilt table testing with the initial test using glyceryl trinitrate provocation in preference to drug free passive tilt. The protocols used in our laboratory for drug free passive head-up tilt table testing and isoproterenol and glyceryl trinitrate provoked tilt testing are described below.

POSITIVITY CRITERIA

The head-up tilt table test is judged to be positive if syncopal or pre-syncopal symptoms *reproducing the patients original symptoms* are accompanied by arterial hypotension, bradycardia (relative or otherwise) or both.[19,111] Heart rate and blood pressure changes in isolation should not prompt a diagnosis of vasovagal syncope.

CLASSIFICATION OF VASOVAGAL SYNCOPE

Prior to 1992, vasovagal syncope was arbitrarily defined as vasodepressor, cardioinhibitory or mixed, depending on the relative contributions of the component haemodynamic changes to the syncopal reaction. The observation of specific patterns of heart rate and blood pressure changes, along with the need for appropriate classification to aid further clinical and pathophysiological investigation and treatment evaluation prompted the development of a proposed classification for tilt induced vasovagal syncope.[114] The classification divides the positive responses to tilting into four subgroups as follows:

Type 1 – Mixed

Heart rate falls during syncope but the ventricular rate is never less than 40 beats per minute (bpm), or falls to less than 40 bpm for less than 10s with or without asystole of less than 3. Blood pressure falls prior to the fall in heart rate.

Type 2A – Cardioinhibitory

Heart rate falls at syncope to a ventricular rate less than 40 bpm for more than 10s or asystole occurs for more than 3s. Blood pressure again falls prior to the heart rate fall.

Type 2B – Cardioinhibitory

Heart rate falls at syncope to a ventricular rate less than 40 bpm for more than 10s or asystole occurs for more than 3s. Blood pressure falls to hypotensive levels (<80 mm Hg systolic) only at or after the onset of rapid and severe heart rate fall as previously defined.

Type 3 – Pure Vasodepressor

Heart rate does not fall more than 10% from its peak at the time of syncope. Blood pressure falls to precipitate syncope.

The main exceptions to this classification include chronotropic incompetence, an excessive heart rate rise (>130 bpm) during tilt, and where carotid sinus hypersensitivity is present.

THE NEWCASTLE PROTOCOLS FOR HEAD-UP TILT TABLE TESTING

Patient preparation, environment and monitoring should proceed as described previously. Termination of the test should occur immediately positivity criteria are achieved, if patient discomfort should supervene, or significant arrhythmia develops. Blood pressure is recorded at one-minute intervals, with ECG recordings made at baseline, every five minutes or continuously during symptoms or haemodynamic derangements. Intra-venous cannulation is avoided in all but the isoproterenol protocols to preclude its adverse effects on specificity.[11,61]

DRUG FREE PASSIVE HEAD-UP TILT

The patient is rested supine for 30 minutes prior to the test, then tilted upright for 40 minutes at an angle of 70°.

ISOPROTERENOL HEAD-UP TILT TEST

The patient remains supine for 30 minutes, and is then tilted to 70° for five minutes. The supine position is again assumed for five minutes reequilibration. Isoproterenol is then infused at a rate of 1 µg/min for five minutes supine, then five minutes at 70° tilt. The infusion is discontinued for two minutes during which the patient remains supine. Isoproterenol is then recommenced at 3 µg/min for five minutes supine and five minutes at 70°. Two minutes further supine rest follows, then further infusion of isoproterenol at 5 µg/min for five minutes supine and at 70°. The infusion should be discontinued if heart rate exceeds 150 bpm, blood pressure exceeds 180/100 or if arrhythmia, chest pain, severe tremulousness, vomiting or other intolerable side effects supervene.

GLYCERYL TRINITRATE HEAD-UP TILT TEST

The patient is rested supine as above, and 200–400 g sublingual glyceryl trinitrate administered. The patient remains supine for a further five minutes and is then tilted to 70° for 30 minutes.

HEAD-UP TILT TABLE TESTING IN PAEDIATRIC PATIENTS.

Children as young as 3 years of age have successfully undergone tilt table testing.[115] Indications, contraindications and methodology are essentially the same as in adults,[116–119] with the proviso that isoproterenol dosage should not exceed 0.03 µg/kg/min where it is indicated.[115] Sensitivities and specificities of head-up tilt testing in children are similar to those reported in adults for both drug-free passive tilt and isoproterenol provoked tilt.[119]

HEAD-UP TILT TABLE TESTING IN THE ELDERLY

Again, few studies have addressed the issue of tilt testing specifically in the elderly, but those studies reported again suggest that there should be no differences in tilting methodology or outcomes in this group.[61,120] The main difference in older subjects lies in the more variable presenting symptoms seen in this group. Elderly patients with vasovagal syncope frequently have sudden onset of syncope with little or no prodrome,[20] hence tilt test positivity criteria need to be viewed with this in mind.

HEAD-UP TILT TABLE TESTING IN THE EVALUATION OF THERAPEUTIC INTERVENTIONS FOR VASOVAGAL SYNCOPE

While the role of head-up tilt table testing in the diagnosis of vasovagal syncope is well established, its place in therapy choice and monitoring is less certain. Inconsistencies in the reproducibility of tilt test results make repeat tests during therapeutic intervention difficult to interpret. High rates of placebo sensitivity,[121] the low statistical power of many of the studies[121] and the lack of control groups[122,123,124,125] serve to cloud the issue further. Nonetheless, several randomised controlled studies of pharmacological interventions have demonstrated the utility of repeated tilt testing in the assessment of therapeutic efficacy.[126,127] Studies on the efficacy of pacing therapy in vasovagal syncope suffer from similar methodological difficulties, though one small study from the Westminster group used tilt table testing on three successive days to evaluate reproducibility and treatment efficacy.[128] 10 patients underwent tilt testing on three successive days – the first two to establish reproducibility of the positive response and the third to evaluate the response to temporary dual chamber pacing.[128] The results of the third tilt test dictated the prescription of permanent pacing, with the subjects assessed prospectively for symptom recurrence.[128] Head-up tilt table testing is currently being used in the European multicentre VASIS (Vasovagal Syncope International Study) trial of dual chamber pacing intervention in the management of vasovagal syncope, both in assessing the reproducibility of positive responses to head-up tilt and in monitoring the results of intervention.[129]

HEART RATE VARIABILITY DURING HEAD-UP TILT TABLE TESTING

The role of heart rate variability in the analysis of autonomic function is well established, but its relevance in the assessment and prediction of the physiologic and pathophysiologic responses to head-up tilt table testing is uncertain. Studies to date report conflicting results, and are methodologically hampered by the small numbers of subjects studied, inconsistencies in the tilt protocols and techniques of heart rate variability analysis used, differences in the duration of heart rate variability measurement and the variable use of control groups.[130–140]

Several studies have shown a fall in parasympathetic tone during tilt in both patients with vasovagal syncope[124] and normal healthy subjects.[131,132] In contrast, others have found no evidence of parasympathetic withdrawal in syncopal patients undergoing head-up tilt.[130,136,139] The issue is further confused by studies reporting an increase in parasympathetic tone during tilt in vasovagal patients.[134,140] Data on the variations in sympathetic tone during tilt are similarly contradictory, although the association of parasympathetic withdrawal with sympathetic excess demonstrated in one study[124] may explain the therapeutic success of anticholinergics noted in several reports.[87,122,124,125] The utility of heart rate variability in predicting responsiveness to head-up tilt table testing is similarly contentious. On the basis of current evidence, heart rate variability is not predictive in adults,[136,138,139] though a recent paper examining a paediatric population found that the low frequency power to high frequency power ratio provided a positive predictive value of 0.93, and a negative predictive value of 0.79 in syncopal children undergoing tilt table testing.[140] Heart rate variability reproducibility has been assessed by one group,[136,138] with high degrees of concordance reported in healthy subjects but extreme variability found in tilt positive patients with syncope.[136,138] Such variability was even more pronounced in those with different responses to two separate tilt table tests.[138]

Thus heart rate variability analysis has yet to provide definitive data on the autonomic cascade culminating in vasovagal syncope. Its utility as a research tool is evident, but its role in the diagnosis and management of the vasovagal syndrome is unclear. Newer spectral analysis techniques promising a more accurate method of studying the transition period from sinus rhythm to arrhythmia and syncope[137] may provide more insights in the future.

OTHER USES OF HEAD-UP TILT TABLE TESTING

POSTURAL ORTHOSTATIC TACHYCARDIA SYNDROME AND ORTHOSTATIC HYPOTENSION

The phenomenon of tilt test induced postural tachycardia and relatively mild hypotension has been recently described, and is thought to represent a mild form of autonomic dysfunction rather than a variant of vasovagal syncope.[141] The syndrome is diagnosed on the basis of a heart rate increase of >30 bpm (or a maximum heart rate of 120 bpm) in the absence of profound hypotension but reproducing symptoms.[141] Symptoms include light-headedness, fatigue, pre-syncope and dizziness.[141] Further clarification of this clinical entity is awaited with interest.

Orthostatic hypotension has traditionally been diagnosed on the basis of a fall in blood pressure during active standing. This definition was recently expanded by the American Autonomic Society and the American Academy of Neurology to include a fall in systolic blood pressure of >20 mm Hg or diastolic blood pressure of >10 mm Hg using a tilt table in the head-up position within 3 minutes and at an angle of 60°.[142]

PSYCHOGENIC AND HYPERVENTILATION SYNCOPE

Psychogenic syncope, or syncope in the absence of heart rate, blood pressure, electroencephalographic or transcranial Doppler abnormalities, has been reliably diagnosed using the head-up tilt table test.[143] Clinical features of the syncopal attack induced by tilt table testing include sudden and dramatic syncope, a prolonged recovery period and mild disorientation after the episode, all features rarely seen in vasovagal attacks.[143] Hyperventilation syncope is considered under the same rubric of psychiatric disorder,[144] with diagnosis established by the demonstration of hyocapnia (which is thought to stimulate cerebral vasoconstriction)and alkalosis during head-up tilt table testing.[144,145] Traditionally, arterial blood gas measurement has provided the gold standard for these values, but a recent study reported on the use of end-tidal PCO_2 as the marker of hyperventilation ('capnography tilt test').[144]

CAROTID SINUS SYNDROME

The diagnosis of carotid sinus syndrome rests on the finding of >3s asystole, >50 mm Hg fall in systolic blood pressure or both during carotid sinus massage.[146] In a sub-group of symptomatic individuals the response cannot be elicited in the supine position and is only present during massage in the head-upright tilt position.[16,146,147] Estimates of the prevalence of this facet of carotid sinus syndrome range from 5%–10%,[147] though recent data on elderly patients with recurrent unexplained falls and carotid sinus hypersensitivity investigated at our laboratory found that 40% had a positive response to carotid sinus massage in the head-upright tilt position only. Where history is suggestive of carotid sinus syndrome and massage is negative in the supine position a repeat in the upright position is recommended.

DIFFERENTIAL DIAGNOSIS OF CONVULSIVE SYNCOPE

Convulsive, tonic-clonic type movements during vasovagal syncope may result in the misdiagnosis of epilepsy, particularly where syncope is accompanied by prolonged asystole.[18] Head-up tilt testing, particularly with concurrent electroencephalography, may be helpful in differentiating such convulsive syncope from epilepsy,[18,148] with characteristically brief tonic-clonic activity accompanying the hallmark hypotension/bradycardia.

REFERENCES

1. Allen SC, Taylor CL, Hall VE. A study of orthostatic insufficiency by the tiltboard method. *Am J Physiol* 1945; **143**:11–19.
2. Hellebrandt FA, Franseen EB. Physiologic study of vertical stance in man. *Physiol Rev* 1943; **23**:220–225.
3. Brigden W, Howarth S, Sharpey-Shafer EP. Postural changes in peripheral blood flow of normal subjects with observations on vasovagal fainting reactions as a result of tilting and the lordotic posture, pregnancy, and spinal anaesthesia. *Clin Sci* 1950; **9**:79–92.
4. Stead EA, Warren JV, Merrill AJ, Brannon ES. The cardiac output in male subjects as measured by the technique of right atrial catheterization: Normal values and observations on the effect of anxiety and tilting. *J Clin Invest* 1945; **24**:326–334.
5. Lamb LE, Johnson RL, Stevens PM. Cardiovascular deconditioning during chair rest. *Aerospace Med* 1964; **35**:646–652.
6. Miller PB, Johnson RL, Lamb LE. The effects of four weeks of absolute bed rest on circulatory function of man. *Aerospace Med* 1964; **35**:1194–2012.
7. Graveline DE, Balke B, McKenzie RE, Hartman B. Psychobiologic effects of water immersion induced hypodynamics. *Aerospace Med* 1961; **32**:387–396.
8. Graybiel A, McFarland RA. The use of the tilt table test in aviation medicine. *J Aviation Med* 1941; **12**:194–203.

9. Gilbert CA, Stevens PM. Comparative effects of posture and lower body negative pressure on peripheral vascular reflexes. 36th Annual Convention, Aerospace Medical Association, New York, April 1965.
10. Staff of NASA Manned Spacecraft Center: Results of 4th Manned Orbital Flight, May 15–16, 1963. NASA SP–45. US Government Printing Office, Washington, DC.
11. Stevens PM. Cardiovascular dynamics during orthostasis and the influence of intravascular instrumentation. *Am J Cardiol* 1966; **17**:211–218.
12. Weissler AM, Warren JV, Estes EH, McIntosh HD, Leonard JJ. Vasodepressor syncope. Factors affecting cardiac output. *Circulation* 1957; **15**:875–882.
13. Warren JV, Brannon ES, Stead EA, Merrill AJ. Effect of venesection and the pooling of blood in the extremities on the atrial pressure and cardiac output in normal subjects, with observations on acute circulatory collapse in three instances. *J Clin Invest* 1945; **24**:337–346.
14. Brigden N, Howarth S, Sharpey-Shafer EP. Postural changes in peripheral blood flow in cases of left heart failure. *Clin Sci* 1950; **9**:93–106.
15. Epstein SE, Stampfer M, Beiser GD. Role of the capacitance and resistance vessels in vasovagal syncope. *Circulation* 1968; **37**:524–533.
16. Hammill SC, Holmes DR, Wood DL. Electrophysiologic testing in the upright position: improved evaluation of patients with rhythm disturbance using a tilt table. *J Am Coll Cardiol* 1984; **4**:65–71.
17. Kenny RA, Ingram A, Bayliss J, Sutton R. Head-up tilt: a useful test for investigating unexplained syncope. *Lancet* 1986; **1**:1352–1354.
18. Grubb BP, Samoil D. Neurocardiogenic syncope. In: Kenny RA (Ed) Syncope in the Older Patient: Causes, Investigations and Consequences of Syncope and Falls. Chapman & Hall, London 1996, 91–106.
19. Benditt DG, Ferguson DW, Grubb BP, Kapoor WN, Kugler J, Lerman BB, Maloney JD, Raviele A, Ross B, Sutton R, Wolk MJ, Wood DL. Tilt table testing for assessing syncope. ACC Expert Consensus Document. *J Am Coll Cardiol* 1996; **28**:263–275.
20. Sutton R. Vasovagal syncope: clinical features, epidemiology and natural history. In: Blanc JJ, Benditt D, Sutton R (Eds) Neurally Mediated Syncope: Pathophysiology, Investigations, and Treatment. Futura, Armonk 1996, 71–76.
21. Kapoor WN, Smith MA, Miller NL. Upright tilt testing in evaluating syncope: A comprehensive literature review. *Am J Med* 1994; **97**:78–88.
22. Hill MB. The influence of the force of gravity on the circulation of the blood. *J Physiol (Lond)* 1895; **18**:15–53.
23. Gowers WR. A lecture on vagal and vasovagal attacks. *Lancet* 1907; **173**:1551–1553.
24. Lewis T. Vasovagal syncope and the carotid sinus mechanism. With comments on Gower's and Nothnagels syndrome. *Brit Med J* 1932; **1**:873–876.
25. Barcroft H, Edholm OG. On the vasodilation in human skeletal muscle during post-haemorrhagic fainting. *J Physiol (Lond)* 1945; **104**:161–175.

26. Barcroft H, Edholm OG, McMichael J. Post-haemorrhagic fainting: study by cardiac output and forearm flow. *Lancet* 1944; **1**:489–490.
27. Almquist A, Goldenberg IF, Milstein S, Chen M-Y, Chen X, Hansen R, Gornick CC, Benditt DG. Provocation of bradycardia and hypotension by isoproterenol and upright posture in patients with unexplained syncope. *N Engl J Med* 1989; **320**:346–351.
28. El-Bedawi KM, Hainsworth R. Combined head-up tilt and lower body suction: a test of orthostatic tolerance. *Clin Auton Res* 1994; **4**:41–47.
29. Benditt DG. Neurally mediated syncopal syndromes: pathophysiological concepts and clinical evaluation. *PACE* 1997; **20**:572–584.
30. Benditt DG, Lurie KG, Adler SW, Sakaguchi S, Schultz J. Pathophysiology of vasovagal syncope. In: Blanc JJ, Benditt D, Sutton R (Eds) Neurally Mediated Syncope: Pathophysiology, Investigations, and Treatment. Futura, Armonk, 1996; 1–24.
31. Weiling W, Smit AAJ, De Jong-De Vos Van Steenwijk CCE, Van Lieshout JJ, Karemaker JM. Pathophysiological mechanisms underlying vasovagal syncope in young subjects. *PACE* 1997; **20**:2034–2038.
32. Van Lieshout JJ, Wieling W, Karemaker JM. Neural circulatory control in vasovagal syncope. *PACE* 1997; **20**:753–763.
33. Waxman MB, Cameron DA, Wald RW. Role of ventricular vagal afferents in the vasovagal reaction. *J Am Coll Cardiol* 1993; **21**:1138–1141.
34. Kaufman H. Neurally mediated syncope and syncope due to autonomic failure: differences and similarities. *J Clin Neurophysiol* 1997; **14**:183–196.
35. Greenfield ADM. An emotional faint. *Lancet* 1951; **1**:1302–1303.
36. Le Doux JE. Emotion and the amygdala. In: Aggleton JP (Ed) The Amygdala: Neurobiological Aspects of Emotion, Memory, and Mental Dysfunction. Wiley-Liss, New York 1992; 339–351.
37. Simone F. Vegetative nervous system and syncopes. *Functional Neurol* 1990; **5**:187–192.
38. Koizumi K, Kollai M. Control of reciprocal and non-reciprocal action of vagal and sympathetic efferents: study of centrally induced reactions. *J Auton Nerv Syst* 1981; **3**:483–501.
39. Verrill PJ, Aellig WH. Vasovagal faint in the supine position. *Br Med J* 1970; **IV**:348.
40. Rowell LB. Human Circulation. Regulation during Physical Stress. Oxford University Press, New York 1986.
41. Quan KJ, Carlson MD, Thames MD. Mechanisms of heart rate and arterial blood pressure control: implications for the pathophysiology of neurocardiogenic syncope. *PACE* 1997; **20**:764–774.
42. Coupland NJ, Bailey JE, Wilson SJ, Horvath R, Nutt D. The effects of clonidine on cardiovascular reponses to standing in healthy volunteers. *Clin Auton Res* 1995; **5**:171–177.
43. Brown GE, Wood EH, Lambert EH. Effects of tetra-ethyl-ammonium chloride on the cardiovascular reactions in man to changes in posture and exposure to centrifuged force. *J Appl Physiol* 1949; **2**:117–132.
44. Abboud FM. Ventricular syncope: is the heart a sensory organ? *N Engl J Med* 1989; **320**:390–392.

45. Mark AL. The Bezold-Jarisch reflex revisited: clinical implications of inhibitory reflexes originating in the heart. *J Am Coll Cardiol* 1983; **1**:90–102.
46. Coleridge HM, Coleridge JCG, Kidd C. Cardiac receptors in the dog, with reference to two types of afferent endings in the ventricular wall. *J Physiol (Lond)* 1964; **174**:323–329.
47. Sharpey-Schafer EP, Hayter CJ, Barlow ED. Mechanism of acute hypotension from fear or nausea. *Brit Med J* 1958; **2**:878–880.
48. Von Bezold A, Hirt L. Uber die physiologischen wirkungen des essigsannen veratrins. *Unters Physiol Lab Wurtzburg* 1867; **1**:73–156.
49. Jarisch A, Zottermann Y. Depressor reflexes from the heart. *Acta Physiol Scand* 1949; **16**:31–51.
50. Oberg B, Thoren P. Increased activity in vagal cardiac afferents correlated to the appearance of reflex bradycardia during severe haemorrhage in cats. *Acta Physiol Scand* 1970; **80**:22A–23A.
51. Oberg B, Thoren P. Increased activity in left ventricular receptors during haemorrhage or occlusion of caval veins in the cat. A possible cause of the vaso-vagal reaction. *Acta Physiol Scand* 1972; **85**:164–173.
52. Ludbrook J, Ventura S. Roles of carotid baroreceptor and cardiac afferents in haemodynamic responses to acute central hypovolaemia. *Am J Physiol* 1996; **270**:H1538–H1548.
53. Shalev Y, Gal R, Tchou PJ, Anderson AJ, Avitall B, Akhtar M, Jazayeri R. Echocardiographic demonstration of decreased left ventricular dimensions and vigorous myocardial contraction during syncope induced by head-up tilt. *J Am Coll Cardiol* 1991; **18**:746–751.
54. Fitzpatrick A, Williams T, Ahmed R, Lightman S, Bloom SR, Sutton R. Echocardiographic and endocrine changes during vasovagal syncope induced by prolonged head-up tilt. *Eur J Card Pacing Electrophys* 1992; **2**:121–128.
55. Yamanouchi Y, Jaalouk S, Shehadeh AA, Jaeger F, Goren H, Fouad-Tarazi FM. Changes in left ventricular volume during head-up tilt in patients with vasovagal syncope: an echocardiographic study. *Am Heart J* 1996; **131**:73–80.
56. Brignole M, Menozzi C, Corbucci G, Garberoglio, Plicchi D. Detecting incipient vasovagal syncope: intraventricular acceleration. *PACE* 1997; **20**:801–805.
57. Fitzpatrick AP, Banner N, Cheng A, Yacoub M, Sutton R. Vasovagal reactions may occur after orthotopic heart transplantation. *J Am Coll Cardiol* 1993; **21**:1132–1137.
58. Scherrer U, Vissing S, Morgan BJ, Hanson P, Victor RG. Vasovagal syncope after infusion of a vasodilator in a heart transplant recipient. *N Engl J Med* 1990; **322**:602–604.
59. Rudas L, Pflugfelder PW, Kostuk WJ. Vasodepressor syncope in a cardiac transplant recipient. A case of vagal re-innervation. *Can J Cardiol* 1992; **8**:403–405.
60. Dickinson CJ. Fainting precipitated by collapse-firing of venous baroreceptors. *Lancet* 1995; **342**:970–972.

61. McIntosh SJ, Lawson J, Kenny RA. Intravenous cannulation alters the specificity of head-up tilt testing for vasovagal syncope in elderly patients. *Age Ageing* 1994; **23**:317–319
62. Seller H. Central baroreceptor reflex pathways. In: Persson PB, Kircheim HR (Eds) Baroreceptor Reflexes. Springer, Berlin 1991, 45–74.
63. Ross RT. Syncope. Saunders, London 1988.
64. Roddie IC. The role of vasoconstrictor and vasodilator nerves to skin and muscle in the regulation of the human circulation. *Ann Roy Coll Surg Engl* 1963; **32**:180–193.
65. Wallin BG, Sundlof G. Sympathetic outflow to muscles during vasovagal syncope. *J Auton Nerv Syst* 1982; **6**:287–291.
66. Vingerhoets AJJM. Biochemical changes in two subjects succumbing to syncope. *Psychosom Med* 1984; **46**:95–103.
67. Sander-Jensen K, Secher NH, Astrup A, Christensen NJ, Diese J, Schwartz TW, Warberg J,Bie P. Hypotension induced by passive head-up tilt: endocrine and circulatory mechanisms. *Am J Physiol* 1986; **251**:R742–R748.
68. Sra JS, Murthy V, Natale A, Jazayeri MR, Dhala A, Deshpaned S, Sheth M, Akhtar M. Circulatory and catecholamine changes during head-up tilt testing in neurocardiogenic (vasovagal) syncope. *Am J Cardiol* 1994; **73**:33–37.
69. Robinson BJ, Johnson RH. Why does vasodilatation occur during syncope? *Clin Sci* 1988; **74**:347–350.
70. Jardine DL, Melton IC, Crozier IG, Bennett SI, Donald RA, Ikram H. Neurohormonal response to head-up tilt and its role in vasovagal syncope. *Am J Cardiol* 1997; **79**:1302–1306.
71. Wallbridge DR, MacIntyre HE, Gray CE, Denvir MA, Oldroyd KG, Rae AP, Cobbe SM. Increase in plasma beta endorphins precedes vasodepressor syncope. *Br Heart J* 1994; **71**:597–599.
72. Abboud FM. Neurocardiogenic syncope. *N Engl J Med* 1993; **328**:1117–1120.
73. Elam RF, Bergman F, Feurstein G. The use of antiserotonergic agents for treatment of acute haemorrhagic shock in cats. *Eur J Pharmacol* 1985; **107**:275–278.
74. Hainsworth R, al Shamma YMH. Cardiovascular responses to upright tilting in healthy subjects. *Clin Sci* 1988; **74**:17–22.
75. Sneddon JF, Counihan PJ, Bashir Y, Haywood GA, Ward DE, Camm AJ. Impaired immediate vasoconstrictor responses in patients with recurrent neurally mediated syncope. *Am J Cardiol* 1993; **71**:72–76.
76. Schvartz E. Reliability of quantifiable tilt table data. *Aerospace Med* 1968; **39**:221–226.
77. Fitzpatrick AP, Theodorakis G, Vardas P, Sutton R. Methodology of head upright tilt table testing in patients with unexplained syncope. *J Am Coll Cardiol* 1991; **17**:125–130.
78. De May C, Enterling D. Assessment of the hemodynamic responses to single passive head-up tilt by non-invasive methods in normotensive subjects. *Methods Find Exp Clin Pharmacol* 1986; **8**:449–457.

79. Raviele A, Menozzi C, Brignole M, Gasparini G, Alboni P, Musso G, Lolli G, Oddone D, Dinelli M, Mureddu R. Value of head-up tilt testing potentiated with sublingual nitroglycerin to asses the origin of unexplained syncope. *Am J Cardiol* 1995; **76**:267–272.
80. Natale A, Akhtar M, Jazayeri M, Dhala A, Blanck Z, Deshpande S, Krebs A, Sra JS. Provocation of hypotension during head-up tilt testing in subjects with no history of syncope or presyncope. *Circulation* 1995; **92**:54–58.
81. Grubb BP, Temesy-Armos P, Hahn H, Elliott L. Utility of upright tilt-table testing in the evaluation and management of syncope of unknown origin. *Am J Med* 1991; **90**:6-10.
82. Fouad FM, Sitthisook S, Vanerio G, Maloney J, Okabe M, Jaeger F, Schluchter M, Maloney JD. Sensitivity and specificity of the tilt table test in young patients with unexplained syncope. *PACE* 1993; **16**:394–400.
83. Fish FA, Strasburger JF, Woodrow Benson D. Reproducibility of a symptomatic response to upright tilt in young patients with unexplained syncope. *Am J Cardiol* 1992; **70**:605–609.
84. De May C, Enterling D. Variant responses impair the usefulness of passive upright tilt in drug research. *Methods Find Exp Clin Pharmacol* 1988; **10**:57–64.
85. Morillo CA, Klein GJ, Zandri S, Yee R. Diagnostic accuracy of a low dose isoproterenol head-up tilt protocol. *Am Heart J* 1995; **129**:901–906.
86. Chen XC, Chen MY, Remole S, Kobayashi Y, Dunnigan A, Milstein S, Benditt D. Reproducibility of head-up tilt-table testing for eliciting susceptibility to neurally mediated syncope in patients without structural heart disease. *Am J Cardiol* 1992; **69**:755–760.
87. Raviele A, Gasparini G, DiPede F, Delise P, Bonso A, Piccolo E. Usefulness of head-up tilt test in evaluating patients with syncope of unknown origin and negative electrophysiologic study. *Am J Cardiol* 1990; **65**:1322–1327.
88. Petersen MEV, Price D, Williams T, Jensen N, Riff K, Sutton R. Short AV interval VDD pacing does not prevent tilt induced vasovagal syncope in patients with cardioinhibitory vasovagal syndrome. *PACE* 1994; **17**:882–891.
89. Blanc JJ, Mansourati J, Maheu B, Boughaleb D, Genet L. Reproducibility of a positive passive upright tilt test at a seven-day interval in patients with syncope. *Am J Cardiol* 1993; 469–471.
90. Sheldon R, Splawinski J, Killam S. Reproducibility of upright tilt-table tests in patients with syncope. *Am J Cardiol* 1992; **69**:1300–1305.
91. Brooks R, Ruskin JN, Powell AC, Newell J, Garan H, McGovern BA. Prospective evaluation of day-to-day reproducibility of upright tilt-table testing in unexplained syncope. *Am J Cardiol* 1993; **72**:1289–1292.
92. Folino AF, Buja G, Martini B, Bassan L, Nava A. Upright tilt test: correlation between results and patient clinical features. *PACE* 1996; **19**:1582–1587.
93. Yao L, Cameron D, Roseman J, Waxman MB. Isoproterenol-induced vasodepressor reaction in man. *Clin Invest Med* 1984; **7**:40 (abstract).
94. Calkins H, Kadish A, Sousa J, Rosenheck S, Morandy F. Comparison of responses to isoproterenol and epinephrine during head-up tilt. *Am J Cardiol* 1991; **67**:207–209.

95. Carlioz R, Graux P, Haye J, Letourneau T, Guyomar Y, Hubert E, Bodart JC, Lequeuche B, Burlaton J-P. Prospective evaluation of high dose and low dose isoproterenol upright tilt protocol for unexplained syncope in young adults. *Am Heart J* 1997; **133**:346–352.
96. Waxman MB, Yao L, Cameron D, Wald RW, Roseman J. Isoproterenol induction of vasodepressor-type reaction in vasodepressor-prone persons. *Am J Cardiol* 1989; **63**:58–65.
97. Sheldon R, Killam S. Methodology of isoproterenol-tilt table testing in patients with syncope. *J Am Coll Cardiol* 1992; **19**:773–779.
98. Blanc J, Victor J, Mansourati J, Le Davay M, Dupuis J-M, Maheu B. Accuracy and mean duration of different protocols of head-up tilt testing. *Am J Cardiol* 1996; **77**:310–313.
99. Kapoor WN, Brant N. Evaluation of syncope by upright tilt testing with isoproterenol. A non-specific test. *Ann Intern Med* 1992; **116**:358–363.
100. Fitzpatrick AP, Lee RJ, Epstein LM, Lesh MD, Eisenberg S, Scheinman MM. Effect of patient characteristics on the yield of prolonged baseline head-up tilt testing and the additional yield of drug provocation. *Heart* 1996; **76**:406–411.
101. Grubb BP, Kosinski D, Temesy-Armos P, Brewster P. Responses of normal subjects during 80 degree head upright tilt table testing with and without low dose isoproterenol infusion. *PACE* 1997; **20**: 2019–2023.
102. Raviele A, Gasparini G, Di Pede F, Menozzi C, Brignole M, Dinelli M, Alboni P, Piccolo E. Nitroglycerine infusion during upright tilt: a new test for the diagnosis of vasovagal syncope. *Am Heart J* 1994; **127**:103–111.
103. McIntosh S, Lawson J, da Costa D, Kenny RA. Use of sublingual glyceryl trinitrate during head-up tilt as a provocative test for vasovagal responses in elderly patients with unexplained syncope. *Cardiol in the Elderly* 1996; **4**:33–37.
104. Aerts A, Dendale P, Strobel G, Block P. Sublingual nitrates during head-up tilt testing for the diagnosis of vasovagal syncope. *Am Heart J* 1997; **133**:504–507.
105. Lurie KG, Dutton J, Mangat R, Newman D, Eisenberg S, Scheinman MM. Evaluation of edrophonium as a provocative agent for vasovagal syncope during head-up tilt table testing. *Am J Cardiol* 1993; **72**:1286–1290.
106. Kaufman H. Head-up tilt, lower body negative pressure, pacemakers and vasovagal syncope. *Clin Auton Res* 1994; **4**:231–232.
107. Hainsworth R, El-Bedawi KM. Orthostatic tolerance in patients with unexplained syncope. *Clin Auton Res* 1994; **4**:239–244.
108. Ruiz GA, Madoery C, Socas AG, Scaglione J, Tentori C. Tolerance to lower body negative pressure exposure: comparison between patients with neurocardiogenic syncope and controls. *PACE* 1997; **20**:706–713.
109. Victor J. Tilt Test: environment, material, patient preparation. In: Blanc JJ, Benditt D, Sutton R (Eds) Neurally Mediated Syncope: Pathophysiology, Investigations, and Treatment. Futura, Armonk 1996; 77–78.
110. Jellema WT, Imholz BPM, Van Goudoever J,Wesseling KH, Van Lieshout J. Finger arterial versus intrabrachial pressure and continuous cardiac output during head-up tilt testing in healthy subjects. *Clin Sci* 1996; **91**:193–200.

111. Mansourati J, Blanc JJ. Tilt test procedure: angle, duration and positivity criteria. In: Blanc JJ, Benditt D, Sutton R (Eds) Neurally Mediated Syncope: Pathophysiology, Investigations, and Treatment. Futura, Armonk 1996, 79–83.
112. Davies R, Slater JDH, Forsling ML. The response of arginine vasopressin and plasma renin to postural change in normal man, with observations on syncope. *Clin Sci* 1976; **51**:267–274.
113. Fitzpatrick A, Sutton R. Tilting towards a diagnosis in recurrent unexplained syncope. *Lancet* 1989; **1**:658–770.
114. Sutton R, Petersen M, Brignole M, Raviele A, Menozzi C, Giani P. Proposed classification for tilt induced vasovagal syncope. *Eur J Card Pacing Electrophys* 1992; **3**:180–183.
115. Grubb BP, Orecchio E, Kurczynski TW. Head-upright tilt table testing in evaluation of recurrent unexplained syncope. *Paediatr Neurol* 1992; **8**:423–427.
116. Thilenius OG, Quinones JA, Huusayni TS, Novak J. Tilt test for diagnosis of unexplained syncope in paediatric patients. *Pediatrics* 1991; **87**:334–338.
117. Lerman-Sagie T, Rechavia E, Strasberg B, Sagie A, Blieden L, Mimouni M. Head-up tilt for the evaluation of syncope of unknown origin in children. *J Pediatr* 1991; **118**:676–679.
118. Strieper JM, Auld DO, Hulse E, Campbell RM. Evaluation of recurrent pediatric syncope: role of tilt table testing. *Pediatrics* 1994; **93**:660–662.
119. Alehan D, Lenk M, Ozme S, Celiker A, Ozer S. Comparison of sensitivity and specificity of tilt protocols with and without isoproterenol in children with unexplained syncope. *PACE* 1997; **20**:1769–1776.
120. Grubb BP, Wolfe D, Samoil D, Madu E, Temesy-Armos P, Hahn H, Elliott L. Recurrent unexplained syncope in the elderly: the use of head-upright tilt table testing in evaluation and management. *J Am Ger Soc* 1992; **40**:1123–1128.
121. Moya A, Permanyer-Miralda G, Sagrista-Sauleda J, Carne X, Rius T, Mont L, Soler-Soler J. Limitations of head-up tilt test for evaluating the efficacy of therapeutic interventions in patients with vasovagal syncope: results of a controlled study of etilefrine versus placebo. *J Am Coll Cardiol* 1995; **25**:65–69.
122. Grubb BP, Temesy-Armos P, Moore J, Wolfe D, Hahn H, Elliot L. Head-upright tilt table test in evaluation and management of the malignant vasovagal syndrome. *Am J Cardiol* 1992; **69**:904–908.
123. Dan D, Grubb BP, Mouhaffel A, Kosinski DJ, Elliott L, Brewster PS. Preliminary observations on the use of nefazodone, a new selective 5-HT receptor blocker in the treatment of neurally mediated syncope. *Eur J Card Pacing Electrophys* 1996; **6**:89–94.
124. Nelson SD, Stanley M, Love CJ, Coyne KS, Schaal SF. The autonomic and hemodynamic effects of oral theophylline in patients with vasodepressor syncope. *Arch Intern Med* 1991; **151**:2425–2429.
125. Milstein S, Buetikofer J, Dunnigan A, Benditt DG, Gornick C, Reyes WJ. Usefulness of disopyramide for prevention of upright tilt induced hypotension-bradycardia. *Am J Cardiol* 1990; **65**:1339–1344.

126. Mahanonda N, Bhuripanyo, Kangkagate C, Wansanit K, Kulchot B, Nademanee K, Chaithiraphan S. Randomized double-blind placebo controlled trial of oral atenolol in patients with unexplained syncope and positive upright tilt table test results. *Am Heart J* 1995; **130**:1250–1253.
127. Biffi M, Boriani G, Sabbatani P, Bronzetti G, Frabetti L, Zannoli R, Branzi A, Magnani B. Malignant vasovagal syncope: a randomised trial of metoprolol and clonidine. *Heart* 1997; **77**:268–272.
128. Fitzpatrick A, Theodorakis G, Ahmed R, Williams T, Sutton R. Dual chamber pacing aborts vasovagal syncope induced by head-up 60 degree tilt. *PACE* 1991; **14**:13–19.
129. Sutton R. Personal communication.
130. Lippman N, Stein KM, Lerman BB. Failure to decrease parasympathetic tone during upright tilt predicts a positive tilt-table test. *Am J Cardiol* 1995; **75**:591–595.
131. Pagani M, Lombardi F, Guzzetti S, Rimoldi O, Furlan R, Pizzinelli P, Sandorne G, Malfatto G, Dell'OrtoS, Piccaluga E, Turiel M, Baselli G, Cerutti S, Malliani A. Power spectral analysis of heart rate and arterial pressure variabilities as a marker of sympatho-vagal balance in man and conscious dog. *Circ Res* 1986; **59**:178–193.
132. Vybrial T, Bryg RJ, Maddens ME, Boden WE. Effect of passive tilt on sympathetic and parasympathetic components of heart rate variability in normal subjects. *Am J Cardiol* 1989; **63**:1117–1120.
133. Lipsitz LA, Mietus J, Moody GB, Goldberger AL. Spectral characteristics of heart rate variability before and during postural tilt. Relations to aging and risk of syncope. *Circulation* 1990; **81**:1803–1810.
134. Pruvot E, Vesin JM, Schlaepfer J, Fromer M, Kappenberger L. Autonomic imbalance assessed by heart rate variability analysis in vasovagal syncope. *PACE* 1994; **17**:2201–2206.
135. Sneddon JF, Bashir Y, Mugatroyd FD, Ward DE, Camm AJ, Malik M. Do patients with neurally mediated syncope have augmented vagal tone? *Am J Cardiol* 1993; **72**:1314–1315.
136. Vardas P, Dochiadakis G, Orfanakis A, Kalaitzakis M, Manios E. Intraindividual reproducibility of heart rate variability before and during postural tilt in patients with syncope of unknown origin. *PACE* 1994; **17**:2207–2210.
137. Jasson S, Medigue C, Maison-Blanche P, Montano N, Meyer L, Vermeiren C, Mansier P, Coumel P, Malliani A, Swynghedauw B. Instant power spectrum analysis of heart rate variability during orthostatic tilt using a time-frequency-domain method. *Circulation* 1997; **96**:3521–3526.
138. Kochiadakis GE, Orfanakis AE, Rombola AT, Chrysostomakis SI, Chlouverakis GI, Vardas PE. Reproducibility of time-domain indexes of heart rate variability in patients with vasovagal syncope. *Am J Cardiol* 1997; **79**:160–165.
139. Kochiadakis GE, Orfanakis AE, Chryssostomakis SI, Manios EG, Kounali DK, Vardas PE. Autonomic nervous system activity during tilt testing in syncopal patients estimated by power spectral analysis of heart rate variability. *PACE* 1997; **20**:1332–1341.

140. Stewart JM, Erb M, Sorbera C. Heart rate variability and the outcome of head-up tilt in syncopal children. *Pediatr Res* 1996; **40**:702–709.
141. Grubb BP, Kosinski DJ, Boehm K, Kip K. The postural orthostatic tachycardia syndrome. A neurocardiogenic variant identified during head-up tilt table testing. *PACE* 1997; **20**:2205–2212.
142. Consensus Committee of the American Autonomic Society and the American Academy of Neurology. Consensus statement on the definition of orthostatic hypotension, pure autonomic failure, and multiple system atrophy. *J Neurol Sci* 1996; **144**:218–219.
143. Petersen MEV, Williams TR, Sutton R. Psychogenic syncope diagnosed by prolonged head-up tilt testing. *Q J Med* 1995; **88**:209–213.
144. Naschitz JE, Gaitini L, Mazov L, Eridzhanyan D, Keren E, Sabo D, Yeshurun D, Hardoff D, Jaffe M. The capnography-tilt test for the diagnosis of hyperventilation syncope. *Q J Med* 1997; **90**:139–145.
145. Roberston DA, Taylor R. Metabolic and endocrine causes of syncope. In: Kenny RA (Ed) Syncope in the Older Patient: Causes, Investigations and Consequences of Syncope and Falls. Chapman & Hall, London 1996, 240–265.
146. Kenny RA, McIntosh SJ. Carotid sinus syndrome. In: Kenny RA (Ed) Syncope in the Older Patient: Causes, Investigations and Consequences of Syncope and Falls. Chapman & Hall, London 1996, 107–122
147. McIntosh SJ, Lawson J, Kenny RA. Clinical characteristics of vasodepressor, cardioinhibitory and mixed carotid sinus syndrome in the elderly. *Am J Med* 1993; **95**:203–208.
148. Grubb BP, Ferard G, Roush K, Temesy–Armos P, Elliott L, Hahn H, Spann C. Differentiation of convulsive syncope and epilepsy with head-up tilt table testing. *Ann Intern Med* 1991; **115**:871–876.

Chapter Five

AUTONOMIC RESPONSES TO BLOCKADES AND PROVOCATIONS

Laurence Keselbrener and Solange Akselrod

Specific autonomic blockades and provocations have been widely used to identify dysfunction in autonomic control. The strength of these tests has lately been enhanced by the availability of novel quantitative approaches to determine the autonomic response, even under transient conditions. The response to autonomic manoeuvres was initially analysed in the time domain, by measures on RR intervals or blood pressure (BP). At a later stage, autonomic function was also analysed in the frequency domain.[1,2,3] However, the stationarity assumption of standard spectral analysis, (based on Fourier transform [FT] or auto-regressive [AR] modelling), proscribes this approach for the evaluation of a brief or transient response to autonomic blockades or provocations. Lately, methods of time-frequency analysis, have been developed in order to overcome this stationarity limitation. Their application to cardiovascular signals has allowed obtaining information on the system, even during strongly time-dependent phases of the intervention.

Indeed, the frequency content has been shown, in steady state conditions, to reflect various branches of autonomic function. If the autonomic activity is sudden, short-lasting or transient, as often happens in blockades and provocations procedures, then one has to be sure to detect (to time localise) exactly the autonomic response. Measures in the time domain provide information on the time localisation of the response (as illustrated in the instantaneous HR during

Malik M (Ed) Clinical guide to cardiac autonomic tests, 101–148.

Oculocardiac Reflex test in Figure 1a). However the time domain does not allow distinguishing quantitatively enough between the various branches of autonomic function. In contrast, steady state spectral tools, which achieve this discrimination based on frequency specificity, will smear the information over the entire time window considered. The user has then to be very careful and apply the appropriate tools to quantitate such autonomic response, as can be achieved by the time-frequency representation of the fluctuations of the cardiovascular signals.[4] In our instance, the relevant signals are generally heart rate (HR), blood pressure (BP), cardiac output (CO), blood flow (BF) or respiratory signals, and our goal is thus to link their response with the specific autonomic branches involved.

When focusing on blockades and provocations, the advantage of a reliable quantitative method for time-dependent frequency decomposition is its ability to specify, to quantitate and to localise as accurately as possible in time, the changes which occur in the frequency control of a signal, as illustrated in the time-frequency spectrum of HR during diving response test, shown in Figure 1b as a 3D graph, and in Figure 1c as a contours plot.

First, we will review the approaches of time-domain estimates and time-frequency decompositions, which have been applied to cardiovascular signals. Next we will discuss general aspects related to the procedures of the autonomic tests and the evaluation of the elicited response: basic requirements, general recommendations, possible confounding factors and guides for interpretation. Then, providing several representative examples, we will review the most frequently applied autonomic tests, describe the required arrangement, their autonomic effects and interpretation.

METHODS FOR QUANTITATIVE ESTIMATES OF AUTONOMIC RESPONSE

TIME-DOMAIN ESTIMATES

The transient nature of the signals during provocations and blockades have for years restrict the analysis of the HR and BP responses to the time domain, mainly due to the stationarity limitation set by standard spectral analysis. Before dealing with advanced methods for time-frequency analysis that allow to overcome this limitation and provide spectral decomposition of non-stationary signals, we briefly

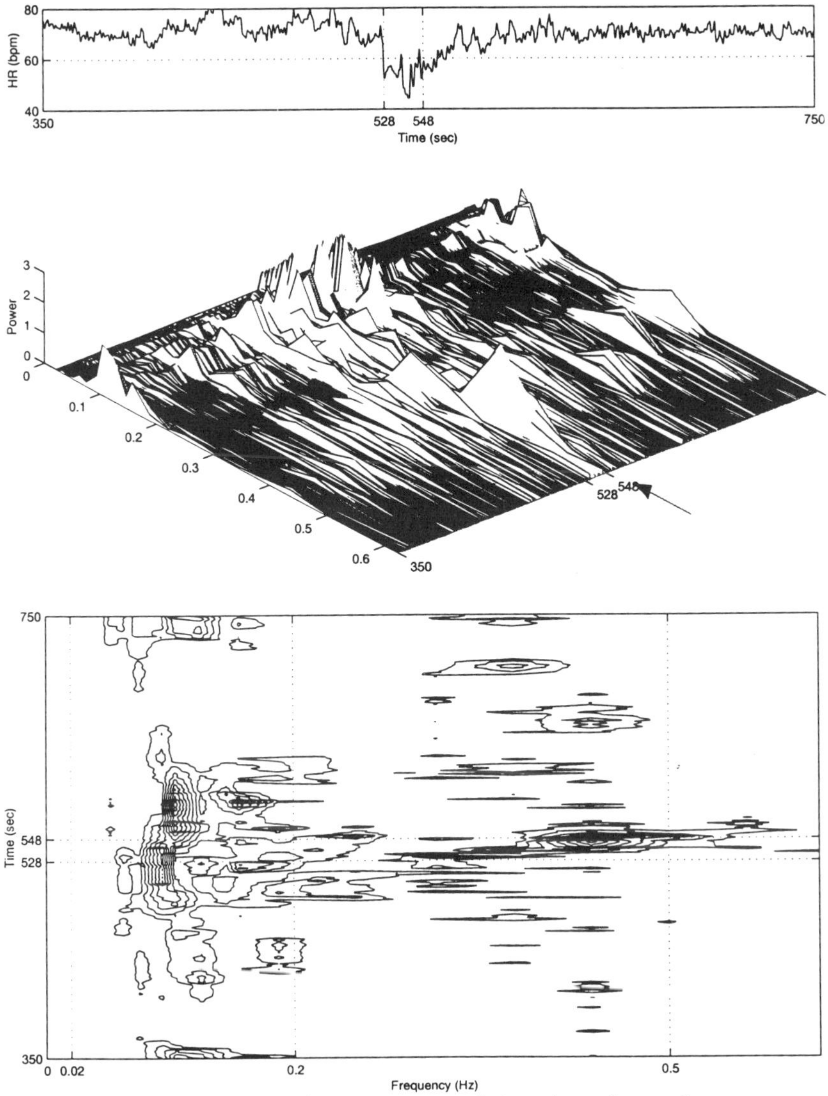

FIGURE 1. Heart rate time series and its time-dependent power spectrum, displayed as a 3D graph and as a contour plot, during a vagal manoeuvre, the diving response (application of an ice bag on the face of a healthy subject) A clear sudden drop in mean HR occurs at the time of the manoeuvre (528–548 s). There is a strong increase in the HF region of the HR power spectrum at this time, which is related to a surge in vagal activity. The increase in the LF range can be related to the vagal surge as well as to an increase in sympathetic activity, due to the cold feeling of the ice on the face.

review the estimates obtained in the time domain. Such estimates do not require stationarity of the signals and can be obtained from very short time windows, and often even from a single beat. They are easily computed and not time consuming. Their main limitation is the lack of discrimination they achieve between the different autonomic branches.

The time-domain estimates are often separated in two classes: a class of variables derived directly from the intervals and a class of variables based on the differences between adjacent cardiac cycles[5].

First Class Estimates

The first class includes: mean HR over a specified period, standard deviation of the heart rate, standard deviation of normal-to-normal intervals (SDNN) and standard deviation of RR intervals (SDRR). These variables are generally calculated from the entire trace length and therefore suffer from the same stationarity limitation as standard spectral analysis does. Other variables are calculated from shorter segments. One such variable is the standard deviation over 5 min intervals (SDANN). Similarly, the mean of the standard deviation of 5-min cycle intervals over the entire recording period (SDNNindx) is often measured. Another interesting estimate takes into account the mean HR, since high levels of HR are associated with a reduced HRV, the SDNN estimate is divided by the mean NN interval. Part of these measures can be applied to the other cardiovascular signals (BP, BF etc), such as the standard deviation, the SDANN and the SDNNindx.

Second Class Estimates

Estimates that belong to the second class include:

The root mean square of successive differences (RMSSD),

The proportion of number of differences between adjacent cycles that exceed an arbitrary limit, such as the proportion of cycles where the difference is greater than 50 ms (pNN50), and the proportion of cycles where the difference is greater than 6.25% of the mean heart period (pNN6.25%),

The number of instances where the difference between adjacent normal-to-normal intervals exceeds an arbitrary limit (e.g. 50 ms, then named NN50) which is reported as "counts".

The first class variables address the HRV without trying to discern any peculiar type of modulation, while the second class variables reflect, almost entirely, vagal activity.

Coumel et al have introduced a method based on time domain estimates, which provides information essentially equivalent to the spectral LF/HF ratio, namely an estimate of sympatho-vagal balance.[6] They propose to detect short oscillations (SO), over 2 to 4 cycles, and long oscillations (LO), over 8 to 15 cycles, by reporting sequences of acceleration or slowing over a preselected number of consecutive RR intervals[7]. The SO/LO ratio highly correlates with the LF/HF ratio.

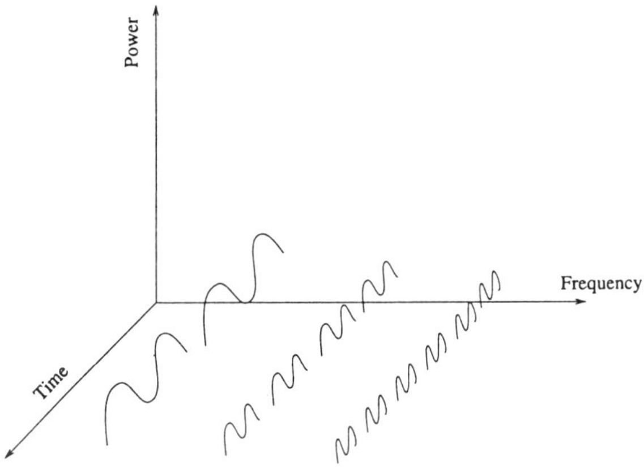

FIGURE 2. The time-frequency plane of the SDA. Each spectral component is calculated from a portion of the time series which includes exactly five times the corresponding time period (for illustration only two periods were drawn). The SDA thus achieves a good time resolution for high frequency components (short time periods) and a lower time resolution for low frequency components (long time periods).

TIME-FREQUENCY DECOMPOSITION OF CARDIOVASCULAR SIGNALS

When considering autonomic provocations or blockades, it was clear that the usual, steady state spectral analysis approach would not fulfil the basic conditions required for a reliable and quantitative determination of the autonomic response. Poor time localisation, and inability to determine whether and when one or more frequency components suddenly changed in frequency or amplitude, were the essential limitations raised.

Several novel approaches were developed and/or implemented to cope with these problems: Our group has developed a method for the time-frequency decomposition based on Fourier transform.[4,8,9] We refer to this method as the *selective discrete Fourier transform algorithm* or *SDA*. For each time and frequency, the algorithm calculates the corresponding spectral component by means of DFT, on a windowed part of the signal, such that the duration of this window is inversely proportional to the analysed frequency (Figure 2). On simulated data, we have shown that the SDA achieves the same results as standard spectral analysis for stationary signals as well as reliably detecting and quantifying abrupt changes in the frequency content for non-stationary signals. When applied during an experiment with voluntarily changing respiration rate, the SDA was able to accurately follow after the location and amplitude of the respiratory peak in the HR spectrum.[4] The SDA also successfully detected and quantified the surge in vagal tone elicited by vagal stimuli[10] (Figure 1). Furthermore, the HR time-dependent power spectrum faithfully displayed the increase in sympathetic activity and the vagal withdrawal which occur on active standing. The SDA for time-frequency decomposition was applied to pathophysiological conditions such as hypertension[11], syncope[12] and autonomic failure.[13]

Novak et al[14,15] applied a *modified Wigner distribution* for the time-frequency mapping variations of RR intervals, BP and respiratory signals. Spurious cross terms are suppressed using a smoothing data window and a Gauss frequency window.[14] The discrete Wigner distribution is easy to implement using the FFT algorithm. This analysis has been applied to head-up tilt and stimulated respiration experiments, in several clinical conditions such as hypertension,[16] postural tachycardia syndrome[15] or Alzheimer disease.[17]

Hayano et al used a *complex demodulation* (CDM) of heart rate and blood pressure variability to evaluate the time dependent responses of the LF and HF amplitudes of the spectra during head-up tilt.[18] The CDM provides a time resolution of less than 15 s and is robust to alterations in the frequency of the components.

Bianchi et al developed a *time-variant algorithm of autoregressive* identification and have applied it to cardiovascular signals.[19] This time variant algorithm permits to obtain a new set of AR parameters whenever a new sample value is available, and thus to obtain a time-dependent spectral description. This is accomplished by updating the previously evaluated sample on the basis of the prediction error and weighting by means of a forgetting factor. This approach was extensively applied to acute cardiac ischaemic events and head-up tilt.[19,20]

Pola et al have compared the evaluation of classical (FFT based), autoregressive and *time-frequency representation* (TFR) based power spectral estimators on artificial data, with patterns typical to the nonstationary series obtained during autonomic tests.[21] Their results documented the superiority of the TFR approach when a sharp time resolution is required. When applied to real RR interval time series obtained during autonomic tests, the TFR achieved a correct representation of the transients in the joint time-frequency domain.

Van Steenis et al applied the *exponential distribution* (ED) to various clinical and experimental human cardiovascular time series and displayed its ability to interpret time-varying frequency components of cardiovascular signals.[22]

Marchesi et al compared the results of the two bilinear distribution, the *Wigner-Ville* and the *exponential distribution*. The ED achieved a good suppression of the cross terms characteristic of bilinear time-frequency distributions, such as the Wigner distribution[22,23]. On the other hand, they claimed that the ED seems to offer less auto terms (or frequency) resolution than WV distribution.

Niccolai et al, applied the *evolutionary Periodogram* (EP), which decomposes the signal by means of amplitude modulated sine waves, to RR series during deep breath and tilt.[24] Compared to the short time Fourier transform (STFT), EP presents a better frequency resolution, while no differences in time resolution were observed between the methods. The same group applied, in 1997, the S-Transform (ST), which uses a scalable gaussian analysis window and produces a *constant-Q-spectral analysis*, in a way similar to the wavelet or SDA (see below).[25]

The *discrete wavelet transform*, using a spline wavelet as the basis function, has lately been applied to cardiovascular signals, during apnea and periodic breathing,[26] providing a powerful technique to detect changes in the signal. In another study, wavelet and cosine packets have been applied to HR signals and compared to STFT, at baseline, during controlled breathing and on standing;[27] however, the results obtained using the STFT were similar to the results obtained using wavelet and cosine packets.

GENERAL RECOMMENDATIONS RELATED TO THE TEST PROCEDURE AND EVALUATION

BASIC REQUIREMENTS FROM AN AUTONOMIC TEST

The autonomic test should be as *non-invasive* as possible. In particular, intravenous instrumentation should be avoided, since its use might affect the outcome of the autonomic test.[28]

The test should have high *sensitivity* in order to detect an abnormality in the great majority of cases, and so, that a normal result provides reassurance that significant autonomic failure is absent. The test should be *specific* (low false positive rate) for detection of autonomic failure. There is typically a trade-off between sensitivity and specificity in the choice of the range of normal response (control).

A test with well-established *reproducibility* usually displays a variation of around 20 percent or less between successive repetitions. In evaluating its validity, a new test should be compared against a well-established test, using an adequate choice and degree of abnormalities.

The autonomic test should not *interfere* with any of the monitored parameters. One should verify whether the test does not have a direct effect (not via autonomic control) on one of the tested variables. An example of such interference is the change in respiratory pattern during mental stress (see below) induced by aloud counting. Subjects should avoid speaking and moving during an autonomic test.

The test should be *minimally dependent* on the co-operation of the subject, and objective criteria should be applied, both during its implementation and during its evaluation.

Stationarity considerations should not be overlooked. Standard spectral analysis requires long periods of stationary data acquisition and was often considered as impractical for routine use.[29] Methods for time-frequency decomposition which perform analysis on non-stationary signals have strongly reduced the stationarity limitation and allow their use in routine tests. They are, in general, *easy to perform* and *not time-consuming*. Most of them can be implemented in real time.

IMPORTANT ENVIRONMENTAL AND PHYSIOLOGICAL FACTORS

Careful *age* consideration or *age matching* is recommended since a strong correlation between HR or HRV responses and age has been established, in most of the autonomic provocations. A reduction with age has been observed in the response to vagal maneuvers,[10,29,30] to deep breathing,[31] to head up tilt (as expressed by the increase in the LF/HF ratio and the decrease in normalised HF power of HRV),[32] to Valsalva manoeuvre,[33,34] or to change of posture.[35,36] Age should thus be taken into account in any protocol designed for the investigation of autonomic responses.

During all tests, *respiration* should be monitored. Any change in the respiratory pattern (rate or depth) should be taken in consideration, when attributing changes in the HF peak of HR or BP spectrum to changes in vagal activity. Controlled respiration at a constant rate might avoid interpretation problems related to changes in breathing rate and depth, however it induces some stress which might interfere with the autonomic influence of the manoeuvre.

The autonomic activity is largely different in supine, sitting or standing position. The *position* of the subject should thus be uniform, during the entire experiment as well as for all investigated subjects.

Temperature should be kept constant at around 22–24°C, since it might interfere with low frequency components of HR and BP control. If thermal investigations are performed (such as thermal entrainment), a valid temperature monitoring procedure is required.[37]

Sex dependence of the response should be verified for each type of manoeuvre. Indeed, no gender effect was found in the study of Gelber et al, on Valsalva manoeuvre in 611 normal subjects.[34] On the contrary, a lower sympathetic response to mental stress was observed in women as compare to men.[38]

A significant *circadian rhythm* was found in responses to autonomic tests, such as ocular compression test.[39] It is thus crucial to perform the test at the same hour of the day when comparing responses. *Postprandial* changes have been observed and a careful protocol should thus avoid meal interference.

A clear disadvantage of most provocations is their strong dependency on the patient's motivation and often requires active co-operation. Also it is often difficult to control accurately the intensity of the trigger such as for the ocular pressure, deep breath test or hand grip. It is therefore recommended to design, when possible, a way to *quantitate the strength of the autonomic provocation.*

GUIDES FOR ANALYSIS AND INTERPRETATION OF RESPONSE

The time-frequency decomposition methods described above permit to assess simultaneously short term as well as long term variability.

Furthermore, such analysis provides information about the *very short term* variability, during the transient stages of an experiment.

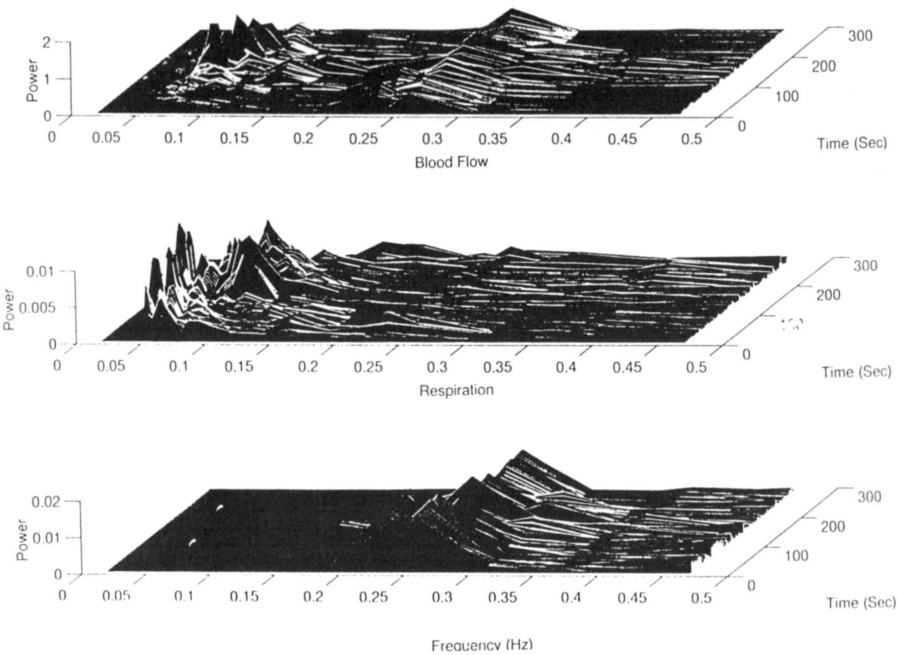

FIGURE 3. Time dependent power spectra of heart rate, blood flow and respiration, obtained by SDA. The signals were sampled from an healthy subject, at supine baseline under spontaneous breathing. The respiratory peak can be clearly defined from the respiration power spectrum and clearly appears in the HR spectrum. Note the LF peak which is present in the HR and BF spectra, and the VLF which is observed only in the BF spectrum. No detrending was applied to the stationary signals obtained at baseline.

Usually during autonomic provocations and blockades, the following signals are monitored: (1) ECG with a subsequently applied R-detection algorithm,[40] (2) blood pressure, preferably non-invasive by Finapres, (3) respiration preferably using a device which provides information about the respiratory volume in addition to rate. Additional optional signals may also be recorded: (4) cerebral blood flow or (5) peripheral blood flow, (6) any direct or indirect measure of cardiac output.

These signals require some *pre-treatment* before being submitted to time-frequency decomposition. RR intervals can be directly analysed or transformed into an equally spaced RR or HR series[41]. Any case of arrhythmia should be corrected as smoothly as possible. Some detrending of most signals should be applied, using for example, a non-linear high pass filter,[4] a non-linear bandpass filter based on the fourth-order polynomial function,[16] or the reconstructed signal after wavelet decomposition up to some specified scale.[27] This detrending removes the long-term modulation and can therefore affect the very low frequency power. One should thus be aware of the lowest frequency that can still be reliably estimated after detrending.

The *time-frequency decomposition* can then be applied to the signal, providing a time-dependent spectrum which can be displayed as *a 3D graph or contour plot* (see Figure 1). In the 3D graph, the x-axis is the time axis, the y-axis is the frequency axis, and the z-axis is the power (or amplitude) axis. In general, in the time-dependent power spectrum of HR and BP fluctuations obtained by SDA algorithm and displayed in Figure 3, two main regions of power are observed, a high frequency (HF) peak located around the respiratory frequency (0.3 Hz) and a low frequency (LF) located around 0.1 Hz, generally related to baroreflex. The very low frequency (VLF) peak which is observed around 0.05 Hz often merges with the LF peak. In the BF time-frequency spectrum (Figure 3), the power is concentrated below 0.2 Hz, in the VLF and LF ranges. *Such time-dependent spectrum allows to follow closely any change in the power or the frequency location of the different peaks.*

The location of the HF peak accurately follows after the respiratory rate.[4,42] The respiratory rate can thus be easily determined and should be taken into account when interpreting results in the HF range. If the respiratory signal is available, its time-dependent power spectrum provides the varying location of the HF peak. One can decide to define the HF range so as to include the entire respiration frequency domain or to define HF according to moving boundaries which accurately follows after the variations in respiratory rate. *Once the boundaries of the spectral peaks are determined, integrals can be calculated and displayed as a function of time.* These integrals provide the quantitative evaluation of the autonomic activity as a function of time. Usually, the integral over the *HF content of HR fluctuations* is used as an estimate of vagal activity, the *LF of HR fluctuations* as an estimate of combined vagal and β-sympathetic activity related to the baroreflex, the *LF of BP fluctuations* as an estimate of α-sympathetic activity and the *VLF of HR and BP fluctuations* as an estimate of humoral activity, related among others to vasomotor control and thermal regulation. The *HF content of BP fluctuations* mostly reflects the mechanical influence of respiration and cannot generally be used as an estimate of vagal tone.[1,43] It strongly depends on the nature of the respiration and may be markedly affected by a change from abdominal to chest respiration, such as occurs during a change of posture or head-up tilt test. The *LF/HF ratio of HR* fluctuations, which can also be calculated and displayed as a continuous function of time, then provides an evaluation of the instantaneous sympatho-vagal balance.

The time-dependent LF and HF contents of HR fluctuations can be calculated *in absolute units*, or *in normalised units* when they are divided by total power.[44] The LF integral normalised by total power provides information on the concentration of power between the sympathetic and parasympathetic branches. It has been shown to be linearly correlated with HR, during graded tilt,[45] and is also (similarly to LF/HF) considered an index of the sympatho-vagal balance.[46] This normalisation is of special importance when total variance varies and may possibly mask the change in one of the spectral components as compared to the other.[44]

CAUTIONS AND LIMITATIONS FOR THE USE OF THE TIME-DEPENDENT SPECTRAL ANALYSIS

The *LF* peak of HR and BP fluctuations is mediated by both branches of the ANS. As for steady state analysis, also during an autonomic provocation or blockade, *the discrimination between these two contributions is often difficult.* An unequivocal conclusion about the sympathetic activity can be obtained only in the extreme cases where the vagal and sympathetic activities have evolved in opposite ways, or if vagal activity (and thus HF) did not change at all. For example, if HF is increased without any marked change in respiration, then it must reflect enhancement of vagal activity. A concomitant decrease in the LF peak can then be ascribed to a decrease in sympathetic activity. However, with a concomitant increase in the LF peak, no clear conclusion can be derived about the sympathetic activity. If HF is reduced or unchanged, than an increase in LF can be interpreted as an increase in sympathetic activity.

The sensitivity of the HF power to changes in breathing rate or in breath volume[47–49] should not be neglected. When applying a time-frequency decomposition to HR fluctuations during triggered breathing at a gradually increasing rate, we have observed a reduction in the HF power, along with its shift in frequency.[4] Care should be taken to verify possible changes in respiratory rate or volume, before interpreting changes in the HF of HR and BP fluctuations as changes in vagal activity. The use of randomised or constant *metronome* breathing rate, might partially overcome this limitation, at the expense of inducing some stress. Transfer function computation from respiration to HR, with randomised breathing, provides a good estimate of vagal function[50]. However these artificial measurement conditions interfere with the autonomic provocation test.

It should be mentioned that the *power of fluctuations mainly reflects variations in neural activity*, without necessarily being proportional to the mean level of neural activity. Usually the concealed assumption is made that the level of fluctuations in neural activity is linearly related to the mean level, which is probably true near midlevel of activity and not near saturation. Indeed, it has been observed that during exercise for example, the fluctuations in HR almost vanish, while sympathetic activity is obviously high, most probably saturated. It is thus crucial to keep in mind that mean HR does reflect in a direct manner the sympatho-vagal balance and conclusions from analysis of variability should always be challenged with the patterns of the change in HR.[6]

Finally, the *varying time-frequency resolutions* should be kept in mind when using constant-Q analysis methods, such as the SDA, wavelet, or S-Transform (see above). Indeed, *the time resolution of the LF power is lower than the time resolution of the HF power* (Figure 2). Any sudden change in autonomic activity will therefore be expressed at slightly different times for each region of power. If the time of interest is positioned at the centre of the window, the change in autonomic activity will appear in the LF region *earlier*, than it will appear in the HF region. The temporal resolution of each spectral peak should be remembered before discussing the time evolution of the autonomic activity.

PHARMACOLOGICAL BLOCKADES

Pharmacological interventions, using agonists and antagonists, are a valid challenge for the evaluation of autonomic control. Obviously, the autonomic limbs which can be directly investigated based on the response of cardiovascular variables are the parasympathetic, β-sympathetic, and α-sympathetic control. Although many of the clinical studies have limited themselves mainly to HR and HRV responses, a much fuller view of autonomic functioning is achieved when considering additional haemodynamic variables.

When focusing on HR and the full spectrum of HR fluctuations, pharmacological agents provide direct quantitative information regarding parasympathetic and β-sympathetic activity, while α-sympathetic control can only be indirectly estimated.[51] When focusing on BP and/or BF signals and their low frequency fluctuations, α-sympathetic agents (agonists or blockers) will allow to probe directly α-sympathetic control.

Obviously, a wide range of drugs which affect haemodynamic parameters, such as Converting Enzyme Inhibitors (captopril, enalapril, quinapril), Calcium blockers (verapamil, nimodiprine), diuretics, a variety of anti-arrhythmic drugs, etc., will inevitably elicit an autonomic response. Such response, although often complex, indirect and more difficult to interpret, can be investigated by the same tools as those applied to agents which affect specific branches of autonomic cardiovascular control.

In most studies which involve pharmacological interventions, effort has been made to achieve steady state conditions of drug concentration, for instance by providing a constant infusion rate following an initial bolus injection. Assuming stability is reached, one can safely compare mean HR (and mean BP) and the pattern of HRV (or BPV), attained in the time *or* frequency domain under the effect of the drug, as compared to baseline.

Limiting ourselves here to the essential pharmacological agents interfering with autonomic cardiovascular control,[52] we will briefly consider how the use of vagal and sympathetic blockades (or activation) may provide additional insight into autonomic activity.

PARASYMPATHETIC BLOCKADES

Parasympathetic blockades are easy to perform and will usually provide clear results in humans. Indeed, in resting conditions, humans present a strong parasympathetic predominance, with low sympathetic activity.[53] Only when vagal function is impaired (inducing a marked reduction in HR fluctuations), will any additional reduction, achieved by partial or full vagal blockade, be difficult to quantitate. On the other hand, this feature in itself, namely a markedly reduced vagolytic response in HR and/or in HRV (more specifically in its HF component), is a valid indication for the diagnosis of vagal neuropathy (such as in diabetes[54]), or of vagal withdrawal (such as in congestive heart failure[55,56]).

In general, vagal blockade achieves a strong increase in HR, with a marked reduction in HRV, gradually reducing, until almost eliminating all HF power[2,53,56,57] and leaving only some of the LF HR variability at full blockade. The degree of reduction in HF fluctuations, under blockade relative to control conditions, can be used to estimate the initial level of vagal activity.[58,59]

Since the time constants of the parasympathetic blockers are quite long, steady state drug levels are easily achieved,[56] even by bolus injection (without necessarily requiring iv. infusion), certainly for short-term measurements (about 5 min). Therefore, the time domain and steady state frequency domain analysis are adequate tools for estimating the effect of vagal antagonists such as atropine or scopolamine. The effect of agonists undergoes a much faster decay, so that steady state can be achieved only by adding a constant rate infusion following the initial bolus injection. Under these conditions, steady state tools of analysis can again be safely used. The estimation of a short term effect, achieved for instance by a bolus of acetylcholine, would require the use of non steady state time-frequency analysis such as our SDA algorithm.

An interesting feature of parasympathetic blockers (atropine or scopolamine) is that, for very low doses, a slight but significant *paradoxical increase* in vagal activity, and thus a small decrease in HR and increase in the HF peak in HR, can be observed (central effect). The low dose vagomimetic effect of atropine[56] or transdermal scopolamine[60,61] has been observed in healthy individuals, in patients with severe coronary disease[62], in patients following acute MI,[63,64] and in patients with advanced congestive heart disease.[65] However, although it seems well accepted that enhanced sympatho-vagal balance is beneficial for cardiac patients, the significant increase in HRV induced by acute, low dose scopolamine did not result in a decreased risk for ventricular fibrillation.[66] Yet, the question whether low dose atropine or scopolamine might have a long-term cardiac protective effect, is important and deserves further research.

At higher doses of parasympathetic blockers, a clear dose-dependent reduction in vagal activity is achieved, reflecting the gradual peripheral muscarinic blockade, as shown in an atropine study performed in healthy individuals.[56] A dose-dependent increase (by about a factor of 2) in mean HR is accompanied by a dose-dependent reduction (of several orders of magnitude) in the HF power of HRV, thus emphasising the sensitivity of the HF fluctuations to the level of vagal activity.

Vagal blockade may have its own clinical applications. One of them is the use of atropine as premedication during thoracic epidural anaesthesia (TEA), in order to reduce its vagotonic effect.[67] Anticholinergic drugs, commonly administered during reversal of neuromuscular blockade, do suppress parasympathetic control (as express by HRV parameters) in the early post-operative period. The effects of glycopyrolate appear to be of shorter duration and therefore preferable to atropine.[68]

SYMPATHETIC BLOCKADE

The effects of sympathetic blockade are much more difficult to quantitate and/or to interpret, since healthy humans at rest are to a large extent under vagal control.[53,69–71] Indeed, β-blockade may have a minor effect on the LF component of HRV, for low initial sympathetic activity, such as in supine volunteers[53] or in conscious dogs.[2,72] Yet, it might have a more marked effect, for high initial activity, such as in migraine patients[51] or in conscious rats.[73,74]

Full sympathetic blockade is more often applied in basic research than in the clinical setup.[75] Interruption of sympathetic activity is recommended only in very specific diseases, in which sympathetic control is clearly involved, such as Raynaud's disease, Burger's disease, reflex sympathetic dystrophy. In some other diseases, such as chronic pain syndrome, in which sympathetic involvement is less clear, sympathetic blockade may still be beneficial. In those cases, full sympathetic blockade may be achieved either by non-invasive means, such as pharmacotherapy (adrenergic blockade), or by invasive means, such as sympathectomy or local injections (local anaesthetics, chemical ablation).

Under physiological conditions, sympathetic activation involves in general the synchronous activation of both the β- and α-sympathetic branches. However, when considering sympathetic blockade, one is obliged to discriminate between blockade of either of these branches.

The most frequently prescribed sympathetic intervention is partial β-blockade in cardiac patients (post MI, ischaemic, arrhythmic, etc.) or in hypertensive subjects. The classic β-blocking agent propranolol, which in addition to its β_1 antagonism, also affects β_2 receptors, causing some increase in vascular resistance, tends to be replaced by more specific agents. The selective β_1 antagonists (atenolol, metoprolol, esmolol, etc.)[76] attenuate sympathetically mediated increase in mean HR and reduce myocardial contractility, without affecting the pressor response mediated by the β_2 vascular receptors.

The most commonly used β-blockers (propranolol, atenolol) are quite long acting (several hours). Following a bolus dose, steady state conditions can be assumed over the time scales required for investigating autonomic activity by spectral analysis.

Since β-blockers (and especially those who are purely β_1 agents) affect directly the SA-node, they clearly interfere with the sympathetic component of HRV, reducing any LF component observed at baseline. Indirectly, via HR, HRV and contractility, β-blockade (even pure β_1) will inevitably affect BP and its pattern of fluctuations. However these effects on BP might still be masked by the various pressor mechanisms, which impinge directly on the blood vessels, such as α-sympathetic activity, renin-angiotensin, vasopressin.

The use of α-sympathetic blockers (prazosin(α_1), phenoxybenzamine(α_1+α_2), phentolamine(α_1+α_2)), which directly affect vasoconstriction, may thus have a marked effect on BP and its LF fluctuations. The possible adverse hypotensive effect of these drugs requires caution in the clinical setup. They are still extensively used in basic research, in particular in the investigation of the mechanisms responsible for hypertension.[73,77] A large number of animal studies have been performed, which have allowed to ascribe at least part of the LF BP fluctuations to α-adrenergic modulation.[73,77,78] The HF fluctuations in systolic BP were markedly increased by α-sympathetic blockade (prazosin).[73,74,78]

Indirectly and reflexively, the use of α-blockers should affect mean HR and its LF fluctuations. Yet this effect is expected to be weaker than for BP, being masked by the uninterrupted input to the SA node from both vagal and β-sympathetic nerve endings. For the same reason, the differences between a normal population and a hypertensive (or pre-hypertensive) one are more strongly reflected in the BP signal than in HR.[11,77]

AUTONOMIC PROVOCATIONS

Autonomic provocations include three kinds of stimulation: (1) sympathetic stimulations, such as mental stress and sustained hand grip; (2) vagal manoeuvres, such as diving response, oculocardiac reflex and deep breath test; and (3) mixed provocations, which involve both sympathetic and parasympathetic responses, such as Valsalva manoeuvre, active change of posture, head up tilt, lower body negative pressure and cold pressor test.

Precedent chapters of this book have dealt with the methodology of some of these tests and their interpretation by standard tools of analysis. In this part, we will discuss the interpretation of the response to these tests by advanced methods of time-frequency decomposition. We present a general description of these tests, as well as a description of their autonomic effects. Examples of results obtained by our SDA method for time-frequency decomposition are provided for several of these tests. The results are presented either as 3D plots or as spectral integrals as a function of time.

SYMPATHETIC STIMULATION

Mental Stress

General Description and Arrangements

Mental stress (Figure 4) can be achieved by mental arithmetic, digits repeated backwards, computer quiz, delayed auditory feedback task, speech stress, reaction-time task, etc. It induces an increase in mean HR, in systolic and diastolic BP, and in cardiac output (CO). It is accompanied by an increase in total peripheral resistance, an thus a decrease in flow (Figure 4a). Mental stress is considered effective when mean HR is increased by at least 15 beats per minute.[79]

Arithmetic test should be performed aloud causing the mental stress to be much stronger. However, if the autonomic response is to

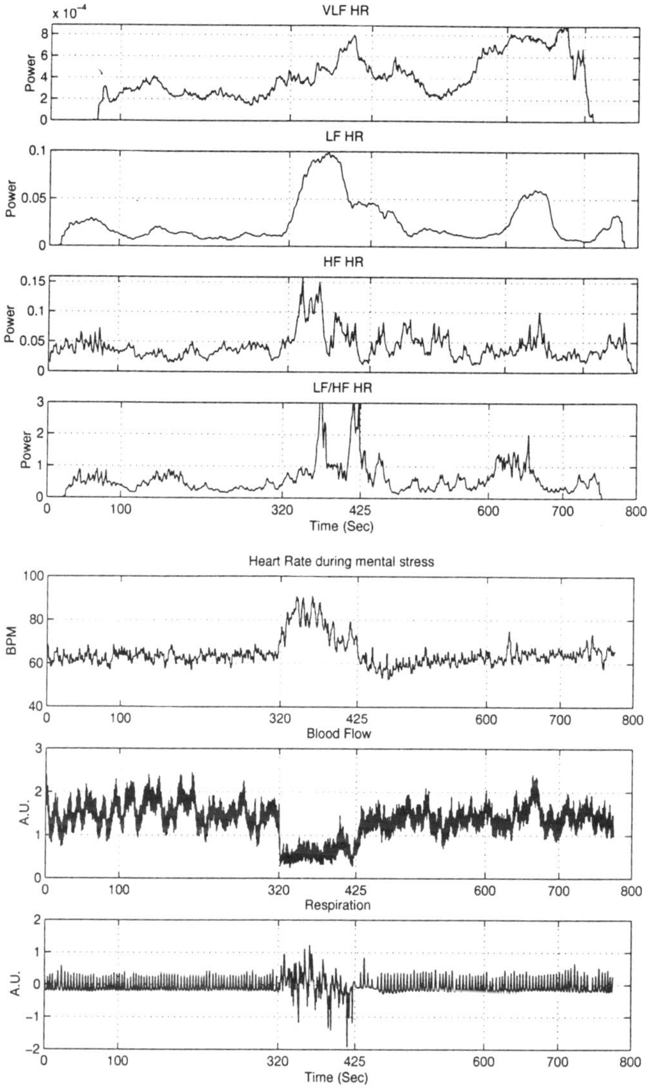

FIGURE 4. Arithmetic Mental stress: Top part shows time series of instantaneous HR, blood flow and respiration. At the time of the stress (320–425 s), a strong increase in HR and a sudden drop in BF are observed. The drop in BF is related to a sudden peripheral vasoconstriction. The change in pattern of the respiration is due to talking. Bottom part shows time-dependent integrals obtained by SDA on HR. The panels are respectively: VLF, LF, HF and the LF/HF ratio. Note the strong increase in the LF region at the time of the stress. The increase in HF results from the change in respiration pattern while counting aloud.

be estimated by means of spectral analysis, mental stress performed by computer quiz seems preferable, since talking affects the respiratory pattern in such a way that the HF peak in the HR spectrum no longer reflects vagal activity.

Autonomic Effects

Invasive studies have displayed an increased sympathetic activity during mental stress. Direct microneurographically measure of muscle sympathetic nerve activity (MSNA) has shown that mental stress mediates the sympathetic nerve activation.[80] Pharmacological blockades have revealed an overall pattern of increased sympathetic control of the heart during mental tasks.[81–83] In heart transplant subjects, mental stress has displayed the participation of humoral factors which also participate in the cardiac chronotropic response.[84]

The parasympathetic response to mental stress is still under debate. Bernston et al[81] reported that mental stress is accompanied by reduced parasympathetic activity, while Jern et al[85] reported that atropine did not affect the magnitude of the HR response to mental stress.

Non-invasive evaluation of the autonomic control by spectral analysis of HRV has also been used during the steady state parts of mental stress experiments. Moriguchi et al have observed enhanced low frequency power and LF/HF power ratio, which are both indices of sympathetic activity.[82] Analysis of 24 hours HRV have shown that psychological stress was significantly associated with an increase in the LF/HF ratio, suggesting an increase in the sympathetic predominance during stressful periods of the day.[86]

Interpretation

The results of a mental stress experiment are presented in Figure 4a, which shows HR, BF and respiration as function of time. Active mental arithmetic was performed between 320 and 393 s, by counting from 200 downward by steps of 7. A strong enhancement in mean HR, and a strong decrease in blood flow, reflecting reduced peripheral resistance, appear at the time of the stress, clearly suggesting a rise in sympathetic activity. The parameters slowly return to normal values after cessation of the mental stress. The respiratory signal displays the change in breathing pattern while counting, thus resulting in a broad-band deeper respiration.

Time-dependent spectral analysis (by SDA[4]) unfolds the time-evolution of the autonomic response to mental stress. Figure 4.b displays the instantaneous VLF, LF, HF and LF/HF of HR fluctuations during mental stress. The strong increase in LF power reflects the sympathetic surge during the mental stress. It is plausible to ascribe the increase in the HF power to the increase in respiration depth (during aloud counting) and not to a vagal increase. Care should be taken in the interpretation of the response to mental stress as reflected in the HF power, since respiration is directly affected by the experimental procedure, thus affecting the HF content of HR fluctuations without correlation to changes in vagal activity.

Sustained Hand Grip

General Description and Arrangement

Sustained muscle contraction causes a rise in systolic and diastolic BP as well as in HR. The stimulus derives from exercising muscle and central command.[87–90] Efferent impulses travel to the muscle and heart, resulting in increased cardiac output, BP, HR and muscle vasoconstriction. There is an effort and time-dependent increase in muscle sympathetic activity.[91] The rise in BP is mediated by an increase in cardiac output and peripheral resistance.

This autonomic manoeuvre has been adopted as a clinical test of sympathetic autonomic function, and 30% of maximal contraction has been recommended for up to 5 min.[92] Nevertheless, many patients are unable to sustain the effort for 5 minutes, though an effort for 3 minutes is probably adequate. A normal response has been suggested to be an increment of the diastolic BP by 16 mm Hg or more, and 11 to 15 mm Hg is considered borderline.[92]

The test is of limited sensitivity and specificity, and confounding variables are not well studied.[33] The early HR increase seems to be due to vagal withdrawal,[87] while the late components are elicited by sympathetic activation.[33]

VAGAL MANOEUVRES

Common vagal manoeuvres include *Diving Response* (apneic facial exposure to an ice water bag), *Valsalva Manoeuvre* (forced expiration against a column of mercury to 40 to 50 mm HG), *Oculocardiac Reflex* (inflation of a Honan balloon against one eye), *Deep Breath Test* and unilateral Carotid Sinus Massage.

Diving Response

General Description and Arrangement

Diving response (Figure 1) is usually elicited by placing an ice-bag on the face for 10–30 s, while breath-holding. Placing an ice-bag on the forehead causes similar cardiac and vascular responses as face immersion. Being technically much easier to perform and not requiring active co-operation of the patient, the ice-bag on the forehead test may be a valid replacement for the full face immersion test accompanied by breath-holding.[93]

The diving response in humans is characterised by breath-holding, slowing of the HR (diving bradycardia), reduction of limb blood flow and gradual rise in mean ABP.[94] The average human diving response consists in a mean HR reduction from 70 to 45 beats/min; a small proportion of healthy subjects develop bradycardia to less than 20 beats/min.[95] The diving response can be modified by many factors but the most important are water temperature, oxygen tension in the arterial blood and emotional factors.[94] The response is stronger at lower water temperatures and with immersion of the whole face than immersion of mouth and noise only.

The diving response is clinically applied as a therapeutic measure to treat paroxysmal supraventricular tachycardia, and as an investigatory tool in autonomic dysfunction.[96,97]

Autonomic Effects

The observed bradycardia results solely from increased parasympathetic stimulation to the cardiac pacemaker and is thus not influenced by β-adrenoceptor blockade (metoprolol).[98] The concomitant reduction in limb blood flow results from vasoconstriction due to increased sympathetic input to arteries in the arms and legs.

The nervous inputs and outputs for the response seem to be coordinated in the brain stem by the respiratory, vasomotor and cardioinhibitory "centres". The diving response is accompanied with an increase in sympathetic activity, not only in vasomotor tone. Multi-unit recordings of muscle nerve sympathetic activity have shown a pronounced activation of MNSA during diving, with a mean increase of 360% compared to baseline. Diving without apnea elicited a significant but weaker increase in MNSA, whereas apnea alone did not influence the sympathetic outflow.[99]

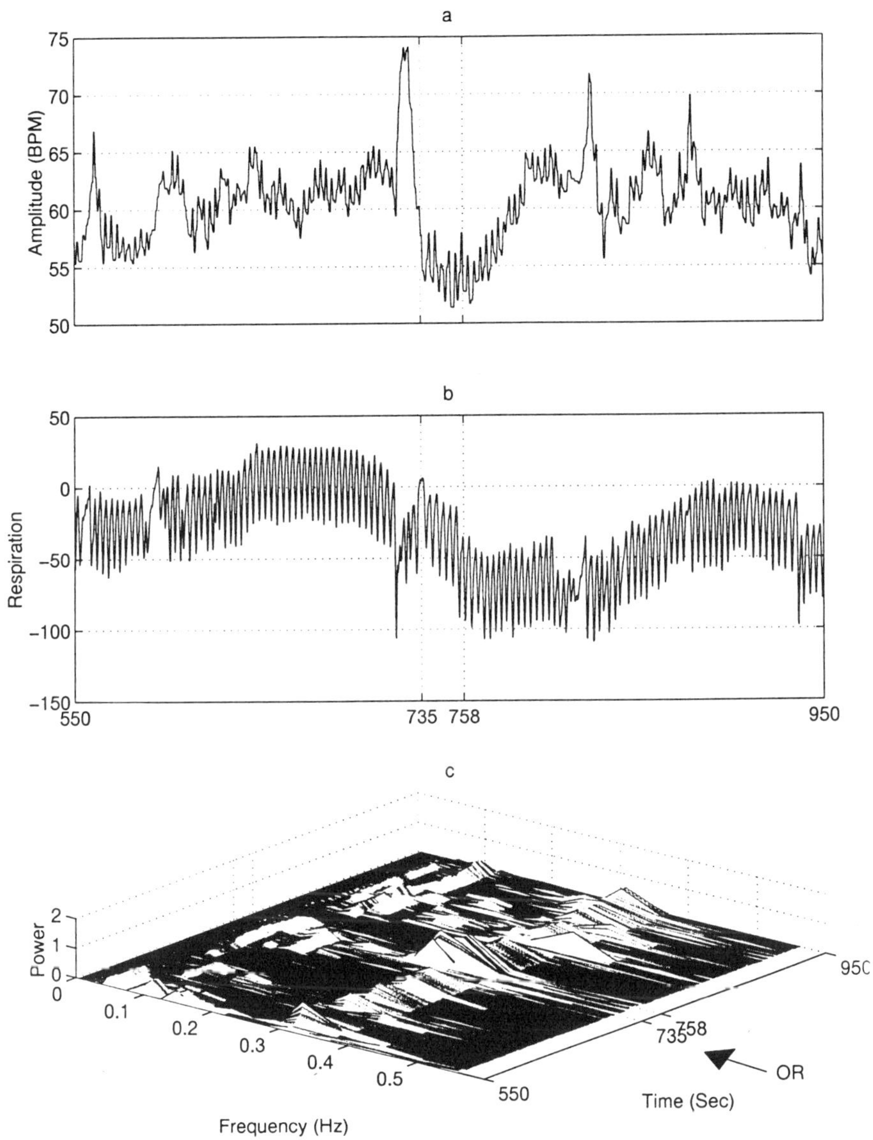

FIGURE 5. Oculocardiac Reflex – pressure on the eyelids of an healthy subject. (a) HR, (b) respiration and (c) HR time-dependent power spectrum. The strong decrease in mean HR is clearly related to the increase in the HF range of the spectrum, thus to a vagal surge at the instant of the manoeuvre (735–758 s). The deep breath just before the manoeuvre induces the surge in mean HR.

Interpretation

Figure 1 displays an example of the mean HR and the time-dependent HR power spectrum (obtained using the SDA algorithm[4,10]) during the application of an ice bag to the face. A sudden increase in the time-dependent HF peak which reflects the sudden surge in vagal activity, is clearly observed. This increase can be measured and provides a quantitative evaluation of the vagal response. The rise in the LF power can be attributed partly to the increase in vagal activity which also contributes to this range, and partly to the expected increase in sympathetic vasomotor activity.

Oculocardiac Reflex

General Description and Arrangement

The ocular compression test (Figure 5) is performed by placing a water-filled pressure sensor (external diameter 18 mm) on each eyelid. The pressure applied to each sensor is around 100 mm Hg, maintained without change for 10 s up to 30 s,[30] and is identical for both eyes.[100]

The oculocardiac (trigemino-vagal) reflex, which was discovered by Aschner,[101] is frequently observed during anaesthesia for paediatric strabismus surgery.[102] It is a well-known entity which may result in vagal depression of the heart when pressure is placed on the eyeball, or, when traction is applied to the extra-ocular muscles. Bradycardia is, by far, the most common cardiac response to these stimuli.[103,104]

Moonie et al[105] suggested that preoperative orbital pressure (inducing oculocardiac reflex) may predict which patients might have bradycardia intraoperatively.

Autonomic Effects

The oculocardiac reflex involves an afferent pathway through the ophthalmic branch of the trigeminal nerve and an efferent pathway involving the motor nucleus of the vagus nerve and its cardiac efferent. Several authors[106,107] proposed assessing cardiac reactivity to vagal stimuli by evaluating the extent of cardiac response to ocular pressure.

Interpretation

Typically, quantitative time-domain measures of the response to ocular pressure are obtained from: (1) the mean RR interval during

60 s of the steady state preceding the ocular compression test; (2) the longest RR interval during the ocular compression; (3) the longest RR interval during the ocular compression divided by the control RR interval[100]. However this does not provide separate information about each limb of the autonomic system. Therefore, spectral analysis of HR fluctuations is recommended, yet this analysis should be a time-dependent analysis due to the short time duration of the vagal response.

The SDA time-frequency decomposition of HR fluctuations obtained during OR (Figure 5) discloses a strong increase in the HF peak, thus reflecting the increase in vagal activity at the time of the stimulus (t=735–758 s). Note that the spectral content reverts to normal immediately with the cessation of the ocular pressure. Such time-dependent evaluation allows to separate vagal from other autonomic components.

The brief increase in HR (Figure 5a) which occurs slightly before the application of the OR is due to a sudden strong inspiration, as observed in the respiratory signal. It is of major importance to verify that the respiratory depth does not vary during the autonomic stimulus, so that the increase in the HF peak of the HR spectrum can be solely attributed to vagal surge.

Deep Breath Test

General Description and Arrangement

During the deep breath test, patients are asked to breathe regularly in and out, six times for 5 seconds each. For each of the six breathing phases, the shortest RR interval is measured each time during every inspiration, and the longest RR interval during every expiration. The shortest and the longest mean RR intervals are calculated, from which the maximal and minimal HR is subsequently determined[31]. Bennett et al claim that a single deep breath is a more potent stimulus for HR changes than repeated deep breaths.[108–111]

Autonomic Effects

The HR response to deep breath is exclusively parasympathetically related. Propranolol, as well as alpha-adrenergetic blockers, do not have any influence on the test, while atropine completely suppresses the HR response[31]. An investigation on cardiac transplants was able to show that humoral factors failed to influence the test.[112]

In our patient's OR example shown in Figure 5a, we incidentally observe the deep-breath linked artificial HR and HF increase, which occurred slightly before the OR test itself.

MIXED AUTONOMIC PROVOCATIONS INVOLVING SYMPATHETIC AND PARASYMPATHETIC RESPONSE

Valsalva Manoeuvre

General Description and Arrangement

The Valsalva manoeuvre (VM), which was originally designed to help clear infected and blocked middle ears, is now better known as an apparently simple test of autonomic function. In practice, however, although the manoeuvre itself is simple, namely to blow against resistance for 10 to 20 s, the reflex pathways involved in the response are complex. To induce the haemodynamic changes accompanying the manoeuvre, forced expiration needs to be continued for at least 7 s. Four phases of the normal circulatory response to the manoeuvre are usually analysed and are discussed in a preceding chapter of this book.

In order to determine the *Valsalva Ratio* (VR), the subject is asked to maintain a column of mercury at 40 mm Hg (not exceeding 50 mm Hg) for 15 s blowing into a bugle with an air leak (to ensure an open glottis). This should be repeated until two similar BP and HR are obtained. Usually up to four manoeuvres are performed.[33] The test is easy to perform but is effort dependent.

The VR is derived from the maximal HR generated by the manoeuvre divided by the lowest HR occurring within 30 s of the peak HR[113]. The pressure of 40 mm Hg is chosen because it seems to yield reproducible results, whereas VR determined at a pressure below 20 mm is inadequate and above 60 mm is less reproducible.[114]

Autonomic Effects

Parasympathetic blockade with atropine abolishes the HR changes,[115] whereas cardiac sympathetic blockade does not.[116] The BP overshoot can only be abolished by combined and autonomic blockades.[114,117] The reflex pathways thus involve both parasympathetic and sympathetic fibres, and, despite their complexity, provide a useful way of evaluating autonomic activity.

Active Change of Posture

A change of posture (CP) from supine to standing (Figure 6) results in a shift of blood away from the chest to the venous system below the diaphragm, usually referred to as *venous pooling*. The test is very simple. The subject is asked to lie quietly on a couch and then to stand up unaided and remain standing quietly for 1 to 5 min. Most subjects can stand up within 2 to 3 s, but even if a subject stands up slowly (over 10 or more seconds), the characteristic responses still occur.[118]

Almost invariably in normal subjects, HR is maximal around the 15th beat after starting to stand up, and a relative bradycardia is reached around the 30th beat, although the actual maxima and minima points vary slightly from individual to individual.[35] The 30/15 ratio, or RR interval at beat 30/RR interval at beat 15, has been recommended as an index of cardiovagal function.[33]

However, time-frequency decomposition of HR and BP signals may provide a much deeper insight in the time course of the autonomic response to CP.

Autonomic Effects

The initial orthostatic reflexes are mediated by the neural regulatory system alone. With standing, the regulatory system increases heart rate, cardiac contractility and vascular tone, by a decrease in vagal outflow and an increase in sympathetic outflow. During prolonged standing posture, additional reflexes of the humoral system become involved in the maintenance of ABP. The presumed mechanism of the initial response to CP consists in (Figure 6a): the immediate increase in HR results from vagal inhibition (eliminated by vagal blockade[119]). A secondary, more gradual, HR increase results from further vagal inhibition and sympathetic enhancement. This part of the response most probably reflects arterial baroreflex compensation for the transient ABP fall.[87,120] The rapid HR decrease at about 20 s is associated with the return to normal ABP levels (in some cases, with overshoot), causing rapid vagal activation of the sinus node. Within 30 s from the onset of standing, stabilisation of the circulation is reached.[11] This response is extremely complex, and each of its stages is important in the fulfilment of an adequate response to CP, especially in the diagnosis and the clinical evaluation of physiopathologies such as the widespread orthostatic hypotension.

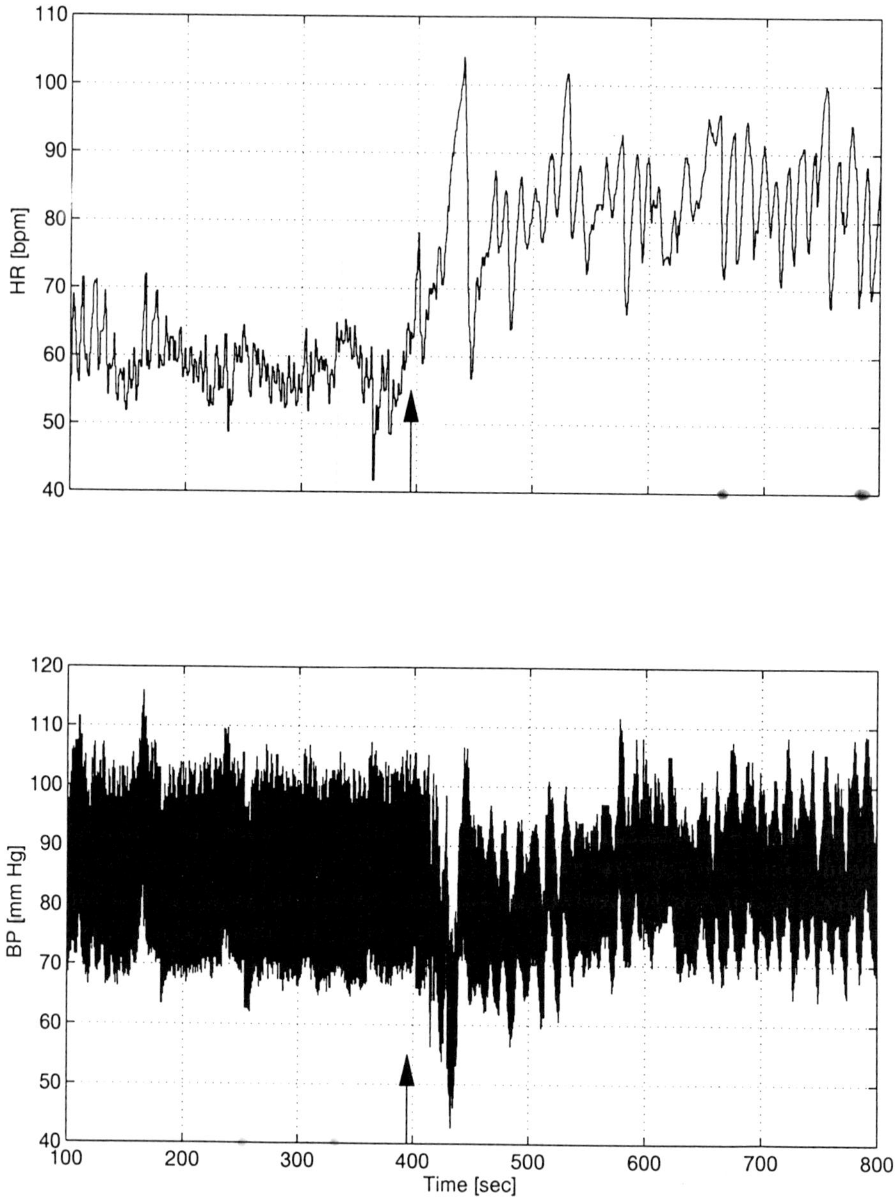

FIGURE 6A. Change of posture: HR and BP during CP (onset of CP is marked by the arrow). The drop in mean BP is corrected by the increase in mean HR. There is then an overshoot in HR and stabilisation to a higher level. The change in the pattern of the fluctuations is clearly observed both in HR and in BP.

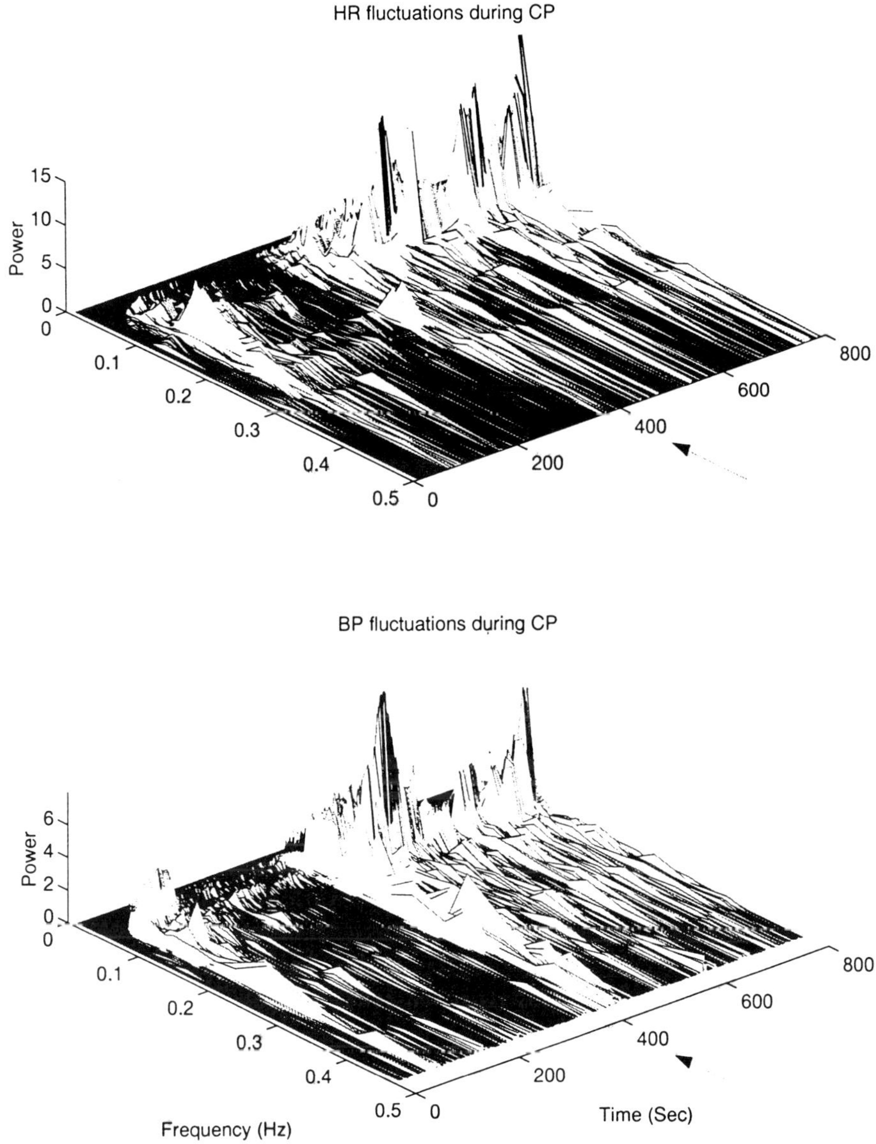

FIGURE 6B. Change of posture: Time-dependent spectra of HR and BP as 3D plots. Note the increase in the LF region of the spectrum. The increase in the HF content in BP spectrum is related to change in the mechanical influence of respiration.

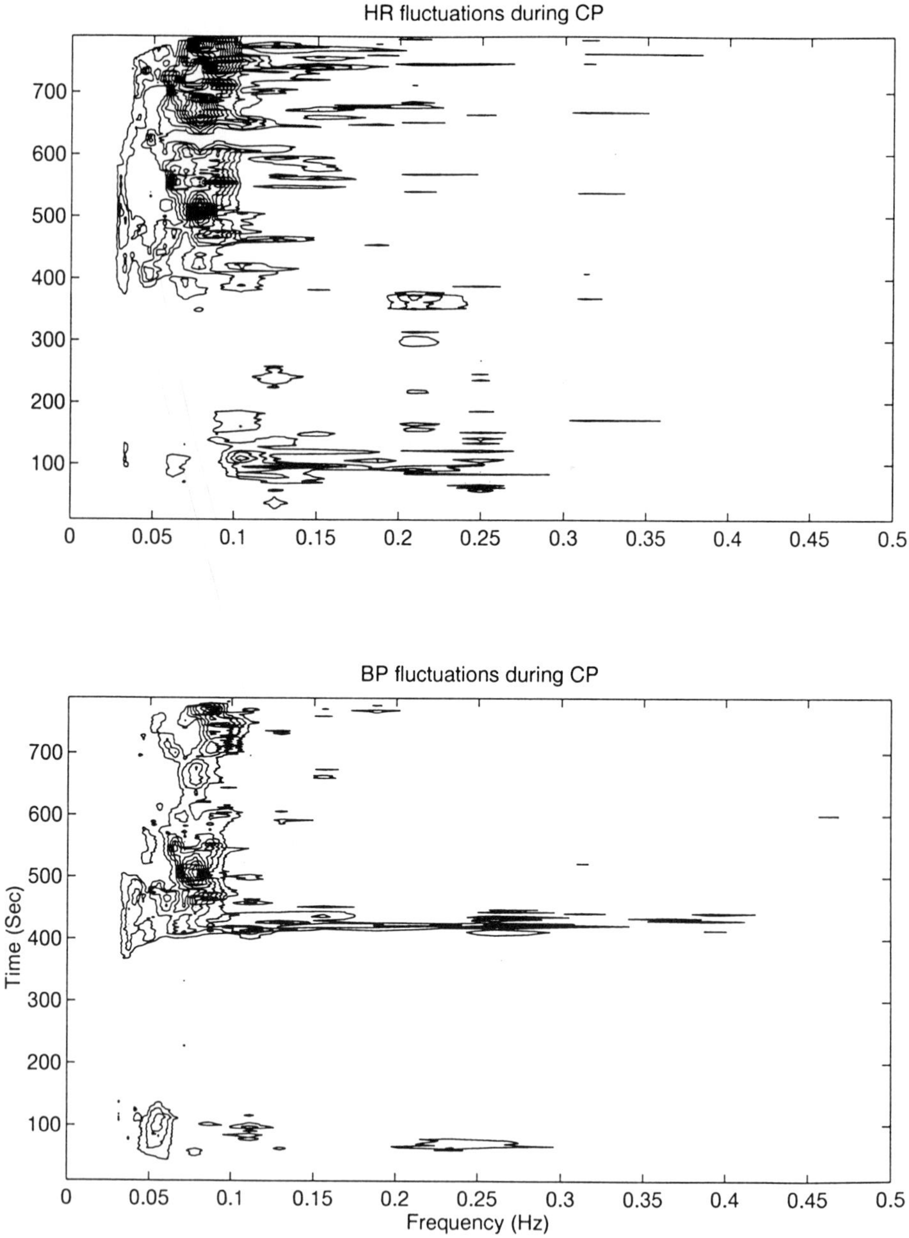

FIGURE 6C. Change of posture: Time-dependent spectra of HR and BP as contour plots. Again note the increase in the LF region of the spectrum.

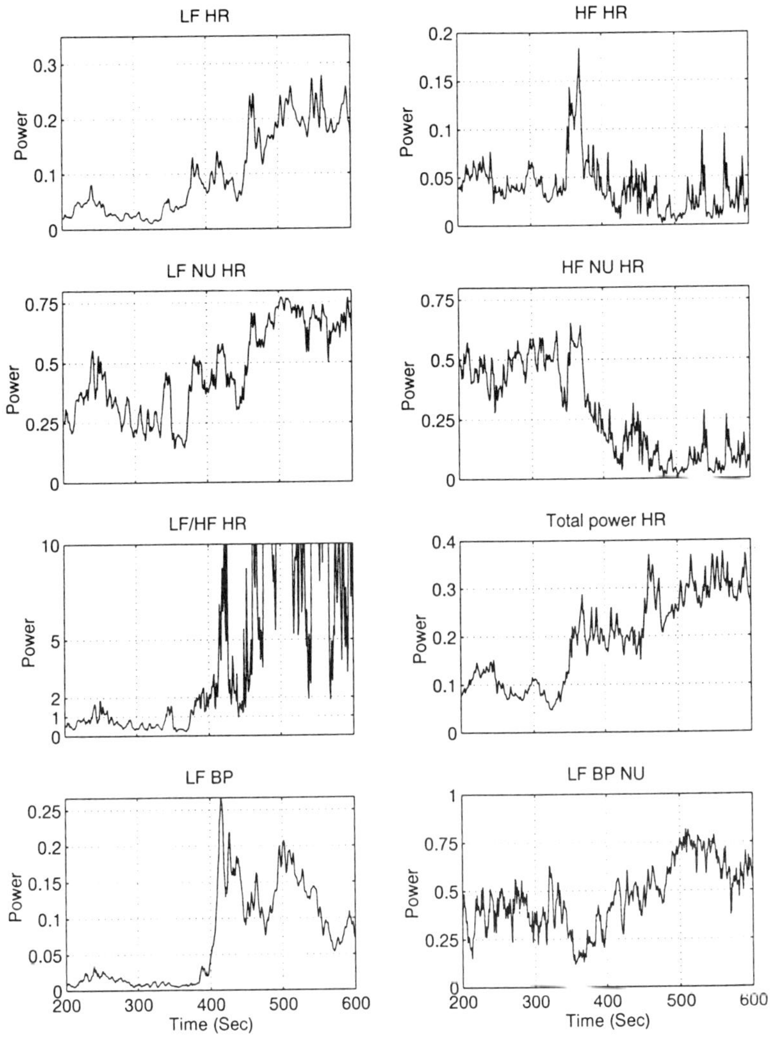

FIGURE 6D. Change of posture: Time-dependent integrals of (from left to right and from top to bottom) LF and HF of HR, in absolute and normalised units, LF/HF ratio and total power of HR fluctuations, LF of BP fluctuations in normalised and absolute units. The increase in sympathetic activity is reflected in the LF content of HR and BP spectra. The decrease in HF is related to an increase in vagal activity. The normalised LF and HF contents of HR fluctuations and the LF/HF ratio display the switch from vagal predominance in supine position (50% HF for 25% LF) to sympathetic predominance on standing (70% LF for 10% HF).

Pharmacological studies have confirmed that the response to CP is mediated by autonomic pathways. Atropine alters the characteristic response leaving only a very slight increase in HR during 30 s after standing. Additional propranolol reduces the response still further, although propranolol alone does not alter the normal HR pattern.[121] Thus the parasympathetic pathways mediate most, but not all, of the HR response.

Steady state spectral analysis of HR fluctuations has shown that the total power and the LF (0.05–0.15 Hz) power in absolute and normalised units are significantly raised on standing as compared to supine position[122,123] (Figure 6d). The normalised HF power is significantly reduced. The coherence between HR and respiration decreased significantly during orthostatic load.[122] All these results reflect an increase in sympatho-vagal balance.

Time-frequency analysis is the ideal tool for CP response analysis and has therefore been applied to HR and BP fluctuations during an active CP. The instantaneous changes revealed in autonomic activity are very clear and can be used to detect pathologies such as in hypertension.[19,11]

Interpretation

An example of the effects of CP on HR and BP is shown in Figure 6, for a normal subject. The subject started standing at t=400 s. The surge in HR overcomes the fall in BP. HR overshoots and then stabilises at some intermediate level (Figure 6a). Mean BP remains constant before and after CP. A change in the fluctuations pattern occurs, both in HR and BP and can be displayed as 3D time-dependent spectra and contour plots (Figures 6b and 6c). The increase in the LF peak of both HR and BP power spectra is obvious. The increase in the HF peak of the BP spectrum is due to the change in nature of respiration from mainly chest to mainly abdominal respiration, and is not directly related to vagal activity.

The time evolution of the activity of both autonomic branches, as well as the sympatho-vagal balance, is displayed in figure 6d. The LF integrals of HR and BP reflect the increase in sympathetic activity, while the HF integral of HR reflects the decrease in parasympathetic activity. Power becomes concentrated in the LF range (Figure 6d). There is a strong increase in the total power of both HR and BP on standing.

Head Up Tilt

General Description and Arrangement

The subject initially lies down on a tilt table in supine position which is then tilted to a 60–80° angle. The tilt table usually has a foot support and an antislip mattress.[124] The use of a 60–80° angle of tilt has been advised to avoid alarm reactions associated with the falling-forward sensation, while at 70 tilt, the gravitational influence on the circulation is only 6% less than in the vertical posture. Ewing et al reported a difference HR response to fast (2–3 s) and slow (12 s) tilt, with a gradual shortening of RR interval and no rebound RR interval lengthening for slow tilt.[119]

Autonomic Effects

Head up tilt induces a gradual increase of the HR with little or no overshoot. It is accompanied by a gradual rise in diastolic pressure and with little change in systolic pressure.[125] The gradual increase in HR is probably due to both diminished parasympathetic and enhanced sympathetic outflow to the heart[126] which in turn, have been attributed to the unloading of carotid sinus and cardiopulmonary receptors.[127,128] The enhanced sympathetic outflow to the vessels which prevents any fall of arterial pressure can be ascribed predominantly to a reduction in the activity of cardiopulmonary receptors.

In normal subjects, passive tilt always leads to a decrease in HF HRV and an increase in LF variability.[12,53,129–131] With tilt, the LF component of HRV becomes predominant (up to 90%) and the LF/HF ratios of HRV and of BPV increase.[132] A strong correlation between the degree of tilt and the LF and HF components of HRV, expressed in normalised units has been displayed.[45,46]

Time-dependent changes of the spectral components of HR fluctuations during tilt have been analysed in several studies by applying various approaches such as by: time-varying auto-regressive modeling[133], smoothed Wigner-Ville distribution,[14,15] SDA,[12] Evolutionary Periodogram[24] and complex demodulation.[18]

Lower Body Negative Pressure

The technique of lower body negative pressure (LBNP), or lower body subatmospheric pressure (LBSP) is used in physiological laboratories to study the effect of cardiopulmonary baroreflexes and orthostatic stress after exposure to microgravity.[134] Application of negative pressure to the lower body impedes venous return, causing a reduction in central blood volume. Adding LBNP to the gravitational shift of blood which occurs on standing exaggerates orthostatic stress. The sequential application of head-up tilt and then, head-up tilt plus LBNP, thus seems a sensitive ideal means of assessing orthostatic tolerance. When the LBNP test is performed 20 min after tilt (LBNP of –40 mm Hg of pressure), 85% of normal subjects faint. Yet, Hainsworth et al reported that the addition of LBNP increases the sensitivity of the tilt test without markedly affecting its specificity.[135]

Autonomic Effects

The method may be used to differentiate between the reflex cardiovascular effects induced by the deactivation of the low-pressure cardiovascular receptors and those produced by the deactivation of the arterial baroreceptors. The former results in an increase in the limb vasculature resistance without a change in HR and BP, while the latter produces a tachycardia, an increase in diastolic BP and a decrease in systolic BP. Cardiovascular effects of –40 mm Hg are similar to those produced by a change of posture from supine to standing.[136]

Cold Pressor Test

General Description and Arrangement

The cold pressor test (CPT) is performed by hand or foot immersion in water at 1 to 5C for 1 to 6 min.[137-139] The CPT has been widely applied since the 1930s, first described by Hines and Brown.[140] It causes a strong sympathetic stimulus[141] and is accompanied, in normal subjects, by an increase in blood pressure, heart rate, total vascular resistance, muscle sympathetic nerve discharge, galvanic skin resistance and a decrease in forearm blood flow.[137,138]

The CPT has been extensively used, mostly for the evaluation of vascular reactivity in healthy subjects.[138,142-146]

Autonomic Effects

The cold pressor test (CPT) is a non-specific sympatho-excitatory stimulus.[147] It induces a rise in cardiac output caused by accelerated heart rate and a positive inotropic effect, which occur through sympathetic activation as well as inhibition of vagal tone.[148] Ebert et al reported a significant increase in muscle sympathetic nerve activity (MSNA) during CPT.[139] Weise et al used standard FFT analysis and reported that the CPT induces an increase in the low frequency components of heart rate and systolic blood pressure fluctuations which reflect increased sympathetic vasomotor activity. They reported no changes in the transfer gain and phase characteristics of the HR modulation by systolic BP.[138]

GENERAL CLINICAL APPLICATIONS

The clinical applications of autonomic manoeuvres can be divided into two distinct domains of investigation: (1) pathologies, diseases and clinical situations in which the autonomic responsiveness is to be evaluated; and (2) clinical and basic research in which some specific part of the cardiovascular system is to be investigated.

AUTONOMIC RESPONSIVENESS IN PATHOLOGIES

Essential Hypertension

Essential hypertension and its various autonomic causes and effects have been extensively investigated. In borderline hypertensive subjects, an attenuated HR response to deep breath test, as well as a hyperactivity in response to mental stress and sustained hand grip have been reported.[149,150,151,152] A reduced response to active standing and to head up tilt has been repeatedly observed in mild-hypertensive subjects.[11,32,130] Furthermore, the altered effects of tilt were significantly correlated with the degree of hypertensive state.[130] Finally, cold pressor test has been used for the evaluation of vascular reactivity.[140,144,153]

Diabetes

Diabetes, with its induced autonomic neuropathy, is an important candidate for autonomic tests. A reduced epinephrine response to mental stress and an attenuated HR response to deep breath test have been observed among diabetic patients with autonomic neuropathy as compared to patients without neuropathy and normal subjects. The responses to change of posture, hand grip and Valsalva have also been used as diagnostic or research tools in diabetes.[92,154–158]

Postural Tachycardia Syndrome

Novak et al used time-frequency analysis of HR and BP fluctuations during tilt test in postural tachycardia syndrome (POTS).[15]

Asthma

Investigation of the autonomic control in asthmatic patients has been performed by means of the response to Valsalva, deep breathing, vagal blockade by atropine, change of posture and hand grip. In general, asthmatics showed heightened cardiovascular responses to tests of parasympathetic function as compared to controls.[159]

Other Conditions

Autonomic response to some of these tests has been estimated in a variety of additional diseases, such as in Parkinson's disease,[160] multiple sclerosis,[161] coronary heart disease,[162–164] congestive heart failure,[123,165] anginal patients,[166] alcoholic patients,[167] autonomic failure patients.[13]

CLINICAL AND BASIC RESEARCH

Head up tilt is widely used to reproduce symptoms of syncope.[12,168–172]

The time-dependent response to an active change of posture during the transition itself allows to establish the autonomic dysfunction in hypertensive patients.[11]

The LBNP test is useful in the aeromedical assessment of apparently normal individuals with low orthostatic tolerance, in the evaluation of the effects of physiological and pharmacological interventions on cardiovascular reflexes, and in the evaluation of patients with autonomic neuropathies. It is also a means of inducing safe, well-controlled syncope in order to study the genesis of this phenomenon.[136]

The CPT has been extensively used, mostly for the evaluation of vascular reactivity in healthy subjects.[142–146] The stimulus response of the LF component could serve as an index of integrity and reactivity of the sympathetic nervous system in patients with autonomic neuropathy and hyper-dynamic circulation, respectively.[138] These tests have also been extensively applied in the basic research concerned with the autonomic side-effects of drugs.

CONCLUSION

Any combination chosen from the set of autonomic tests described above can be applied for the challenging of autonomic responsiveness in pathologies or for the investigation of their basic mechanisms and function in health or disease. The recent insight added by time-frequency decomposition techniques have enlarged the possibilities of the analysis of autonomic tests. The transient response to such provocations can reveal autonomic alteration or dysfunction, otherwise masked at steady-state.

The interpretation of the response may gain important added value if the quantification procedure is reliably adapted to the short-term, non steady state nature of the provocation and its response. Procedures such as time-frequency decomposition seem to provide the rigorous insight required.

REFERENCES

1. Kitney RI, Fulton T, Mac Donald R, Linkern DA. Transient interaction between blood pressure, respiration and heart rate in human. *J Biomed Engin* 1985; **7**:217–224.
2. Akselrod S, Gordon D, Ubel FA, Shannon DC, Barger AC, Cohen RJ. Power spectrum analysis of heart rate fluctuation: a quantitative probe of beat-to-beat cardiovascular control. *Science* 1981; **213**:220–222.
3. Hyndman BW, Kitney RI, Sayers BA. Spontaneous rhythms in physiologic control systems. *Nature* 1971; **341**:233–239.
4. Keselbrener L, Akselrod S. Selective discrete fourier transform algorithm for time-frequency analysis: method and application on simulated and cardiovascular signals. *IEEE Trans Biomed Eng* 1996; **43**:789–802.
5. Kleiger RE, Stein PK, Bosner MS, Rottman JN. Time-domain measurements of heart rate variability. In: Malik M, Camm AJ (Eds) Heart Rate Variability. Futura, Armonk 1995, 33–46.
6. Coumel P, Maison-Blanche P, Catuli D. Heart rate and heart rate variability. In: Malik M, Camm AJ (Eds) Heart Rate Variability. Futura, Armonk 1995, 207–222.
7. Coumel P, Hermida JS, Wennerblom B, Leenhardt A, Maison-Blanche P, Cauchemez B. Heart rate variability in myocardial hypertrophy and heart failure, and the effects of beta-blocking therapy. A nonspectral analysis of heart rate oscillations. *Eur Heart J* 1991; **12**:412–422.
8. Baharav A, Keselbrener L, Oz O, Lerhman-Sagie T, Akselrod S. The autonomic control in vasovagal syncope investigated by spectral and time-dependent analysis of heart rate fluctuations. In: Computers in Cardiology 1993, 269–272.
9. Keselbrener L, Baharav A, Akselrod S. Selective windowed time-frequency analysis for the quantitative evaluation of non-stationary cardiovascular signals. In: Computers in Cardiology 1994, 5–8.
10. Keselbrener L, Baharav A, Akselrod S. Estimation of fast vagal response by time-dependent analysis of heart rate variability in normal subjects. *Clin Auton Res* 1996; **6**:321–327.

11. Akselrod S, Oz O, Grinberg M, Keselbrener L. Dynamic autonomic response to change of posture investigated by time-dependent heart rate variability among normal and mild-hypertensive adults. *J Auton Nerv Syst* 1997; **64**:33–43.
12. Keselbrener L, Baharav A, Ben-Dov Y, Diamant S, Berger A, Sivan Y, Akselrod S. Time-frequency decomposition of heart rate and blood pressure variability in reflex syncope. In: Computers in Cardiology 1997, 219–222.
13. Keselbrener L, Korzets A, Akselrod S. Failure of erythropoietin treatment in a case with primary autonomic failure: an evaluation by spectral analysis of heart rate variability. *Clin Auton Res* 1997; **7**:333–336.
14. Novak P, Novak V. Time/frequency mapping of the heart rate, blood pressure and respiratory signals. *Med Biol Eng Comput* 1993; **31**:103–10.
15. Novak V, Novak P, Opfer-Gehrking TL, Low PA. Postural tachycardia syndrome: time-frequency mapping. *J Auton Nerv Syst* 1996; **61**:313–320.
16. Novak V, Novak P, De Champlain J, Nadeau R. Altered cardiorespiratory transfer in hypertension. *Hypertension* 1994; **23**:104–113.
17. Novak P, Novak V, Li Z, Remillard G. Time-frequency analysis of slow cortical activity and cardiovascular fluctuations in a case of alzheimer's disease. *Clin Auton Res* 1994; **4**:141–148.
18. Hayano J , Sakakibara Y, Yamada A, Yamada M, Mukai S, Fujinami T, Yokoyama K, Watanebe Y, Takata K. Accuracy of assessment of cardiac vagal tone by heart rate variability in normal subjects. *Am J Cardiol* 1991; **67**:199–204.
19. Bianchi AM, Mainardi L, Petrucci E, Signorini MG, Mainardi M, Cerutti S. Time-variant power spectrum analysis for the detection of transient episodes in hrv signal. *IEEE Trans Biomed Eng* 1993; **40**:135–144.
20. Bianchi A, Mainardi L, Meloni C, Chierchia S, Cerutti S. Continuous monitoring of the sympatho-vagal balance through spectral analysis. *IEEE Eng Med Biol* 1997; **16**:64–73.
21. Pola S, Macerata A, Emdin M, Marchesi C. Estimation of the power spectral density in nonstationary cardiovascular time series: assessing the role of the time-frequency representation (TFR). *IEEE Trans Biomed Eng* 1996; **43**:46–59.
22. Van Steenis HG, Tulen JH. The exponential distribution applied to nonequidistantly sample cardiovascular time series. *Comput Biomed Res* 1996; **29**:174–193.
23. Marchesi C, Venturi M, Pola S, Conforti F, Macerata A, Varanini M, Emdin M. Sequential estimation of the power spectrum for the analysis of variability of non stationary cardiovascular signals. *Proc IEEE* 1991; **13**:578–579.
24. Niccolai M, Varanini M, Macerata A, Pola S, Emdin M, Cipriani M, Marchesi C. Analysis of non stationary heart rate series by evolutionary periodogram. In: Computers in Cardiology 1995; 449–452.
25. Varanini M, De Paolis G, Emdin M, Macerata A, Pola S, Cipriani M, Marchesi C. Spectral analysis of cardiovascular time series by the s-transform. In: Computers in Cardiology 1997, 383–386.

26. Figliola A, Serrano E. Analysis of physiological time series using wavelet transforms. *IEEE Eng Med Biol* 1997; **16:**74–79.
27. Wiklund U, Akay M, Niklasson U. Short-term analysis of heart-rate variability by adapted wavelet transforms. *IEEE Eng Med Biol* 1997; **16:**113:118.
28. Bootsma M, Swenne CA, Lenders JWM, Jacobs MC, Bruschke AVG. Intravenous instrumentation alters the autonomic state. *Eur J Appl Physiol* 1996; **73:**113–116.
29. Low PA, Pfeifer MA. Standardization of clinical tests for practice and clinical trials. In: Low PA (Ed) Clinical Autonomic Disorders. Rochester 1993, 287–296.
30. Arnold RW, Dyer JA, Gould AB, Low PA. Sensitivity to vasovagal maneuvers in normal children and adults. *Mayo Clinic Proc* 1991; **66:**797–804.
31. Andresen D, Bruggemann T, Behrens S, Ehlers C. Heart rate response to provocative maneuvers. In: Malik M, Camm AJ (Eds) Heart Rate Variability. Futura, Armonk 1995, 267–274.
32. Yo Y, Nagano M, Nagano N, Iiyama K, Higaki J, Mikami H, Ogihara T. Effect of age and hypertension on autonomic nervous regulation during passive head-up tilt. *Hypertension* 1994; **23**(Suppl I): I82–I86.
33. Low PA. Laboratory evaluation of autonomic function. In: Low PA (Ed) Clinical Autonomic Disorders. Rochester 1993, 169–196.
34. Gelber DA, Pfeifer M, Dawson B, Schumer M. Cardiovascular autonomic nervous system tests: determination of normative values and effect of confounding variables. *J Auton Nerv Syst* 1997; **62:**40–44.
35. Ewing DJ, Campbell IW, Murray A, Neilson JM, Clarke BF. Immediate heart-rate response to standing: simple test for autonomic neuropathy in diabetes. *Br Med J* 1978; **1:**145–147.
36. Braune S, Schulte-Monting J, Schwerbrock S, Lucking CH. Retest variation of cardiovascular parameters in autonomic testing. *J Auton Nerv Syst* 1996; **60:**103–107.
37. Shefi O, Davidson S, Maayan A, Akselrod S. The effect of thermal stimulation on the heart rate variability in neonates. *Early Human Dev*. In Press.
38. Madden K, Savard GK. Effects of mental state on heart rate and blood pressure variability in men and women. *Clin Physiol* 1995; **15:**557–569.
39. Ramet J, Hauser B, Waldura J, De Prins J. Circadian Rhythm of Cardiac Responses to Vagal Stimulation Tests. *Ped Neurol* 1992; **8:**91–96.
40. Keselbrener L, Keselbrener M, Akselrod S. Nonlinear high pass filter for R-wave detection in ECG signal. *Med Eng Phys* 1997; **19:**481–484.
41. Berger RD, Akselrod S, Gordon D, Cohen RJ. An efficient algorithm for spectral analysis of heart rate variability. *IEEE Trans Biomed Eng* 1986; **33:**900–904.
42. Akselrod S. Components of heart rate variability: basic studies. In: Malik M, Camm AJ (Eds) Heart Rate Variability. Futura, Armonk 1995, 147–163.

43. Malliani A, Pagani M, Lombardi F, Rimoldi O. Low frequency oscillation of heart rate and arterial pressure variabilities as a marker of sympathetic modulation of cardiovascular function. In: Di Rienzo M, Mancia G, Parati G, Pedotti A, Zanchetti A (Eds) Blood Pressure and Heart Rate variability, IOS Press, 1993, 215–229.
44. Task Force of the European Society of Cardiology and the North American Society of Pacing and Electrophysiology. Heart rate variability, standards of measurement, physiological interpretation and clinical use. *Circulation* 1996; **93:**1043–1065.
45. Bootsma M, Swenne CA, Van Bolhuis HH, Chang PC, Cats VM, Bruschke AVG. Heart rate and heart rate variability as indexes of sympathovagal balance. *Am J Physiol* 1994; **266:**H1565–H1571.
46. Montano N, Ruscone TG, Porta A, Lombardi F, Pagani M, Malliani A. Power spectrum analysis of heart rate variability to assess the changes in sympathovagal balance during graded orthostatic tilt. *Circulation* 1994; **90:**1826–1831.
47. Angelone A, Coulter NA. Respiratory and sinus arrhythmia: a frequency dependent phenomena. *J Appl Physiol* 1964; **19:**479–482.
48. Hirsh JB, Bishop B. Respiratory sinus arrhythmia in humans: how breathing pattern modulates heart rate. *Am J Physiol* 1981; **241:**H620–H629.
49. Brown TE, Beightol LA, Koh J, Eckberg DL. Important influence of respiration on human rr interval power spectra is largely ignored. *J Appl Physiol* 1993; **75:**2310–2317.
50. Berger RD, Saul JP, Cohen RJ. Assessment of autonomic response by broad band respiration. *IEEE Trans Biomed Eng* 1989; **36:**1061–1065.
51. Appel S, Kuritxky A, Zahavi I, Zigelman M, Akselrod S. Evidence for instability of the autonomic nervous system in patients with migraine headache. *Headache* 1992; **32:**10–17.
52. Gilman AG, Rall TW, Nies AS, Taylor P. Goodman and Gillman's The Pharmacological Basis of Therapeutics, Pergamon Press, New York 1990.
53. Pomeranz B, Macauley RJB, Caudill MA, Kutz I, Adam D, Gordon D, Kilborn KM, Barger AC, Shannon DC, Cohen RJ. Assessment of autonomic function in humans by heart spectral analysis. *Am J Physiol* 1985; **248:**H151–H153.
54. Lishner M, Akselrod S, Mor-Avi V, Oz O, Divon M, Ravid M. Spectral analysis of heart rate fluctuations: a non-invasive method for early diagnosis of diabetic neuropathy. *J Auton Nerv Syst* 1987; **19:**119–125.
55. Binkley PF, Nunziata E, Haas GJ, Nelson SD, Cody RJ. Parasympathetic withdrawal is an integral component of autonomic imbalance in congestive heart failure: demonstration in human subjects and verification in a paced canine model of ventricular failure. *J Am Coll Cardiol* 1991; **18:**464–472.
56. Izraeli S, Alcalay M, Benjamini Y, Wallach-Kapon R, Tochner Z, Akselrod S. Modulation of the dose-dependent effects of atropine by low-dose pyridostigmine: quantification by spectral analysis of heart rate fluctuations in healthy human beings. *Pharmacol Biochem Behav* 1991; **39:**613–617.

57. Ali Melkkila T, Kaila T, Antila K, Halkola L, Iisalo E. Effects of glycopyrrolate and atropine on heart rate variability. *Acta Anaesthesiol Scand* 1991; **35**:436–441.
58. Hayano J, Taylor AJ, Yamada A, Mukai S, Hori R, Asakawa T, Yokoyama K, Watanabe Y, Takata K and Fujinami T. Continuous assessment of hemodynamic control by complex demodulation of cardiovascular variability. *Am J Physiol* 1993; **264**:H1229–H1238.
59. Uusitalo AL, Tahvanainen KU, Uusitalo AJ, Rusko HK. Non-invasive evaluation of sympathovagal balance in athletes by time and frequency domain analyses of heart rate and blood pressure variability. *Clin Physiol* 1996; **16**:575–588.
60. Vybiral T, Bryg RJ, Maddens ME, Bhasins SS, Cronin S, Boden WE, Lehmann MH. Effects of transdermal scopolamine on heart rate variability in normal subjects. *Am J Cardiol* 1990; **65**:604–608.
61. Vesalainen RK, Tahvanainen KU, Kaila TJ, Kantola IM, Kuusela TA, Eckberg DL. Effects of low-dose transdermal scopolamine on autonomic cardiovascular control in healthy young subjects. *Clin Physiol* 1997; **17**:135–148.
62. Kochiadakis GE, Rombola AT, Kanoupakis EM, Zuridakis EG, Skalidis EI, Vardas PE. Effect of transdermal scopolamine on heart rate variability in patients with severe coronary heart disease. *Pacing Clin Electrophysiol* 1996; **19**:1867–1871.
63. Pedretti R, Colombo E, Sarzi-Braga S, Caru B. Influence of transdermal scopolamine on cardiac sympathovagal interaction after acute myocardial infarction. *Am J Cardiol* 1993; **72**:384–392.
64. Casadei B, Pipilis A, Sessa F, Conway J, Sleight P. Low doses of scopolamine increase cardiac vagal tone in the acute phase of myocardial infarction. *Circulation* 1993; **88**:353–357.
65. La Rovere MT, Mortara A, Pantaelo P, Maestri R, Cobelli F, Tavazzi L. Scopolamine improves autonomic balance in advanced congestive heart failure. *Circulation* 1994; **90**:838–843.
66. Hull SS, Vanoli E, Adamson PB, De-Ferrari GM, Foreman RD, Schwartz PJ. Do increases in markers of vagal activity imply protection from sudden death? The case of scopolamine. *Circulation* 1995; **91**:2516–2519.
67. Kawamoto M, Matsumoto C, Yuge O. Atropine premedication attenuates heart rate variability during high thoracic epidural anesthesia. *Acta Anaesthesiol Scand* 1996; **40**:1132–1137.
68. Parlow JL, Van Vlymen JM, Odell MJ. The duration of impairment of autonomic control after anticholinergic drug administration in humans. *Anesth Analg* 1997; **84**:155–159.
69. Langewitz W, Ruddel H, Schachinger H, Lepper W, Mulder LJ, Veldman JH, Van Roon A. Changes in sympathetic and parasympathetic cardiac activation during mental load: an assessment by spectral analysis of heart rate variability. *Homeost Health Dis* 1991; **33**:23–33.
70. Ahmed MW, Kadish AH, Parker MA, Goldberger JJ. Effect of physiologic and pharmacologic adrenergic stimulation on heart rate variability. *J Am Coll Cardiol* 1994; **24**:1082–1090.

71. Jokkel G, Bonyhay I, Kollai M. Heart rate variability after complete autonomic blockade in man. *J Auton Nerv Syst* 1995; **51**:85–89.
72. Akselrod S, Gordon D, Madwed JB, Snidman NC, Shannon DC, Cohen RJ. Hemodynamic regulation: investigation by spectral analysis. *Am J Physiol* 1985; **249**:H867–H875.
73. Japundzic N, Grichois ML, Zitoun P, Laude D, Elgozhi JL. Spectral analysis of blood pressure and heart rate in conscious rats: effects of autonomic blockers. *J Auton Nerv Syst* 1990; **30**:91–100.
74. Murphy CA, Sloan RP, Myers MM. Pharmacological responses and spectral analyses of spontaneous fluctuations in heart rate and blood pressure in shr rats. *J Auton Nerv Syst* 1991; **36**:237–250.
75. Niv D, Chayen MS. Sympathetic block and sympathectomy. In: Korczyn AD (Ed) Handbook of Autonomic Nervous System Dysfunction, Marcel Dekker, New York 1995, 399–412.
76. Airaksinen KE, Niemela MJ, Huikuri HV. Effect of beta-blockade on baroreflex sensitivity and cardiovascular autonomic function tests in patients with coronary artery disease. *Eur Heart J* 1994; **15**:1482–1485.
77. Akselrod S, Eliash S, Oz O, Cohen S. Hemodynamic regulation in the spontaneously hypertensive rat: investigation by spectral analysis. *Am J Physiol* 1987; **253**:H176–183.
78. Almog Y, Eldor S, Oz O, Akselrod S. Beat to beat fluctuations in the bp related signals in rats: can it contribute to the understanding of the development of hypertension? *J Auton Nerv Syst*, In Press.
79. Folino AF, Buja G, Turrini P, Oselladore L, Nava A. The effects of sympathetic stimulation induced by mental stress on signal-averaged electrocardiogram. *Int J Cardiol* 1995; **48**:279–285.
80. Matsuwaka T, Sugiyama Y, Mano T. Increased muscle sympathetic nerve activity during delayed auditory feedback in humans. *Jpn J Physiol* 1995; **45**:905–911.
81. Bernston GG, Cacioppo JT, Binkley PF, Uchino BN, Quigley KS, Fieldstone A. Autonomic cardiac control. III. Psychological stress and cardiac response in autonomic space as revealed by pharmacological blockades. *Psychophysiology* 1994; **31**:599–608.
82. Moriguchi A, Otsuka A, Kohara K, Mikami H, Katahira K, Tsunetoshi T, Higashimori K, Ohishi M, Yo Y, Ogihara T. Spectral change in heart rate variability in response to mental arithmetic before and after the beta-adrenoceptor blocker, carteolol. *Clin Auton Res* 1992; **2**:267–270.
83. Albus M, Zellner A, Bondy B, Muller-Spahn F, Engel R, Ackenheil M. Influence of cpg 361a, propanolol and diazepam on autonomous reactions to different stressors. *Prog Neuropsychopharmacol Biol Psychiatry* 1989; **13**:87–97.
84. Shapiro PA, Sloan RP, Horn EM, Myers MM, Gorman JM. Effect of innervation on heart rate response to mental stress. *Arch Gen Psychiatry* 1993; **50**:275–279.
85. Jern S, Pilhall M, Jern C. Effect of cholinergic blockade on heart rate, blood pressure and plasma catecholamine responses to mental stress in normal subjects. *Clin Auton Res* 1991; **1**:225–231.

86. Sloan RP, Shapiro PA, Bagiella E, Boni SM, Paik M, Bigger JT, Steinman RC, Gorman JM. Effect of mental stress throughout the day on cardiac autonomic control. *Biol Psychol* 1994; **37**:89–99.
87. Borst C, Hollander AP, Bouman LM. Cardiac acceleration elicited by voluntary muscle contractions of minimal duration. *J Appl Physiol* 1972; **32**:70–77.
88. Seals DR, Chase PB, Taylor JA. Autonomic mediation of the pressor responses to isometric exercise in humans. *J Appl Physiol* 1988; **64**:2190–2196.
89. Ewing DJ, Bellavere F, Espi F, McKibben BM, Buchanan KD, Riemersma RA, Clarke BF. Correlation of cardiovascular and neuroendocrine tests of autonomic function in diabetes. *Metabolism* 1986; **35**:349–353.
90. Vissing SF and Victor RG. Central motor command activates sympathetic outflow to the skin during static exercise. *Acta Physiol Scand* 1989:136(suppl):44.
91. Saito M, Mano T, Abe H, Iwase S. Responses in muscle sympathetic nerve activity to sustained hand-grips of different tensions in humans. *Eur J Appl Physiol* 1986; **55**:493–498.
92. Ewing DJ, Irving JB, Kerr F, Wildsmith JA, Clarke BF. Cardiovascular responses to sustained handgrip in normal subjects and in patients with diabetes mellitus: a test of autonomic function. *Clin Sci Mol Med* 1974; **46**:295–306.
93. Trouerbach J, Duprez D, De Buyzere M, De Sutter J, Clement D. Cardiovascular responses elicited by different simulated diving manoeuvres. *Eur J Appl Physiol* 1994; **68**:341–344.
94. Gooden BA. Mechanism of the human diving response. *Integr Physiol Behav Sci* 1994; **29**:6–16.
95. Arnold RW. Extremes in human breath hold, facial immersion bradycardia. *Undersea Biomed Res* 1985; **12**::183–190.
96. Gooden BA. The diving response in clinical medicine. *Aviat Space Environ Med* 1982; **53**:273–276.
97. Aydin M, Baysal K, Kucukoduk S, Cetinkaya F, Yaman S. Application of ice water to the face in initial treatment of supraventricular tachycardia. *Turk J Pediatr* 1995; **37**:15–17.
98. Olsson G, Rehnqvist N, Ekman AM. Responsiveness to diving reflex after acute myocardial infarction. *Acta Med Scand* 1988; **223**:431–435.
99. Fagius J, Sundlof G. The diving response in man: effects on sympathetic activity in muscle and skin nerve fascicles. *J Physiol Lond* 1986; **377**:429–433.
100. Ramet J, Hauser B, Dehan M, Curzi-Dascalova L, Gaultier C. Effect of state of alertness on the hr response to ocular compression in human infants. *Biol Neonate* 1995; **68**:270–275.
101. Aschner B. Uber einen bisher noch nicht beschriebenen Reflex von Auge auf Kreislauf and Atmung: Verschwinden des Radialispulses bei Druck auf das Auge. *Wien Klin Wochenschr* 1908; **21**:1521–1530.
102. Yamashita M. Oculocardiac reflex and the anesthesiologist. *Middle East J Anesthesiol* 1986; **8**:399–415.

103. Gold RS, Pollard Z, Bucheald IP. Asystole due to oculocardiac reflex during strabismus surgery: a report of two cases. *Ann Opthalmol* 1988; **20**:473–475.
104. Lang S, Lanigan DT, Van der Wal M. Trigeminocardiac reflexes: maxillary and mandibular variants of the oculocardiac reflex. *Can J Anaesth* 1991; **38**:757–760.
105. Moonie GT, Rees DL, Elton D. The oculocardiac reflex during strabismus surgery. *Can Anaesth Soc J* 1964; **11**:621–632.
106. Kahn A, Riazi J, Blim D. Oculocardiac reflex in near miss for sudden infant death syndrome infants. *Pediatrics* 1983; **71**:49–52.
107. Blanc VF, Hardy JF, Milot J, Jacob JL. The oculocardiac reflex: a graphic and statistical analysis in infants and children. *Can Anaesth Soc J* 1983; **30**:360–369.
108. Bennett T, Farqujar IK, Hosking DJ, Hampton JR. Assessment of methods for estimating autonomic nervous control of the heart in patients with diabetes mellitus. *Diabetes* 1978; **27**:1167–1174.
109. Abraham RR, Abraham RM, Wynn V. Autonomic and electrophysiological studies in patients with signs or symptoms of diabetic neuropathy. *Electroencephalogr Clin Neurophysiol* 1986; **63**:223–230.
110. O'Brien IA, O'Hare JP, Lewin IG, Corrall RJ. The prevalence of autonomic neuropathy in insulin-dependent diabetes mellitus: a controlled study based on heart rate variability. *Q J Med* 1986; **61**:957–967.
111. Mitchell EA, Waelthall SR, Elliott RB. Tests for autonomic neuropathy in diabetic children. *Aust Paediatr J* 1985; **21**:105–109.
112. Ewing DJ. Cardiac Autonomic Neuropathy. In: Jarret RJ (Ed) Diabetes and Heart Disease. Elsevier, Amsterdam 1984, 99–132.
113. Low PA, Zimmerman BR, Dyck PJ. Comparison of distal sympathetic with vagal function in diabetic neuropathy. *Muscle Nerve* 1986; **9**:592–596.
114. Korner PI, Tonkin AM, Uther JB. Reflex and mechanical circulatory effects of graded valsalva maneuvers in normal man. *J Appl Physiol* 1976; **40**:434–440.
115. Leon DF, Shaver JA, Leonard JJ. Reflex heart rate control in man. *Am Heart J* 1970; **80**:729–739.
116. Spodick DH, Meyer MB, Quarry-Pigott VM. Effect of β-adrenergic blockade on beat-to-beat response to valsalva manoeuvre. *Br Heart J* 1974; **36**:1082–1086.
117. Cudkowitcz L. Normal, pharmacological and disordered systemic circulatory responses to the valsalva maneuver. *Respiration* 1968; **25**:81–115.
118. Ewing DJ. Noninvasive evaluation of heart rate: the time domain. In: Low PA (Ed) Clinical Autonomic Disorders, Rochester 1993, 297–314.
119. Ewing DJ, Hume L, Campbell TW, Murray A, Neilson JMM, Clarke BF. Autonomic mechanisms in the initial heart rate response to standing. *J Appl Physiol* 1985; **49**:809–814.
120. Sprangers RLH, Wesseling KH, Imholtz ALT, Weiling W. Initial blood pressure fall upon stand up and onset of exercise explained by changes in total peripheral resistance. *J Appl Physiol* 1991; **70**:523–530.

121. Ewing DJ, Hume L, Campbell IW, Murray A, Neilson JM, Clarke BF. Autonomic mechanisms in the initial heart rate response to standing. *J Appl Physiol* 1980; **49**:809–814.
122. Weise F, Heydenreich F, Runge U. Contributions of sympathetic and vagal mechanisms to the genesis of heart rate fluctuations during orthostatic load. a Spectral Analysis. *J Auton Nerv Syst* 1987; **21**:127–134.
123. Sanderson JE, Yeung DT, Kay RL, Tomlinson B, Critchley JA, Woo KS, Bernardi L. Impact of changes in respiratory frequency and posture on spectral analysis of heart rate and systolic blood pressure variability in normal subjects and patients with heart failure. *Clin Sci Colch* 1996; **91**:35–43.
124. Wieling W, Wesseling KH. Importance of reflexes in the circulatory adjustments to postural changes. In Hainsworth R, Mark AL (Eds) Cardiovascular Reflex Control in Health and Disease, London 1993, 35–63.
125. Borst C, Van Brederode JF, Wieling W, Van Montfrans GA, Dunning AJ. Mechanisms of initial blood pressure response to postural change. *Clinic Sci 1984;* **67**:321–327.
126. Marin Neto JA, Gallo L, Manco JC, Rassi A, Amorin DS. Mechanisms of tachycardia on standing: studies in normal individuals and in chronic chagas heart patients. *Cardiovasc Res* 1980; **14**:541–550.
127. Donald DE, Shepherd JT. Reflexes form the heart and lungs:physiological curiosities or important regulatory mechanisms? *Cardiovasc Res* 1978; **12**:449–469.
128. Wolthuis RA, Bergman SA, Nicogossian AE. Physiological effects of locally applied reduced pressure in man. *Physiol Rev* 1974; **54**:566–595.
129. Saul JP. Heart rate variability during congestive heart failure: observations and implications. In: Di Rienzo M, Mancia G, Parati G, Pedotti A, Zanchetti A (Eds) Blood Pressure and Heart Rate Variability: Computer Analysis, Methodology and Clinical Applications. IOS Press, Amsterdam 1993, 266–275.
130. Malliani A, Pagani M, Lombardi F, Cerutti S. Cardiovascular neural regulation explored in the frequency domain. *Circulation* 1991; **84**:482–492.
131. Saul JP, Berger RD, Chen MH, Cohen RJ. Transfer function analysis of autonomic regulation. II. Respiratory sinus arrhythmia. *Am J Physiol* 1989; **25**:H153–H161.
132. Pagani M, Lombardi F, Guzzetti S, Rimoldi O, Furlan R, Pizzinelli P, Sandrone G, Malfatto G, Dell'Orto S, Piccaluga E. Power spectral analysis of heart rate and arterial pressure variabilities as a marker of sympathovagal interactions in man and conscious dog. *Circ Res* 1986; **59**:178–193.
133. Cerutti S, Baselli G, Bianchi AM, Mainardi LT, Signorini MG, Malliani A. Cardiovascular variability signals: from signal processing to modeling complex physiological interactions. *Automedica* 1994; **16**:45–69.
134. Kaufman H. Head-up tilt, lower body negative pressure, pacemaker and vasovagal syncope. *Clin Auton Res* 1994; **4**:231–232.
135. Hainsworth R, El-Bedawi KM. Orthostatic tolerance in patients with unexplained syncope. *Clin Auton Res* 1994; **4**:239–244.

136.Dikshit MB. Lower body suction and cardiovascular reflexes: physiological and applied considerations. *Indian J Physiol Pharmacol* 1990; **34**:3–12.
137.Singh PI, Khurana I. Cardiovascular responses to cold pressor test: a test for autonomic functions. *J Indian Med Assoc* 1991; **89**:229–231.
138.Weise F, Laude D, Girard A, Aitoun P, Siche JP, Elghozi JL. Effects of the cold pressor test on short-term fluctuations of finger arterial blood pressure and heart rate in normal subjects. *Clin Auton Res* 1993; **3**:303–310.
139.Ebert TJ, Monhanty PK, Kampine JP. Lidocaine attenuates efferent sympathetic responses to stress in humans. *J Cardiothorac Vasc Anesth* 1991; **5**:437–443.
140.Hines EA, Brown GE. The cold pressor test for measuring the reactibility of the blood pressure: data concerning 571 normal and hypertensive subjects. *Am Heart J* 1936; **11**:1–9.
141.Sayinalp S, Sozen T, Ozdogan M. Cold pressor test in diabetic autonomic neuropathy. *Diabetes Res Clin Pract* 1994; **26**:21–28.
142.Feldr R, Wenstrand DEW, Milwnkee WS. The cold pressor test in subjects with normal blood pressure. *Am Heart J* 1942; **23**:766–771.
143.Godden JO, Roth GM, Hines EA. The change in intra-arterial pressure during the immersion of the hand in ice-cold water. *Circulation* 1955; **12**:963–973.
144.Parati G, Pomidossi G, Casadei R, Ravogli A, Groppelli A, Cesana B, Mancia G. Comparison of the cardiovascular effects of different laboratory stressors and their relationship with blood pressure variability. *J Hypertens* 1988; **6**:481–488.
145.Victor RG, Leimbach WN, Seals DR, Wallin BG, Mark AL. Effects of the cold pressure test on muscle sympathetic nerve activity in humans. *Hypertension* 1987; **9**:429–436.
146.Fagius J, Karhuvaara S, Sundolf G. The cold pressor test: effects on sympathetic nerve activity in human muscle and skin nerve fascicles. *Acta Physiol Scand* 1989; **137**:325–334.
147.Herd JA. Cardiovascular response to stress. *Physiol Rev* 1991; **71**:305–330.
148.Scharf M, Korczyn AD. Cold pressor test. In: Korczyn AD (Ed) Handbook of Autonomic Nervous System Dysfunction, Marcel Dekker, New York 1995.
149.Johnston LC. The abnormal heart response to a deep breath in borderline labile hypertension: a sign of autonomic nervous system dysfunction. *Am Heart J* 1980; **99**:487–493.
150.Weder AB, Takiyyuddin M, Sekkarie MA, Julius S. Behavior and hypertension: a pathophysiological puzzle. *J Hypertens* 1989; **Suppl 7:**:S13–S17.
151.Schulte W, Ruddel H, Jacobs U, Von Eiff AW. Hemodynamic abnormalities in borderline hypertension during mental stress. *J Cardiovasc Pharmacol* 1986; 8(suppl 5):S128–S130.
152.Mahajan SK, Mahajan KK, Gupta JP. Autonomic reflexes in borderline hypertensives. *J Assoc Physicians India* 1990; **38**:784-786.
153.Cuddy RP, Samlyan H, Keighley JF, Markason CR, Eich RH, Syracuse NY. Haemodynamic and catecholamine changes during a standard cold pressor test. *Am Heart J* 1966; **71**:446–465.

154. Agarwal A, Anand IS, Sakhuja V, Chugh KS. Effect of dialysis and renal transplantation on autonomic dysfunction in chronic renal failure. *Kidney Int* 1991; **40**:489–495.
155. Stojceva–Taneva O, Masin G, Polenakovic M, Stojcev S, Stojkovski L. Autonomic nervous dysfunction and volume responsive hypotension in hemodialysis patients. *Am J Nephrol* 1991; **11**:123–126.
156. Brown FM, Brink SJ, Freeman R, Rabinowe SL. Anti-sympathetic nervous system autoantibodies. Diminished catecholamines with orthostasis. *Diabetes* 1989; **38**:938–941.
157. Weinrauch LA, D'Elia JA, Gleason RE, Keough J, Mann D, Kennedy FP. Autonomic function in type I Diabetes Mellitus complicated by nephropathy. *Am J Hypertens* 1995; **8**:782–789.
158. Spallone V, Bernardi L, Maiello MR, Cicconetti E, Ricordi L, Fratino P, Menziger G. Twenty-four-hour pattern of blood pressure and spectral analysis of heart rate variability in diabetic patients with various degrees of autonomic neuropathy. Comparison to Standard Cardiovascular Tests. *Clin Sci Colch* 1996; **91**(Suppl):105–107.
159. Gupta S, Dolwani S. A study of autonomic status and its effect on ventilatory functions in bronchial asthma. *India J Chest Dis Allied Sci* 1996; **38**:147–156.
160. Bordet R, Benhadjali J, Destee A, Hurtevent JF, Bourriez JL, Guieu JD. Sympathetic skin response and RR interval variability in multiple system atrophy and idiopathic Parkinson's disease. *Mov Disord* 1996; **11**:268–272.
161. Frontoni M, Fiorini M, Strano S, Cerutti S, Giubilei F, Urani C, Bastianello S, Pozzilli C. Power spectrum analysis contribution to the detection of cardiovascular dysautonomia in multiple sclerosis. *Acta Neurol Scand* 1996; **93**:241–245.
162. Raizner AE, Chahine RA, Ishinori T, Verani M, Zacca M. Provocation of coronary artery spasm by the cold pressor test. *Circulation* 1980; **62**:925–932.
163. Manyari DE, Nolewajka AJ, Purves P, Doman A, Kostuk WJ. Comparative value of the cold pressor test and supine bicycle exercise to detect subjects with coronary artery disease using radionuclide ventriculography. *Circulation* 1983; **64**:571–579.
164. Findlay IN, Gillen G, Cummingham AD, Elloot AT, Atchinson T, Dargie HJ. A comparison of isometric exercise, cold pressor stimulation and dynamic exercise in patients with coronary heart disease. *Eur Heart J* 1988; **9**:657–664.
165. Oren RM, Roach PJ, Schobel HP, Berg WJ, Ferguson DW. Sympathetic responses of patients with congestive heart failure to cold pressor test stimulus. *Am J Cardiol* 1991; **67**:993–1001.
166. Ellestad MH, Thomas LA, Bortolozzo TL, Abate JA, Greenberg PS. Autonomic responses in chest pain syndromes as compared to normal subjects. *Cardiology* 1987; **74**:35–42.
167. Duncan G, Johnson RH, Lambie DG, Whiteside EA. Evidence of vagal neuropathy in chronic alcoholics. *Lancet* 1980; **2**:1053–1057.

168. Bernardi L, Salvucci F, Leuzzi S, Radaelli A, Calciati A, Valle F, Savasta S, Scaramuzza A, Lorini R. Cardiovascular reflex changes preceding episodes of vasovagal syncope in paediatric subjects. *Clin Sci* 1996; **91**(Suppl):25–27.
169. Morillo CA, Klein GJ, Jones DL, Yee R. Time and frequency domain analyses of heart rate variability during orthostatic stress in patients with neurally mediated syncope. *Am J Cardiol* 1994; **74**:1258–1262.
170. Bloomfield DM, Bigger JT, Pavri BB, Han J, Fleiss JL, Rolnitsky LM, Steinman RC. Vagal modulation of RR intervals during head-up tilt and the infusion of isoproterenol. *Am J Cardiol* 1995; **75**:1145–50.
171. Rubin AM, Rials SJ, Marinchak RA, Kowey PR. The head-up tilt table test and cardiovascular neurogenic syncope. *Am Heart J* 1993; **125**:476–482.
172. Lippman N, Stein KM, Lerman BB. Failure to decrease parasympathetic tone during tilt predicts a positive tilt-table test. *Am J Cardiol* 1995; **75**:591–595.

CHAPTER SIX

SHORT-TERM MEASUREMENT OF HEART RATE VARIABILITY

Juha EK Hartikainen, Kari UO Tahvanainen and Tom A Kuusela

The heart is an organ with rich innervation from the parasympathetic and sympathetic limbs of the autonomic nervous system. Although the heart is capable of intrinsic regulation of cardiac rhythm, electrical conduction and contractility, those functions are largely under the control of autonomic nervous system. Particularly, autonomic nervous system is responsible for rapid regulation of cardiac rhythm and pump function in order to match cardiac output with the body needs during various exogenic stimuli of daily life – physical and mental stress, posture changes, etc. Sino-atrial node and atrioventricular node receive both parasympathetic and sympathetic efferent innervation, whereas in the ventricles efferent neural connections are almost solely of sympathetic origin.[1] In addition to efferent innervation, the heart has afferent autonomic receptors, which subserve various reflexes originating from the heart.[2]

The easiest way to demonstrate the effects of autonomic modulation on the heart is to monitor the function of the sino-atrial node, i.e. changes in heart rate. Increased efferent vagal activity is characterised by reduced heart rate and variability of heart rate, whereas sympathetic stimulation increases heart rate and decreases variability of heart rate.[3] Due to the close link between autonomic nervous

Malik M (Ed) Clinical guide to cardiac autonomic tests, 149–176.

system and the function of the sino-atrial node, heart rate and its fluctuations reflect changes in cardiac autonomic control. This neural link creates the basis of assessment of cardiac autonomic regulation through measurement of heart rate variability (HRV).

HEART RATE VARIABILITY

The term "heart rate variability" has become the conventionally accepted term to describe variation of both instantaneous heart rate and RR intervals. Also other terms like "cycle length variability", "heart period variability" and "RR variability" have been used to describe oscillations in consecutive cardiac cycles.

Assessment of HRV is performed from analysis of intervals between adjacent cardiac beats or from instantaneous heart rate values for each cardiac beat from continuous electrocardiogram (ECG) recordings (more precisely intervals between successive fiducial points of R peaks of QRS complexes). In fact, HRV does not measure the rhythms of the sino-atrial node *per se*, because the analysis is not based on measurement of PP intervals (atrial depolarisations) but RR intervals (ventricular depolarisations). Thus, HRV represents not only fluctuations in the rhythm of the sino-atrial node but also fluctuations in atrioventricular conduction (PQ interval). However, it has been shown recently that during sinus rhythm the direct effect of autonomic regulation on atrioventricular conduction is counteracted by the indirect influence of heart rate on atrioventricular node.[4] As a result, during sinus rhythm HRV assessed form RR intervals reflects very accurately the variability of the rhythm of the sino-atrial node.

Most commonly HRV analysis is based on long-term measurement of HRV; the subjects carry an ambulatory ECG recorder during their daily activities, usually for 24 hours. It is generally recommended that long-term analysis of HRV contains at least 18 hours of recording that includes the whole night and morning hours.[5] Indeed, major variations in HRV during long-term recordings are caused by day-night difference in heart rate and HRV. In addition, challenges of daily life, e.g. changes in body posture, physical and mental activity influence substantially HRV measurements based on long-term ambulatory ECG recordings. Although the response of an individual to these challenges likely contains important information about the subject's cardiovascular regulation, the differences in the life style of individuals and in the surrounding environment make it difficult, if not impossible, to standardise the recording conditions throughout the entire long-term recording.

Short-term analysis of HRV is based on shorter, usually 5 minute ECG recordings. An advantage of short-term recordings is that due to restricted duration they can rather easily be performed in very strictly standardised conditions; in a laboratory with constant temperature, free of noise and other exogenic stimuli which might influence cardiovascular autonomic status. In most cases this is performed at rest in the supine, free standing or in head-up tilted position. Alternatively, short-term recordings are suitable for studying the response of cardiac autonomic regulation (response of HRV) to standardised manoeuvres influencing cardiovascular autonomic control (standing up, tilt to an upright position, static or dynamic exercise, mental loading, deep breathing, Valsalva manoeuvre or pharmacological agents).

Traditionally, short-term measurements of autonomic nervous system have been based on changes in heart rate or RR interval and blood pressure following some stimulus, such as Valsalva manoeuvre, deep breathing, orthostatic stimulus and isometric hand grip test[6] However, in this chapter these tests are not reviewed, but we introduce more recent methods used to measure, not absolute changes in heart rate, but variability of heart rate, i.e. HRV – HRV during stable conditions and HRV response to standardised manoeuvres.

ECG RECORDING AND ANALYSIS

The beat-to-beat heart rate signal after QRS complex detection must meet several conditions that are prerequisites for meaningful power spectral analysis; the signal must be sufficiently long, random, stationary and free of arrhythmias and noise.

"Sufficiently long" means that, assuming constant stationarity, the consistency of power spectral estimates improves with an increase in the duration of the signal.[7] The term "random" is defined as sequences that cannot be determined or defined by a unique mathematical expression or rule. Accordingly, the term "stationarity" means that the probability function of the signal does not change over time.

DURATION OF THE RECORDING

The duration of the recording is determined by the nature of the method used for analysis of HRV (time-domain or frequency-domain methods) and the aim of the study.

Time-domain methods are preferred for long-term measurement of HRV and frequency-domain methods for short-term measurement of HRV. However, if time-domain methods are used to assess HRV from short-term measurements, at least 5-minute recordings are recommended.[5]

For frequency-domain measurements it is recommended that the duration of the recording is at least 10 times the wavelength of the lowest frequency boundary of the spectral component under investigation. On the other hand, to ensure the stationarity of the signal (see further), the duration of the recording should not be extended too much. According to the above, the assessment of HF component of heart rate variability (lowest frequency boundary 0.15 Hz) requires a recording of approximately 1 minute duration. Correspondingly, a 4-minute recording is needed for the assessment of LF component (lowest frequency boundary 0.04 Hz) and 1-hour recording for the assessment of VLF component (lowest frequency boundary 0.003 Hz). A recording of 1 hour, however, cannot be considered any more as a short-term recording. In addition, it is difficult to fulfil the stationarity criteria during such a long recording. Thus, in order to give a standardised guideline it is recommended that approximately 5-minute recordings of stationary signal are used for the short-term measurement of HRV and that the interpretation is based on HF and LF spectral components of HRV.

In addition, one has to realise that although long-term or short-term ECG recordings can be analysed with the same mathematical algorithms, there are substantial differences in the physiological background and interpretation between those recordings. Long-term and short-term recordings can not be considered as surrogates for each other just with different duration. Even though the nature of the ECG signal would remain unchanged, total HRV and the proportion of low frequency variations increase with the length of the recording.[8] Thus, it is inappropriate to compare HRV measures obtained from recordings of different duration with each other (Figure 1).

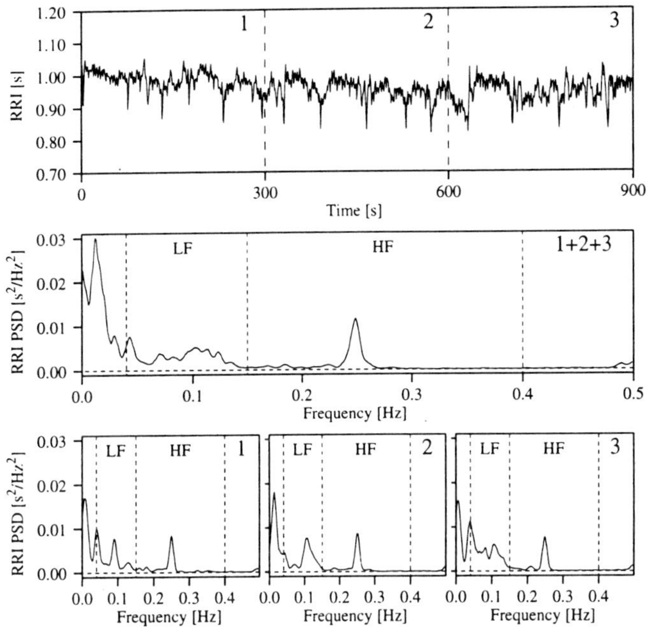

Method: FFT Window type: Hanning Smoothing: triangular. 5 point		Segment of time series 1+2+3	1	2	3
RR interval	(ms)	960	984	952	945
SDNN	(ms)	38	32	34	37
TP	(ms²)	1171	1024	924	1031
VLF	(ms²)	601	437	393	352
LF	(ms²)	327	345	320	473
LFnu	(%)	57	59	60	70
HF	(ms²)	217	221	197	181
HFnu	(%)	38	38	37	27
LF/HF	(%)	151	156	162	262

FIGURE 1. Effect of duration of the recording on frequency-domain measurement of HRV. Recording is performed at supine rest with quiet breathing at 0.25 Hz for 15 min. HRF analysis is performed by FFT from the whole 15-min segment (1+2+3) and from 3 separate non-overlapping 5-min segments (1, 2, and 3, hatched lines cursors in the beat-to-beat time series). The frequency band boundaries at 0.04, 0.15, and 0.4 Hz are indicated in the spectral representation (hatched vertical lines). Note that the VLF power increases when the analysis is extended from 5 to 15 minutes. See Tables 1 and 2 for abbreviations.

TIME AND AMPLITUDE RESOLUTION OF THE RECORDING

The sampling frequency and amplitude resolution of ECG recording for HRV analysis should be properly chosen. The sampling frequency determines the accuracy of the fiducial point recognition of QRS complexes and therefore determines also the accuracy of the measurement of duration between successive RR intervals. A too low sampling rate may produce a jitter in the estimation of the R wave fiducial point. If the fiducial point alters substantially, this fluctuation is measured and it introduces an error in the HRV spectrum that increases with frequency. The optimal sampling frequency is in the range of 250-500 Hz, perhaps even higher.[9] In modern digital Holter systems, sampling frequencies up to 1024 Hz are used. Whether this improves the accuracy of HRV measurement compared to systems utilising lower sampling frequencies, however, has not been studied. A lower sampling frequency may be satisfactory only if some interpolation algorithm is used to define the R wave fiducial point.[10,11] In any case, the sampling frequency should not be less than 100 Hz.

It should be also noted that the amplitude resolution (the number of bits) of the analogue-to-digital conversion must be adequate. Normally 12-bits conversion are used, but the effective resolution is generally significantly lower because the full dynamical range of input signal in analogue-to-digital converters can not be usually utilised in order to prevent saturation or clipping effects in the case of large ECG signals and excessive baseline wandering. If the amplitude of the ECG signal is very small the estimation of the R wave fiducial point is difficult due to low effective resolution.

STATIONARITY OF THE RECORDING

It is essential for the frequency-domain measurements of HRV that during the recording heart rate remains stationary. Only in this case the power of a certain spectral component represents true modulation of HRV at that frequency. If heart rate follows some trend or oscillation caused by endogenous or exogenous stimuli irrelevant for the investigation, the frequency and power of that heart rate change is calculated and incorporated into the HRV analysis. A typical problem of nonstationarity is encountered if the subject is not allowed to rest and heart rate has not stabilised before the recording is started. Namely, when the subject assumes the supine position heart rate starts to decline and results in a trend and nonstationarity of the recording (Figure 2).

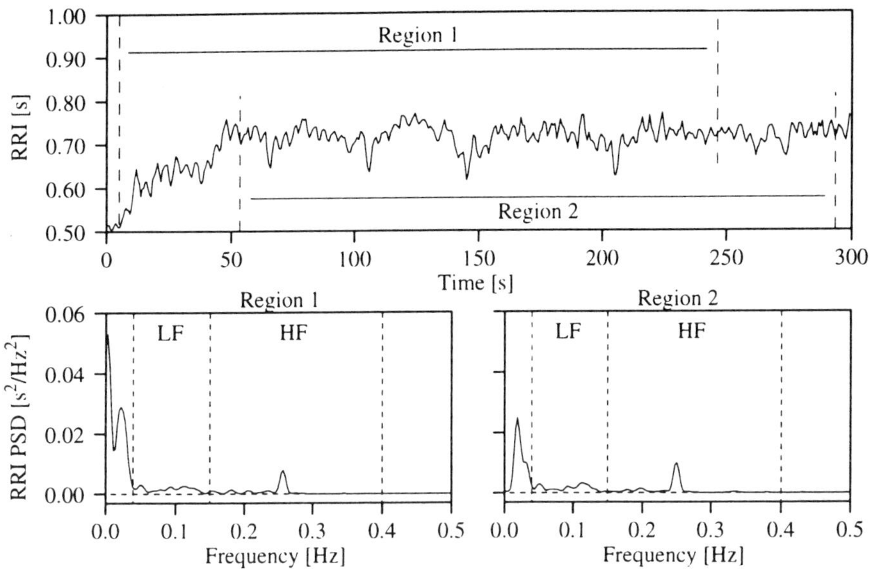

Method: FFT Window type: Hanning Smoothing: triangular, 5 point		Region 1	Region 2
RR interval	(ms)	698	717
SDNN	(ms)	49	26
TP	(ms^2)	1310	797
VLF	(ms^2)	953	411
LF	(ms^2)	193	191
LFnu	(%)	54	50
HF	(ms^2)	160	187
HFnu	(%)	45	48
LF/HF	(%)	121	102

FIGURE 2. Effect of nonstationarity on HRV. Heart rate decreases after the start of the recording and this nonstationarity produces increased variability in the very low frequency (VLF) component (first 4 minutes, region 1). HRV from a stationary segment during the same recording (region 2) demonstrates a drastically decreased total and VLF power. See Tables 1 and 2 for abbreviations.

To overcome the problem of nonstationarity, the subjects should be allowed to rest in the laboratory in the supine position (or possible other body position used in the HRV study) for at least 15 minutes to obtain steady-state haemodynamics before the commencement of the recording. In addition, after a stimulus (e.g. tilting) the HRV analysis should not be performed from a period that includes the early change in heart rate in response to the stimulus, but from a moment after heart rate has stabilised to a new level. For the same reason, the laboratory should be free of any external disturbances, such as unpleasant temperature, noise, bright lights, odours, etc., which might distract the subject. In order to check the stationarity of the signal, normal statistical tests, e.g. those comparing the distributions of RR intervals, may be employed.[12]

If nonstationarity exists, the trend should be removed from the beat-to-beat data. However, this is practically possible only with a linear trend as approximations of non-linear trends, e.g. those related to provocative stimuli, may create a significant bias. Inappropriate trend removal (if nonstationarity exists) results in the dominance of variability in the lowest frequencies and renders the interpretation on the analysis very difficult.

ECTOPIC BEATS, ARRHYTHMIC EVENTS AND NOISE

HRV analysis is based on the assessment of variability of sinus rhythm. Thus, QRS complexes originating outside the sino-atrial node - such as ectopic beats or salvos of arrhythmias - missed beats and noise introduce an error into the HRV analysis. This is due to two reasons. First, ectopic beats are usually premature and produce a very short RR interval followed by a compensatory delay and therefore a prolonged RR interval. Correspondingly, missed beats result in erroneously prolonged RR intervals. These alterations of RR intervals produce a technical error into HRV statistics. Second, ectopic beats and particularly salvos of arrhythmias often result in reduced stroke volume and cardiac output leading to abnormal stimulation of arterial and ventricular baroreceptors. This can activate autonomic reflexes and induce true changes in efferent autonomic activity.[13] The resulting change in HRV analysis is a physiological and not only a statistical phenomenon. Therefore, it is very difficult, if not impossible, to correct.

Arrhythmias and noise can be detected rather easily when the geometric methods are employed for the HRV measurements. Thus, casual editing of the data seems to be sufficient for the assessment of HRV with the geometric methods. On the other hand, other time-domain and particularly frequency-domain measurements are more sensitive to inaccuracies caused by ectopic beats, salvos of arrhythmias and noise. Therefore, careful manual editing of the recorded signal is mandatory to verify possible arrhythmias and to correct identification of QRS complexes. Automatic filtering can be used to help in identifying ectopic beats and artefacts. However, it should not replace manual editing but also in that case visual confirmation is needed.[14] The visual verification and correction of long-term recordings can be extremely time-consuming. Due to the restricted duration of recordings, this, however, is not a major problem in the case of short-term analysis of HRV.

There is no clear consensus how ectopic beats, other arrhythmias and noise should be treated. Perhaps the best option is to select a recording free of ectopic beats, arrhythmias and noise. However, accepting of a short recording for analysis may introduce a significant selection bias. If a proper recording is not available, cleaning of the data by proper interpolation based on the preceding and the successive QRS intervals is recommended[15] (Figure 3). If interpolation is applied, the number of omitted and interpolated QRS complexes should be quoted.

MEASUREMENT OF HEART RATE VARIABILITY

Before any calculations of HRV are performed the recording must be edited and corrected for ectopic beats, arrhythmias, noise and trends. Thereafter, the intervals between successive normal-to-normal QRS complexes (RR intervals resulting from sino-atrial node driven rhythm) or instantaneous heart rate values for each cardiac cycle are determined. After this, the calculation of HRV measures can be done. Several methods have been introduced for the calculation of HRV. According to the mathematical processing they can be divided into two classes; time-domain and frequency-domain methods. Time-domain methods can further be divided into statistical methods and geometric methods. Correspondingly, frequency-domain methods are divided into parametric and nonparametric methods.

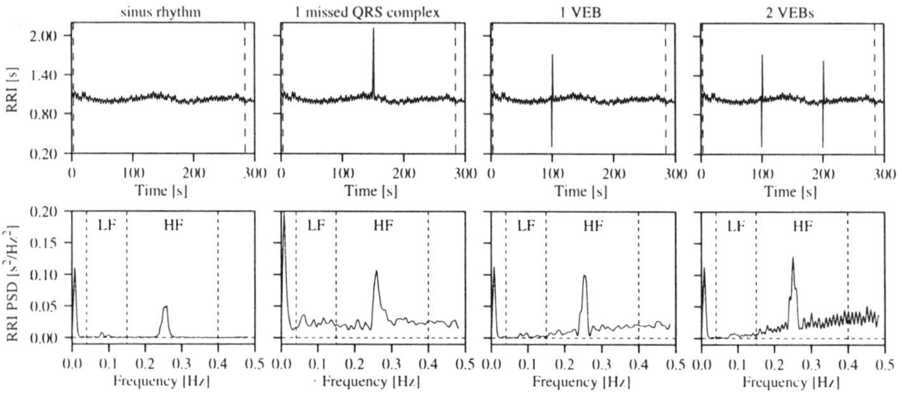

Method: FFT Window type: Hanning Smoothing: triangular, 3 point		sinus rhythm	1 missed QRS complex	1 VEB	2 VEBs
RR interval	(ms)	1028	1032	1028	1028
SDNN	(ms)	45	80	75	94
TP	(ms^2)	2313	13817	7637	12027
VLF	(ms^2)	998	2458	1011	1025
LF	(ms^2)	193	2540	307	506
LFnu	(%)	15	22	5	5
HF	(ms^2)	1081	6930	4596	7396
HFnu	(%)	82	61	69	67
LF/HF	(%)	18	37	7	7

FIGURE 3. Effect of one missed beat (missed QRS detection) and of premature ventricular ectopic beats (VEB) on HRV. The missed beat produces a prolonged RR interval (sum of 2 consecutive RR intervals) and a VEB produces a short RR interval followed by a long RR interval. HRV spectrum is affected markedly. The missed beat and VEBs have been added artificially for demonstration purposes. See Tables 1 and 2 for abbreviations.

TIME-DOMAIN MEASUREMENT OF HRV

Statistical time-domain measurement of HRV

The simplest and most commonly used statistical time-domain measures of HRV that can be derived from direct measurements of RR intervals are the mean RR interval and the standard deviation of RR intervals over the recorded period (SDNN)[16] (Table 1). SDNN represents the overall HRV and encompasses all frequency components responsible for HRV in the period of recording. In long-term recordings, the low frequency variations contribute major proportion of HRV, also SDNN. As the recording shortens, SDNN estimates higher and higher frequency components. In addition, the overall HRV decreases with the shortening of the recording.[8] Thus, the quantity of SDNN depends largely on the duration of the recording. For this reason, SDNN values derived from recordings of different duration should not be compared.

TABLE 1.
Time-domain measures for short-term assessment of heart rate variability

Variable	Unit	Description
SDNN	ms	Standard deviation of RR intervals over the selected time interval
RMSSD	ms	The square root of the mean of sum of the squares of differences between adjacent RR intervals
SDSD	ms	Standard deviation of differences between adjacent RR intervals
NN50 count	n	Number of pairs of adjacent RR intervals differing by more than 50 ms during the selected time interval
pNN50	%	Percentage of NN50 count of all RR intervals, i.e. (NN50 count / Total count of RR intervals) × 100

Other measures that can be derived directly from measurements of RR intervals are the standard deviation of the 5-minute mean RR intervals over the entire recording (SDANN) and the mean of standard deviation of the 5-minute RR intervals over the entire recording (SDNN index). However, because these measures are based on calculations from serial of successive short 5-minute periods, they estimate changes in heart rate caused by cycles longer than 5 minutes and are therefore suitable for the measurement of HRV from long-term recordings but not from short-term recordings.

The most commonly used measures derived from differences between RR intervals include the square root of the mean of sum of the squares of differences between adjacent RR intervals (RMSSD), the number of pairs of adjacent RR intervals differing by more than 50 ms (NN50 count), percentage of NN50 count of all RR intervals (pNN50) and the standard deviation of differences between adjacent RR intervals (SDSD) (Table 1). All these measures reflect high frequency variations of heart rate and are virtually independent of long-term trends, such as diurnal variations.[17] RMSSD, NN50 count and pNN50 are very highly and positively correlated with each other (correlation coefficients > 0.9) and therefore they can be considered as surrogates for each other.[18,19] Although they all are useful for the assessment of HRV, RMSSD is preferred to NN50 and pNN50 because as a continuous variable it has somewhat better statistical properties than NN50 count and pNN50 that are dichotomised variables.

Geometric time-domain measurement of HRV

In the geometric methods the series of RR intervals are converted to construct certain geometric patterns. The construction of the pattern can be performed by (1) constructing a sample density histogram of RR intervals - e.g. HRV triangular index (HRVi)[20] and triangular interpolation of normal to normal RR intervals (TINN),[21] (2) constructing a sample density histogram of differences between successive RR intervals[22] or (3) plotting the duration of each RR interval against the duration of the immediately preceding RR interval (Poincare maps and Lorenz plots).[23] The assessment of HRV from these patterns is performed based on the size, some specific dimension or the shape of the pattern. The mathematical methods for quantification of geometric patterns into measures of HRV are presented elsewhere in this book.

The major advantage of geometric methods is that they are quite insensitive to errors caused by ectopic beats, salvos of bradyarrhythmias or tachyarrhythmias and noise. Therefore, casual editing of the data seems to be sufficient for the assessment of HRV when geometric methods are used. Because of the same reason, geometric methods can often be used for the assessment of HRV from recordings when the quality of data does not permit the use of statistical time-domain or frequency-domain methods.

A disadvantage of geometric measures is that because of their approximate nature they are not as precise measurements as the statistical time-domain and frequency-domain measures. As comes to short-term assessment of HRV, even more important limitation is that a substantial number of RR intervals are needed to construct a representative geometric pattern. Indeed, it is recommended that when using geometric time-domain methods, the duration of the recording should be at least 20 minutes, but preferable 24 hours or even longer.[24] For this reason, geometric methods are inappropriate for the measurement of short-term changes of HRV.

FREQUENCY-DOMAIN MEASUREMENT OF HRV

Cyclic changes in heart rate and several hemodynamic parameters (e.g. blood pressure, stroke volume) have been known for centuries. Already in 1744 Hales provided a well-documented report between respiratory cycle, blood pressure and RR interval.[25] More recently, it has been shown that in addition to the overall variability of heart rate, HRV is composed of certain well-defined rhythms which contain information about the contribution of different regulatory mechanism of cardiovascular control. To evaluate the dynamics of these rhythms and to understand how overall HRV is distributed into different rhythms, the variation of heart rate must be divided into frequency components.[26–28] Although some time-domain measures contain information about heart rate oscillations of certain frequency variations, in first place time-domain measures must be considered as markers of overall HRV rather than markers of certain frequency components of HRV. In order to provide deeper insight to the dynamics and components of HRV more advanced second order statistics, such as power spectral density (PSD) analysis, have to be applied.

Power spectral analysis, by definition, decomposes the heart rate signal into its frequency components and quantifies them in terms of their relative intensity, termed "power". It provides estimates of the power spectrum density function of heart rate, i.e. information how overall HRV (variance of RR intervals or heart rates) is distributed as a function of frequency into different components. The heart rate spectrum can be analysed with two methods; (1) nonparametric Fast Fourier Transformation (FFT) spectra and (2) parametric autoregressive (AR) modelling, directly to beat-to-beat time series or to resampled data.

Fast Fourier transformation or autoregressive modelling

The Fourier transformation is based on the fact the all periodic functions can be expressed as a (infinite) sum of sines (and cosines), e.g. so called harmonic components of the Fourier series. In the case of finite, discretely sampled signal the FFT algorithm produces finite set of amplitudes (and phases) corresponding to the each harmonic component. These amplitudes can be used to estimate the power spectral density function describing the characteristics of HRV from either the equidistantly re-sampled data after interpolation, or from the original beat-to-beat time series data. The number of various frequency components depends on the number of data points in the signal under analysis; from longer data set more frequency components can be computed. In practice this means that the frequency resolution increases.

In the AR methods (several sequential and block algorithms) the original time series is described by an autoregressive model. Each RR interval is assumed to be linear function of M previous values, where M is the model order. Model order selection should be preferably based on the calculation of either Final Prediction Error (FPE) or Akaike Information Criteria (AIC) function.[29] After the determination of M, the power spectral density function is calculated. Statistical tests (e.g. PEWT, prediction error whiteness test) are recommended to yield information on the goodness of the fitted model to the beat-to-beat time series data. Because AR methods are based on underlying model there are no restrictions for maximal number of frequency components.

Discrete peaks for the several components are characteristic to the unsmoothed FFT spectra, whereas the several alternative AR models result in continuous smooth spectra (Figure 4). The windowing in the time domain (Hanning, Welch, Partzen, etc.), and especially smoothing operations in the frequency domain in FFT methods modify the spectral shape. The smaller the smoothing, the more details can be observed in the FFT spectrum, the higher is the smooth-

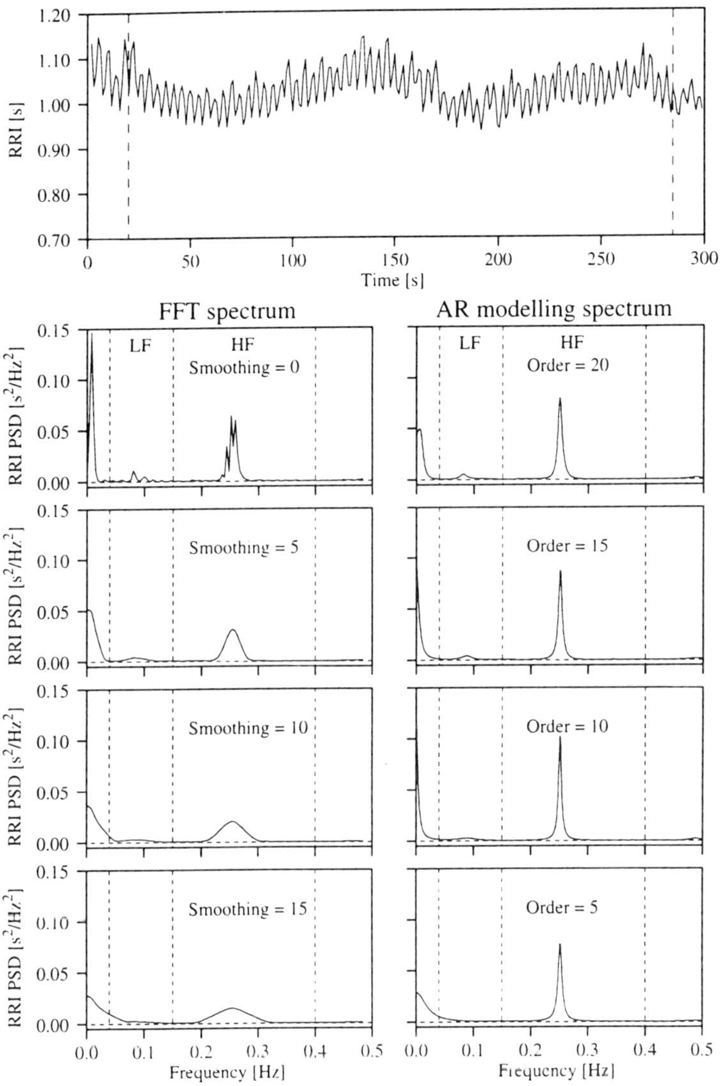

FIGURE 4. Comparison of frequency-domain HRV analysis performed with fast Fourier transformation (FFT) (left) and autoregressive (AR) modelling (right). In addition, the effect of smoothing on FFT spectra and model order on AR spectra is demonstrated. The smaller the smoothing, the more details can be observed in the FFT spectra, the higher the smoothing, the smoother are the spectra. Accordingly, the smaller the model in AR modelling, the smoother the AR spectra.

ing, the smoother is the spectrum (Figure 4). By smoothing the spectrum, it is possible to increase the statistical relevance of the power spectral densities but, of course, the price is lower frequency resolution. Accordingly, the smaller the model in AR modelling the smoother is the AR spectrum. On the other hand, the higher the model order, the more details can be seen in the AR spectrum. A very high model order can produce spurious spectral peaks, whereas a small model order has a tendency to erroneously shift the positions of spectral peaks. The normal range for the model order is 10–20.

The main advantage of the FFT methods is their computationally efficient implementation when compared with AR methods. On the other hand, AR methods are considered to provide smoother and more easily interpretable spectral shapes, better spectral resolution in case of short time series and a possibility to decomposition of the spectrum (division of spectrum into its root components) without the need for predefined spectral bands. In general, both FFT and AR methods provide very comparable results.[30]

Frequency components of HRV

Total power (TP) of the heart rate signal – equivalent to variance of HRV (SDNN squared) – is represented by the area under the power spectral density curve. In short-term recordings three spectral components can be normally distinguished in the power spectral density curve; high frequency (HF, 0.15–0.4 Hz), low frequency (LF, 0.04–0.15 Hz) and very low frequency (VLF, 0–0.04 Hz) components[5] (Table 2). In addition, in long-term recordings a fourth frequency component, ultra low frequency (ULF, 0–0.003 Hz) component, can be identified. The physiological explanation of VLF component has not been well defined. In addition, it represents very low frequency oscillations that need a relatively long recording (about 1 hour) to be assessed reliably. Finally, the nonharmonic component that does not have coherent properties and which is affected by algorithms of baseline and trend removal (necessary prior to spectral estimation) is considered to be included into the VLF component. Thus, VLF component has been considered as a dubious measure that should be avoided when interpreting short-term HRV measurements.

The magnitude of HRV of each frequency component is expressed as power, i.e. the area of power spectral density curve under the corresponding component (Figure 5). The magnitude of each frequency

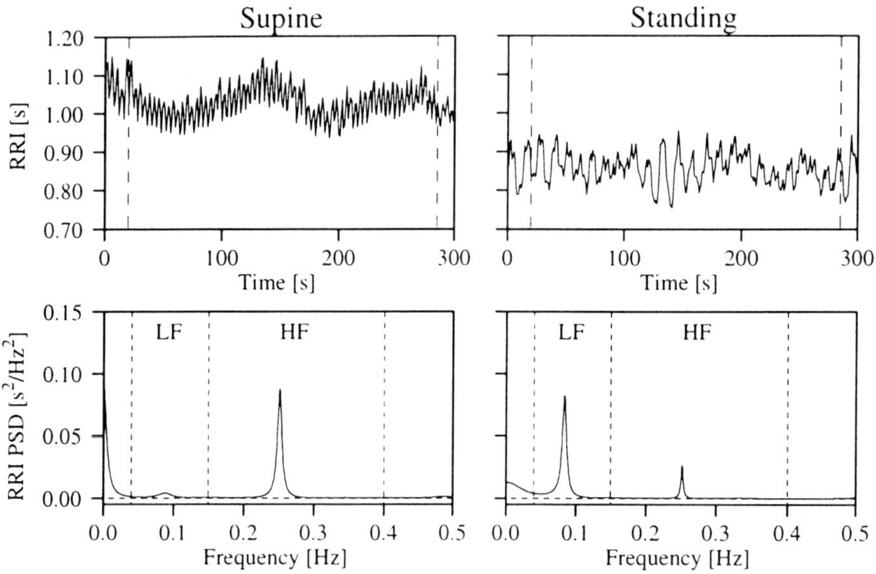

Method: AR modelling Model order: 15		Body position	
		Supine	**Standing**
RR interval	(ms)	1027	856
SDNN	(ms)	45	40
TP	(ms^2)	1961	1547
VLF	(ms^2)	694	321
LF	(ms^2)	170	1028
LFnu	(%)	13	84
HF	(ms^2)	1045	186
HFnu	(%)	82	15
LF/HF	(%)	16	552

FIGURE 5. Frequency-domain assessment of HRV calculated with autoregressive modelling at supine rest and in standing position at breathing frequency of 0.25 Hz. Standing is associated with a decrease in total, VLF and HF power and an increase in LF power expressed in absolute units (ms^2), and in normalised units (nu). See Tables 1 and 2 for abbreviations.

component is usually expressed in absolute power (integral of power spectral density curve within the frequency bounds of the corresponding component) as well as in normalised units (n.u.). The term normalised unit expresses the power centered in the frequency of interest (HF or LF) divided by TP less VLF power (Table 2). In addition, the ratio of power of LF component to power of HF component (LF/HF ratio) has been introduced to describe sympatho-vagal balance (see further).[31,32] However, one has to notice that normalised

TABLE 2.
Frequency-domain measures for short-term assessment of heart rate variability.

Variable	Unit	Frequency range	Description
Total power	ms^2	< 0.4 Hz	The variance of RR intervals over the selected time interval
VLF power	ms^2	≤ 0.04 Hz	Power of very low frequency component
LF power	ms^2	0.04–0.15 Hz	Power of low frequency component
LFnu	n.u.		LF power in normalised units, i.e. LF power / (Total power–VLF power) × 100
HF power	ms^2	0.15–0.4 Hz	Power of high frequency component
HFnu	n.u.		HF power in normalised units, i.e. HF power / (Total power–VLF power) × 100
LF/HF			Ratio of LF power to HF power, i.e. LF power / HF power

units and LF/HF ratio are not accurate measures, but they represent the relative power of the frequency components (LF and HF components). Therefore, the data of power under HF and LF components should not be presented only as normalised units or as LF/HF ratio, but they should be quoted together with the absolute values of HF and LF powers.

INTERPRETATION OF THE FREQUENCY-DOMAIN MEASURES OF HRV

HF component comprises of heart rate variations in the frequency range of 0.15–0.40 Hz. HF component, also termed as respiratory component, corresponds to the heart rate variations related to the respiratory cycles and/or respiratory sinus arrhythmia. HF fluctuations in heart rate are thought to be mediated almost exclusively by fluctuations of efferent parasympathetic activity. This concept is supported by several experimental and human studies. Animal studies have shown that after vagus stimulation the maximum response on heart rate occurs very fast, within only 400 milliseconds after stimulation.[33] On the other hand, the effect of sympathetic stimulation on heart rate starts after a latency of approximately 5 seconds and the maximum response is reached within 20–30 seconds.[34] Thus, only parasympathetic modulation is capable of rapid, high frequency adjustment of heart rate. In another animal study, the effects of stimulation of efferent vagal and sympathetic nerves were investigated. Although vagal stimulation increased the power of HRV of all frequencies, the effect was most pronounced in the HF component.[35] Correspondingly, stimulation of an efferent sympathetic nerve (left inferior cardiac nerve) with different frequencies resulted in only slight but nonsignificant increase of the power of HF component. Finally, in humans HF fluctuations of heart rate are virtually abolished by large dose of atropine.[27,36,37] However, although parasympathetic tone is the major modulator it is not the only factor that influences HF component of HRV. HF component (respiratory component) decreases markedly as breathing volume decreases. In addition, the power of the respiratory peak increases as its frequency decreases. However, if breathing frequency decreases to < 0.15 Hz, it becomes measured not as HF component but as LF component[38,39] (Figure 6). Thus, HF component can be used as a measure of parasympathetic tone only in situations in which breathing frequency and volume are carefully controlled. This is not possible in HRV analysis based on long-term recordings, but usually can be done in short-term recordings performed in laboratory settings.

LF component represents HRV between frequencies 0.04 and 0.15 Hz. The physiological background of this component is somewhat controversial. Animal studies have demonstrated that after the parasympathetic activity has been blocked with high-dose atropine, subsequent sympathetic blockade with β-blocking agents eliminates all heart rate variability remaining in LF component.[27,40] This suggests

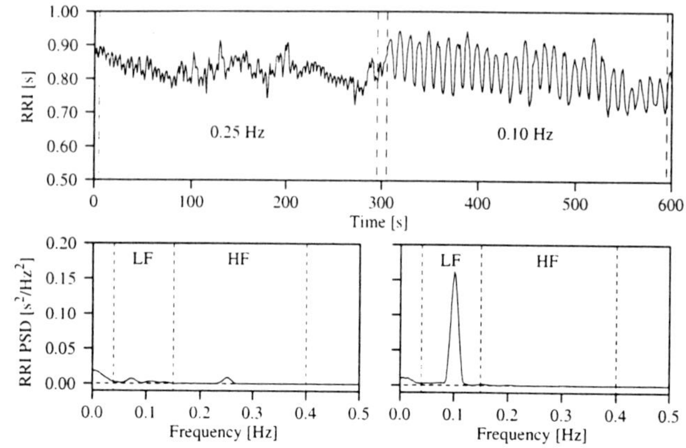

Method: FFT
Window type: Hanning
Smoothing: triangular, 5 points

		Breathing frequency	
		0.25 Hz	**0.10 Hz**
RR interval	(ms)	826	818
SDNN	(ms)	32	64
TP	(ms²)	1040	3276
VLF	(ms²)	470	298
LF	(ms²)	328	2867
LFnu	(%)	58	96
HF	(ms²)	233	102
HFnu	(%)	41	3
LF/HF	(%)	141	2815

FIGURE 6. Effect of respiration on frequency-domain assessment of HRV. RR interval time series are shown during quiet breathing at 0.25 Hz for the first 5 minutes (left) and at 0.1 Hz (right) for the last 5 minutes. Note that slow breathing (without deep breathing) increases total power of HRV. In addition, it shifts the respiratory peak from the HF band to the LF spectral band and results in an increase in LF power and a decrease in HF power. See Tables 1 and 2 for abbreviations.

that after elimination of the parasympathetic modulation, the residual heart rate variation of the LF component is of sympathetic origin. However, the interpretation of the physiology of LF component is more difficult in the case where parasympathetic modulation is intact. Sympathetic stimulation has been shown to increase HRV, particularly HRV of the LF component.[35] However, in the same study, the power of LF component was reduced significantly when vagal stimulation was added on top of sympathetic stimulation. In addition, parasympathetic blockade with vagotomy[41] or high-dose atropine[42,43] has been shown to abolish HRV not only of HF component but also of LF component. Furthermore, Saul et al. and Kingwell et al. reported no significant correlation between myocardial norepinephrine spillover - the golden standard for efferent cardiac sympathetic nerve activity - or peroneal sympathetic nerve activity and the power of LF component of HRV.[44,45] Thus, although it is clear that sympathetic nervous system does contribute to HRV of LF component, LF component is also markedly influenced by parasympathetic nervous system. Therefore the quantitative measurement of sympathetic tone may be very difficult.

A fact that makes the interpretation of the status of cardiac autonomic regulation more difficult is that a change in heart rate is accompanied by a change in HRV and its components. For example, tachycardia due to sympathetic stimulation results in a decrease in TP as well as in the powers of HF and LF components. On the other hand, in this case the power of LF component reduces less than the power of HF component. This can be detected by expressing the powers of LF and HF components in normalised units or as LF/HF ratio (Figure 5). However, using of normalised units or LF/HF ratio does not take into account the fact that LF frequency component is not a pure indicator of sympathetic efferent activity but that parasympathetic stimulation is also capable of producing variations of heart rate in the frequency range of LF component, and the fact that changes in sympathetic and parasympathetic tone can be parallel.[46] Thus, normalised units (LFnu and HFnu) as general markers of sympathetic and parasympathetic tone, respectively, as well as LF/HF ratio as a marker of sympatho-vagal balance should be interpreted with extreme caution.[47] This seems to be justified only in situations in which the behaviour of autonomic nervous system is well documented, the changes in autonomic tone are reciprocal and the breathing pattern

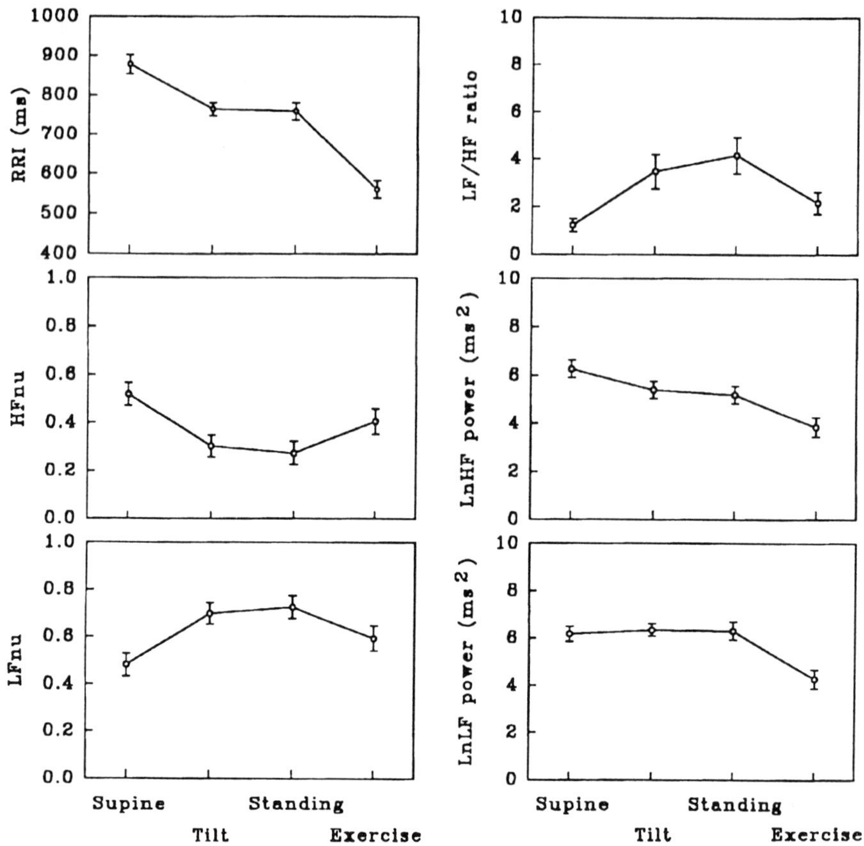

FIGURE 7. Effect of head-up tilt (60°), standing and exercise (stage 1 Bruce protocol) on HRV in 15 healthy subjects. A decrease in the power of HF and LF components expressed in absolute units (LnHF and LnLF) were observed during tilt, standing and exercise. However, when expressed in normalised units, HF power (HFnu) decreased whereas LF power (LFnu) and LF/HF ratio increased during tilt and standing. This is indicative of increased sympathetic tone and/or decreased parasympathetic tone. Although physical exercise is known to result in sympathetic activation, during exercise an increase in HF power (HFnu) and a decrease in LF power (LFnu) and LF/HF ratio were observed. This was due to an increase in breathing frequency and tidal volume, which shifted the respiratory component further to higher frequencies and increased the relative power of respiratory component (HFnu increased).

does not change. Thus, normalised units of HF and LF and LF/HF ratio can be used for estimation of sympatho-vagal balance e.g. during tilt test but not e.g. during physical exercise (Figure 7).

The physiological interpretation of VLF (0.003–0.04 Hz) and ULF (0–0.003 Hz) components are still unknown and warrant further investigation. They have been suggested to account for the long-term regulatory mechanisms related to thermoregulation, renin-angiotensin system, or to other humoral factors.[48] It has been also proposed that physical activity is the main determinant of VLF component[49] and some investigators suggest that all power < 0.04 Hz should be considered as noise.[50]

Finally, it has to be remembered that measures of HRV – irrespectively whether time-domain or frequency-domain measures are considered – do not represent the exact level of activity of autonomic nerves or autonomic tone, but a combination of the level and alterations in efferent autonomic (sympathetic and parasympathetic) nerve activity and their interaction.

CONCLUSION

Measurement of HRV provides an useful tool for assessment of status of the cardiac autonomic regulation. Usually, HRV analysis is based on measurements of RR intervals of consecutive cardiac cycles. It can be performed from long-term or short-term ECG recordings. Long-term measures of HRV represent the responses of cardiac autonomic regulation to challenges of daily life. However, a limitation is that these challenges are not controlled, the recordings are not stationary and the recording conditions cannot be standardised. These problems can be solved with assessment of short-term HRV. Short-term HRV measurements are performed from 5-minute ECG recordings in strictly standardised laboratory conditions, free of exogenic stimuli that might influence the autonomic nervous system. Short-term analysis of HRV is especially suitable for investigation the responses of cardiac autonomic regulation to standardised manoeuvres, such as standing up, tilting, static or dynamic exercise, mental loading, deep breathing, Valsalva manoeuvre, pharmacological interventions and in experimental research.

A short-term ECG recording must meet several conditions. The sampling frequency and amplitude resolution of the recording analogue-to-digital conversion must be adequate. In addition, the recording should be stationary, random and it must be carefully edited and corrected for arrhythmias and noise.

HRV analysis can be performed either by using time-domain or frequency-domain methods. The frequency-domain methods should be preferred to the time-domain methods for short-term assessment of HRV. On the other hand, time-domain methods are recommended for the measurement of long-term HRV. Frequency-domain analysis can be performed with nonparametric (fast Fourier transformation, FFT) or parametric (autoregressive, AR) modelling. The advantages of FFT modelling are simpler algorithm and faster processing, whereas AR has better resolution in the case of short recordings and provides a possibility of decompositioning of the spectra. Practically, both provide very comparable results. If time-domain analyses are used, it is recommended that statistical measures of SDNN and RMSSD are used to describe short-term HRV.

In the short-term frequency-domain analysis, TP and the powers of three spectral components are calculated; HF (0.15–0.4 Hz), LF (0.04–0.15 Hz) and VLF (< 0.04 Hz) components. HF component represents parasympathetic activity upon the heart. However, because breathing (frequency and tidal volume) has significant effect on HF component, a prerequisite for using HF power as an indicator of parasympathetic activity is that breathing remains unchanged. The interpretation of LF component is more controversial. Although sympathetic nervous system is the main contributor of LF component, it is influenced also by the parasympathetic efferent activity. The physiologic basis of VLF component is not known. In addition, short-term recordings are usually too short to allow proper statistic of VLF component. Therefore, VLF component is recommended to be discarded from interpretation of short-term analysis. Thus, interpretation of cardiac autonomic status is based on TP and powers of HF and LF components.

The power of spectral components should be presented in absolute values (ms^2). In addition, it is recommended to express the powers of HF and LF components in normalised units and as LF/HF ratio.

REFERENCES

1. Randall WC, Ardell JL. Functional anatomy of the cardiac efferent innervation. In: Kulbertus HE, Franck G (Eds) Neurocardiology. Futura, Mount Kisco 1988, 3–24.

2. Hainsworth R. The control and physiologic importance of heart rate. In: Malik M, Camm AJ (Eds) Heart rate variability. Futura, Armonk 1995, 3–20.
3. Berger RD, Saul JP, Cohen RJ. Transfer function analysis of autonomic regulation I. Canine atrial rate response. *Am J Physiol* 1989; **256**:H142–H152.
4. Kautzner J, Fiala M, Hartikainen JEK, Hnatkova K, Rowland E, Camm AJ, Malik M. Autonomic modulation of the atrioventricular conduction and its influence on the measurement of heart rate variability. *Submitted.*
5. Task Force for The European Society of Cardiology and The North American Society of Pacing and Electrophysiology. Heart rate variability. Standards of measurement, pysiological interpretation, and clinical use. *Circulation* 1996; **93**:1043–1065.
6. Ewing DJ, Martyn CN, Young RJ, Clarke BF. The value of cardiovascular autonomic function tests: 10 years experience in diabetes. *Diabetes Care* 1985; **8**:491–498.
7. Öri Z, Monir G, Weiss J, Sayhouni X, Singer D. Heart Rate Variability: Frequency Domain Analysis. *Ambulatory Electrocardiography* 1992; **10**:499–537.
8. Saul JP, Albrecht P, Berger RD, Cohen RJ. Analysis of long term heart rate variability: methods, 1/f scaling and implications. In: Computers in Cardiology 1987. IEEE Computer Society press, Washington 1988, 419–422.
9. Pinna GD, Maestri R, Di Cesare A, Colombo R, Minuco G. The accuracy of power spectrum analysis of heart rate variability from annotated RR list generated by Holter systems. *Physiol Meas* 1994; **15**:163–179.
10. Merri M, Farden DC, Mottley JG, Titlebaum EL. Sampling frequency of the electrocardiogram for the spectral analysis of heart rate variability. *IEEE Trans Biomed Eng* 1990; **37**:99–106.
11. Bianchi AM, Mainardi LT, Petrucci E, Signorini MG, Mainardi M, Cerrutti S. Time-variant power spectrum analysis for the detection of transient episodes in HRV signal. *IEEE Trans Biomed Eng* 1993; **40**:136–144.
12. Bendat JS, Piersol AG. Measurement and analysis of random data. New York: Wiley, 1966.
13. Welch WJ, Smith ML, Rea RF, Bauernfeind RA, Eckberg DL. Enhancement of sympathetic nerve activity by single premature ventricular beats in humans. *J Am Coll Cardiol* 1989; **13**:69–75.
14. Malik M, Cripps T, Farrel, Camm J. Prognostic value of heart rate variability after myocardial infarction – a comparison of different data processing methods. *Med Biol Eng Comput* 1989; **27**:603–611.
15. Kamath MV, Fallen EL. Correction of the heart rate variability signal for ectopics and missing beats. In: Malik M, Camm AJ (Eds) Heart rate variability. Futura, Armonk 1995, 75–85.
16. Kleiger RE, Miller JP, Bigger JT, Moss AJ, and the Multicenter post-infarction research group. Decreased heart rate variability and its association with increased mortality after acute myocardial infarction. *Am J Cardiol* 1987; **59**:256–262.

17. Heslegrave RJ, Ogilvie JC, Furedy JJ. Measuring baseline-treatment differences in heart rate variability: variance versus successive difference mean square and beats per minute versus interbeat intervals. *Psychophysiology* 1979; **16**:151–157.
18. Bigger JT, Albrecht P, Steinman RC, Rolnitsky LM, Fleiss JL, Cohen RJ. Comparison of time and frequency-domain based measures of cardiac parasympathetic activity in Holter recordings after myocardial infarction. *Am J Cardiol* 1989; **64**:536–538.
19. Kleiger RE, Bigger JT, Bosner MS, Chung MK, Cook JR, Rolnitzky LM, Steinman R, Fleiss JL. Stability over time of variables measuring heart rate variability in normal subjects. *Am J Cardiol* 1991; **68**:626–630.
20. Malik, Farrell T, Cripps T, Camm AJ. Heart rate variability in relation to prognosis after myocardial infarction – selection of optimal processing techniques. *Eur Heart J* 1989; **10**:1060–1074.
21. Farrell T, Bashir Y, Cripps, Malik M, Poloniecki J, Bennett ED, Ward DE, Camm AJ. Risk stratification for arrhythmic events in post-infarction patients based on heart rate variability and signal averaged ECG. *J Am Coll Cardiol* 1991; **18**:687–697.
22. Björkander I, Held C, Forslund L, Eriksson S, Billing E, Hjemdahl P, Rehnqvist N. Heart rate variability in patients with stable angina pectoris. *Eur Heart J* 1992; **13**:379.
23. Woo MA, Stevenson WG, Moser DK. Complex heart rate variability and serum norepinephrine levels in patients with advanced heart failure. *J Am Coll Cardiol* 1994; **23**:565–569.
24. Malik M. Geometrical methods for heart rate variability assessment. In: Malik M, Camm AJ (Eds) Heart rate variability. Futura, Armonk 1995, 47–61.
25. Hales S. Haemasticks. In: Hales S (Ed) Statistical Essays. Innys and Manby, London 1933, 1–86.
26. Penaz J, Roukenz J, Ban der Waal HJ. Spectral analysis of some spontaneous rhyhms in the circulation. In: Drischel H, Tiedt N (Eds) Biokybernetik Bd. I. Karl Marx University, Leipzig 1968, 233.
27. Axelrod S, Gordon D, Ubel FA, Shannon DC, Barger AC, Cohen RJ. Power spectrum analysis of heart rate fluctuation: a quantitative probe of beat to beat cardiovascular control. *Science* 1981; **213**:220–222.
28. Sayers B, Mc A. Analysis of heart rate variability. *Ergonomics* 1973; **16**:85–97.
29. Akaike H. Statistical prediction identification. *Ann Inst Statist Math* 1970; **22**: 203–217.
30. di Rienzo M, Castiglioni P, Mancia G, Parati G, Pedotti A. 24 h sequential spectral analysis of arterial blood pressure and pulse interval in free-moving subjects. *IEEE Trans Biomed Eng* 1989; **36**:1066–1075.
31. Pagani M, Lombardi F, Guzzetti S, Rimoldi O, Furlan R, Pizzinelli P, Sandrone G, Malfatto G, Dell'Orto S, Piccaluga E, Turiel M, Baselli G, Cerrutti S Malliani A. Power spectral analysis of beat-to-beat heart rate and blood pressure variability as a possible marker of sympatho-vagal interaction in man and conscious dogs. *Circ Res* 1986; **59**:178–193.

32. Malliani A, Pagani M, Lombardi F, Cerutti S. Cardiovascular neural regulation explored in the frequency domain. *Circulation* 1991; **84**:482–492.
33. Levy MN, Martin PJ, Iano T. Effects of single vagal stimuli on heart rate and atrioventricular conduction. *Am J Physiol* 1970; **218**:1256–1262.
34. Furnival CM, Linden RJ, Snow HM. Chronotropic and inotropic effects on the dog heart of stimulating the efferent cardiac sympathetic nerves. *J Physiol* 1973; **230**:137–153.
35. Hedman AE, Tahvanainen KUO, Hartikainen JEK, Hakumäki MOK. Effect of sympathetic modulation and sympathovagal interaction on heart rate variability in anaesthetized dogs. *Acta Physiol Scand* 1995; **155**:205–214.
36. Katona PG, Jih F. Respiratory sinus arrhythmia: noninvasive measure of parasympathetic cardiac control. *J Appl Physiol* 1975; **39**:801–805.
37. Eckberg DL. Human sinus arrhythmia as an index of vagal cardiac outflow. *J Appl Physiol* 1983; **54**:961–966.
38. Hirsch JB, Bishop B. Respiratory sinus arrhythmia in humans. How breathing pattern modulates heart rate. *Am J Physiol* 1981; **241**:H620–H6219.
39. Brown TE, Beightol LA, Koh J, Eckberg DL. Important influence of respiration on human R-R interval power spectra is largerly ignored. *J Appl Physiol* 1993; **75**:2310–2317.
40. Saul JP, Berger RD, Albrecht P, Stein SP, Chen MH, Cohen RJ. Transfer function analysis of the circulation: unique insights into cardiovascular regulation. *Am J Physiol* 1991; **261**:H1231–H1245.
41. Hedman AE, Hartikainen JEK, Tahvanainen KUO, Hakumäki MOK. Power spectral analysis of heart rate and blood pressure variability in anesthetized dogs. *Acta Physiol Scand* 1992; **146**:155–164.
42. Pomeranz B, Macaulay RJB, Caudill MA, Kutz I, Adam D, Gordon D, Kilborn KM, Barger AC, Shannon DC, Cohen RJ, Benson H. Assessment of autonomic function in humans by heart rate spectral analysis. *Am J Physiol* 1985; **248**:H151–H153.
43. Koh J, Brown TE, Beightol LA, Ha CY, Eckberg DL. Human autonomic rhythms: vagal cardiac mechanisms in tetraplegic subjects. *J Physiol Lond* 1994; **474**:483–495.
44. Saul JP, Rea RF, Eckberg DL, Berger RD, Cohen RJ. Heart rate and muscle sympathetic nerve variability during reflex changes of autonomic activity. *Am J Physiol* 1990; **258**:H713–H721.
45. Kingwell BA, Thompson JM, Kaye DM, McPherson GA, Jennings GL, Esler MD Heart rate spectral analysis, cardiac norepinephrine spillover, and muscle sympathetic nerve activity during human sympathetic nervous activation and failure. *Circulation* 1994; **90**:234–240.
46. Fagius J, Sundlöv G. The diving response in man: effects on sympathetic activity in muscle and skin nerve fascicles. *J Physiol Lond* 1986; **377**:429–443.
47. Eckberg DL. Sympathovagal balance. A critical appraisal. *Circulation* 1997; **96**:3224–3232.

48. Kitney RI, Rompelman O. Analysis of the human blood pressure and thermal control systems. In: Perkins J (Ed) Biochemical Computing. Pitman Medical, London 1977, 49–50.
49. Bernardi L, Valle F, Coco M. Physical activity is a major determinant of heat rate variability, its 'very low' frequency component and 1/f (chaotic) distrubution. *Eur Heart J* 1994; **15**:242.
50. Furlan R, Guzetti S, Crivellaro W, Dassi S, Tinelli M, Baselli G, Cerutti S, Lombardi F, Pagani M, Malliani A. Continuous 24-hour assessment of the nerual regulation of systemic arterial pressure and RR variabilities in ambulatory patients. *Circulation* 1990; **81**:537–547.

CHAPTER SEVEN

PHYSIOLOGICAL INTERPRETATION OF SPECTRAL COMPONENTS OF HEART RATE VARIABILITY

Alberto Malliani

Heart rate, under physiological conditions, is the result of sinus node pacemaker rhythmicity and, mainly, of sympathetic and vagal activities.

These two neural outflows are tonically and phasically modulated by the interaction, in closed loop conditions, of at least three major factors: the central integration, peripheral inhibitory reflex mechanisms (with negative feedback characteristics), and peripheral excitatory reflex mechanisms (with positive feedback characteristics).[1] One of the fundamental keys to understand the broad dynamics of heart rate neural regulation is the concept of sympatho-vagal balance, which corresponds to the observation that, in most physiological conditions, the activation of either outflow is accompanied by the inhibition of the other.[2,3] It is well known that reciprocal responses in cardiac vagal and sympathetic neurons are reflexly elicited from the sino-aortic baroreceptors. Stimulation of these receptors (mediating negative feedback reflexes) *excites* cardiac vagal neurons and *inhibits* cardiac sympathetic neurons. This reciprocal relationship was also evident in experimental conditions in which the reflex effects of sympathetic or vagal afferent stimulation were tested on the discharge of single sympathetic or vagal efferent fibres isolated from the same

Malik M (Ed) Clinical guide to cardiac autonomic tests, 177–193.

cardiac nerve impinging on the heart: in these experimental conditions the stimulation of, e.g., afferent sympathetic fibres excited the sympathetic efferent and inhibited the vagal efferent discharge (and vice versa for the stimulation of afferent vagal fibres).[2] Thus, in contradistinction to sino-aortic reflexes, the afferent cardiac sympathetic pathway (mediating positive feedback reflexes) reflexly *inhibited* cardiac vagal fibres and *excited* cardiac sympathetic fibres. Although this reciprocity cannot be, at the moment, quantified and is likely to be non-linear and composite of numerous sub-systems, it surely reflects a well-determined biological strategy, more aimed at subserving complex synergistic patterns (including excitation or inhibition of cardiac performance) rather than single reflexes.

One of the main achievements during the last years has been to progressively demonstrate the hypothesis that this reciprocal interaction can be broadly explored in the frequency domain of heart rate variability (HRV).[1] In fact, a conclusive evidence supports these general principles:

1. The respiratory rhythm of HRV, defined as high-frequency (HF) spectral component, is a marker of vagal modulation.[1,4]
2. The rhythm corresponding to vasomotor waves and present in both heart rate and arterial pressure variabilities, defined as low-frequency (LF) component is a marker of sympathetic modulation.[1,5]
3. A reciprocal relation exists between these two rhythms which is similar to that characterising the sympatho-vagal balance.[1,5]

METHODOLOGY

Numerous exhaustive accounts have been recently published on power spectrum analysis of HRV.[1,6,7] In this context suffice it to say that in our studies we have mainly used autoregressive algorithms that indicate the existence in both the heart period and arterial pressure variabilities of three spectral components, respectively centered around 0. Hz (the very low frequency component, VLF), 0.1 Hz (the LF component) and 0.20–0.35 Hz (the HF component). Usually only the LF and HF components can be adequately studied in time series of 200–500 events. With this approach each spectral component is assessed in terms of its centre frequency and of its absolute power (i.e. the area), expressed in squared units (e.g., msec^2).

THE NORMALISATION PROCEDURE

In view of the physiological model[1] of a reciprocal relation between LF and HF components, their estimation independent of total power has proven to be crucial.

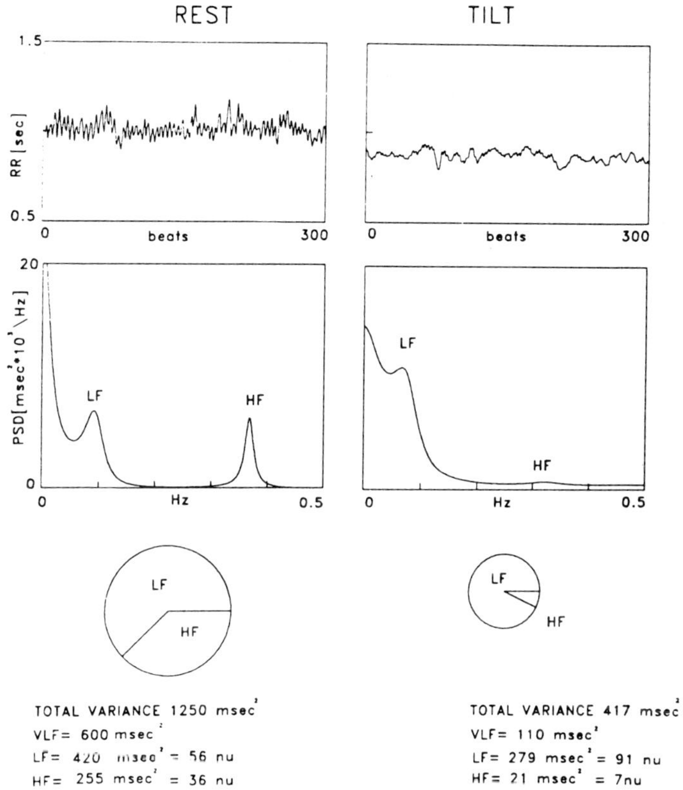

FIGURE 1. Spectral analysis of HRV in a young subject at rest and during 90° tilt. The RR interval time series (tachograms) is illustrated in the top panels. The middle panels contain the autospectra (PSD – power spectral density), with two major components (LF – low frequency, HF – high frequency). During tilt, the LF component becomes predominant. In this example, the total variance is markedly reduced during tilt and, consequently, both LF and HF powers are decreased when expressed in absolute units. The use of normalised units clearly indicates the altered relation between LF and HF during tilt. The pie charts show the relative distribution, together with the absolute power of the two components (represented by the area). (From Malliani et al.[31])

This can be easily obtained by computing their values in normalised units (nu), which are obtained by dividing the power of a given component by the total power (from which VLF has been subtracted) and multiplying by 100. Usually the sum of LF and HF falls short of 100 because of the frequent presence of some small noise components. As an example of the importance of the normalisation procedure for the interpretation of data, one should consider that the sympathetic excitation that leads to tachycardia is often accompanied by a reduction in total power of HRV: this can cause a paradoxical reduction of LF in absolute units, while this component appears clearly increased if expressed in nu (Figure 1).

STRATEGIES FOR VALIDATING THE SYMPATHO-VAGAL MODEL

An observational approach attempting to correlate the effects of graded physical stimuli, known to modify the sympatho-vagal balance, with the graded changes in the HRV spectral profile has provided the most convincing evidence in favour of the proposed model. In a recent study assessing the effects of graded tilt[8] in healthy volunteers, a strong correlation was found (Figure 2) between the degree of tilt incline and LF and HF components expressed in nu ($r=0.78$ and -0.72, $p<0.001$). It was quite obvious that the absolute values of LF underwent contrasting influences during tilt: they tended to be decreased by the reduction of variance (which was significant at 90°), but they also tended to be increased by the greater concentration of power in the LF component, as reflected by its constant rise in nu.

The relationship was also examined[9] between power spectral components of the variabilities in RR interval, intra-arterial pressure, and muscle sympathetic nerve activity (MSNA) obtained by microneurography. In healthy human volunteers these variables, together with respiration, were measured during blood pressure reductions induced by nitroprusside and during blood pressure increases induced by phenylephrine. Both LF and HF components were detected in MSNA variability. Increasing levels of MSNA were associated with a shift of the spectral power toward its LF component; conversely, decreasing levels of MSNA were associated with a shift of MSNA spectral power toward the HF component. Over the range of pressure changes examined, the LF component of MSNA variability

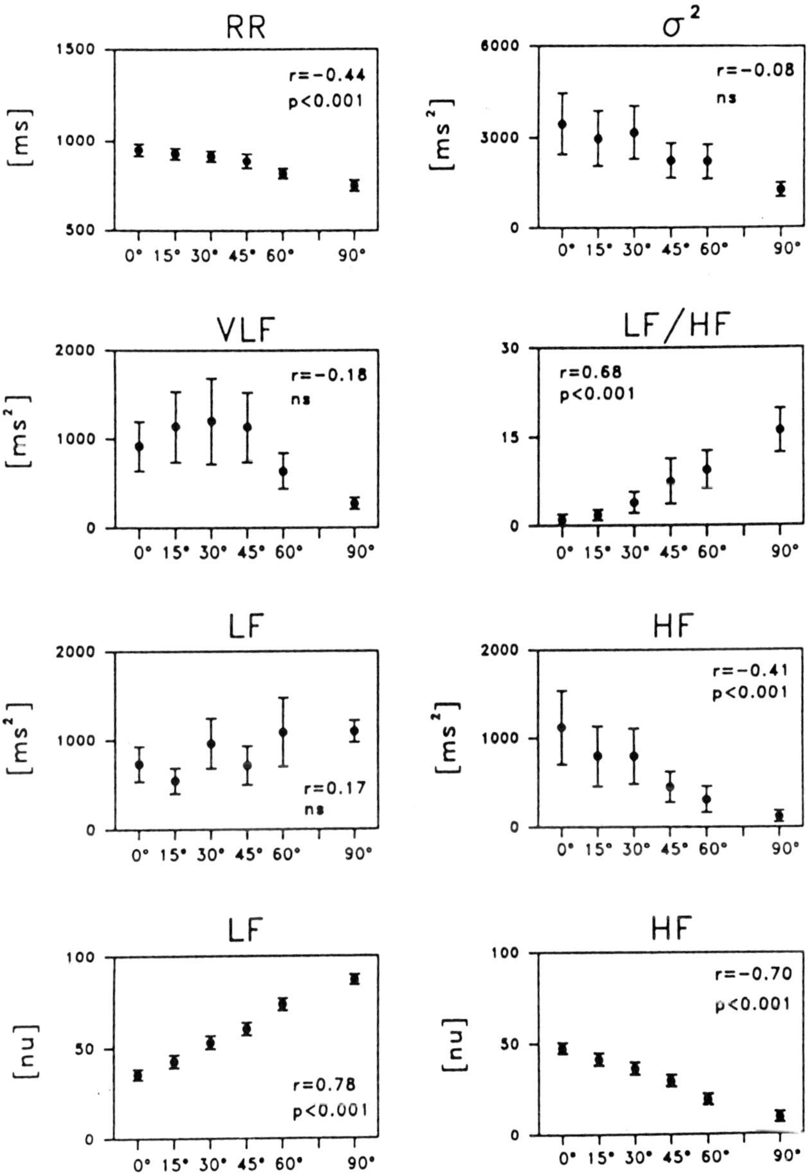

FIGURE 2: Plots of overall relations between tilt incline (in degrees) and measures of RR interval variability. σ^2 indicates variance; VLF – very low frequency; LF – low frequency; HF – high frequency; LF/HF – ratio of low frequency to high frequency; nu – normalised units; and r – Spearman correlation coefficient. (From Montano et al.[8])

was positively and tightly correlated with LF components of RR interval (in nu; $p<10^{-6}$) and of systolic arterial pressure variability (both in millimetres of mercury squared and in nu). Incidentally, these findings seem to counteract the claim that LFnu of RR variability is an invalid marker of sympathetic modulation.[10]

In addition, the smoothed pseudo Wigner-Ville transformation was used[11] in order to obtain an instant spectral power of HRV during tilt. With this time-frequency domain method it was observed that the *transient* effect of tilt was better characterised by a shift of spectral power towards lower frequencies than by changes in amplitude. It was also found, that during the passage to head-up position, the vagal withdrawal and the sympathetic activation occurred nearly simultaneously as revealed by the amplitude of HF and LF oscillations. Therefore, although spectral analysis of HRV requires, as a rule, some degree of stationarity, there are specific algorithms that can attempt to quantify the transients. As a result of this investigation it is clear that LF and HF oscillations should be considered also for the instant changes in their centre frequency and not only for their average centre frequencies for the whole spectrum of a time series of 200–500 events.

Further quite convincing evidence in favour of the biological soundness of this conceptual approach is furnished by the study[12] summarised below.

Ten spectral variables were considered: RR interval, total power (or variance), LF, HF, and the very-low-frequency component (VLF <0.03 Hz) in absolute units $msec^2$, and the centre frequencies (in Hz) of LF and HF; in addition, the LF/HF ratio and LFnu and HFnu were calculated. In 350 healthy subjects ECG tracings were recorded in both supine and upright positions (700 patterns).

The tracings were first ordered consecutively and subsequently assigned alternatively to a training or to a test set (each consisting of 175 cases, providing 350 patterns considered to be independent). A forecasting linear method estimated a normalised activation index (AI ranging from −1 for supine to +1 for upright) that concentrated the information derived from spectral variables.

During the training set, the algorithm had to match the target, i.e. the posture, which was classified by the experimenter, with the information that could be extracted from the interaction of the variables of interest. A pattern was correctly discriminated when target and AI signs were identical. During the test set, as well, a negative value of the AI was intended to recognise the supine and a positive value the upright position.

Such a blind forecasting on the test set was capable of correctly assigning 83.4% (146 of 175) of features to the supine group and 86.3% (151 of 175) to the upright group when 10 variables were evaluated simultaneously.

Three variables (RR, LFnu, and HFnu) were found to hold almost all the information content and could recognise 294 of 350 (84.0%) patterns overall, with a similarly good performance in both supine and upright groups. When one of these three variables was not considered, the forecasting provided inconsistent results. On the other hand, other combinations required at least four variables to furnish good discriminations and recognitions – in some cases, however, with more unbalanced results. In particular, adding the variance to RR, LFnu, and HFnu did not improve the recognition. The use of the absolute values of LF and HF caused an unbalanced forecasting, impairing the recognition of supine group patterns. Finally, when further information was added to the model based on 10 variables, for example the age of the subjects, the recognition power of AI in the test set was slightly increased to 84.3% in the supine group and to 87.6% in the upright group.

It is known[1,6,7] that power spectrum variables are characterised by large inter-subject and intra-subject variations probably reflecting the dynamics of physiological control mechanisms over time, even during steady-state conditions. Therefore, previous studies had addressed only the differences between given data sets. Conversely, this study reports, for the first time, that short ECG time series contain an intrinsic information that can be used to recognise, individual by individual, two different postures and hence two different autonomic profiles. This accomplishment was allowed by combining an appropriate spectral methodology, an adequate physiological model and a forecasting analysis.

The findings above summarised should indicate the heuristic value of this type of observational approach as compared to the more questionable subtractive reasoning, based, for example, on pharmacological interventions, whose limits have been already pointed out in previous reviews.[1,13]

CONTINUOUS 24-HOUR ANALYSIS OF HRV

When LF was analysed in nu over the 24-hour period it displayed a clear circadian pattern with a marked nocturnal decrease corresponding to the reduction of sympathetic activity known to occur during the night.[1,14–16] These changes were mirrored by simultaneous increases in HF component, as an expression of the enhanced vagal activity accompanying the largest portion of sleep.

Interestingly, the circadian oscillation of LF oscillation was lost in patients with essential hypertension,[15] in diabetic patients with or without autonomic neuropathy[16] and markedly blunted in patients 4 weeks after myocardial infarction.[17] However, in all these disturbances a reduced responsiveness of sympatho-vagal balance was also observed when the patients were moved from supine to upright position.[1]

In short, multifarious pathophysiological conditions seem to be characterised by absent or blunted oscillations of sympatho-vagal balance, spontaneous or induced by various manoeuvres.

PATHOPHYSIOLOGICAL MODELS

A number of pathophysiological conditions have already been studied in human subjects and experimental models, as already summarized.[1,13] However, to take an example particularly interesting for what will be discussed below, in the conscious dog we performed transient occlusions of a coronary artery without eliciting behavioural reactions.[18] In these conditions, and in the absence of significant changes in arterial pressure, the coronary occlusion was accompanied by a drastic reduction in variance and by a marked change in spectral profile (Figure 3), consisting in the appearance of ample LF components in both RR and systolic arterial pressure (SAP) variability. It should be noticed that these components were practically absent in control conditions, as it often occurs, especially in the spectrum of RR variability, due the marked vagotonia characterising the conscious dog.

Such a marked shift of sympatho-vagal balance towards sympathetic excitation was likely to reflect an excitatory cardio-cardiac sympatho-sympathetic reflex, as described in acute experiments about thirty years ago.[19]

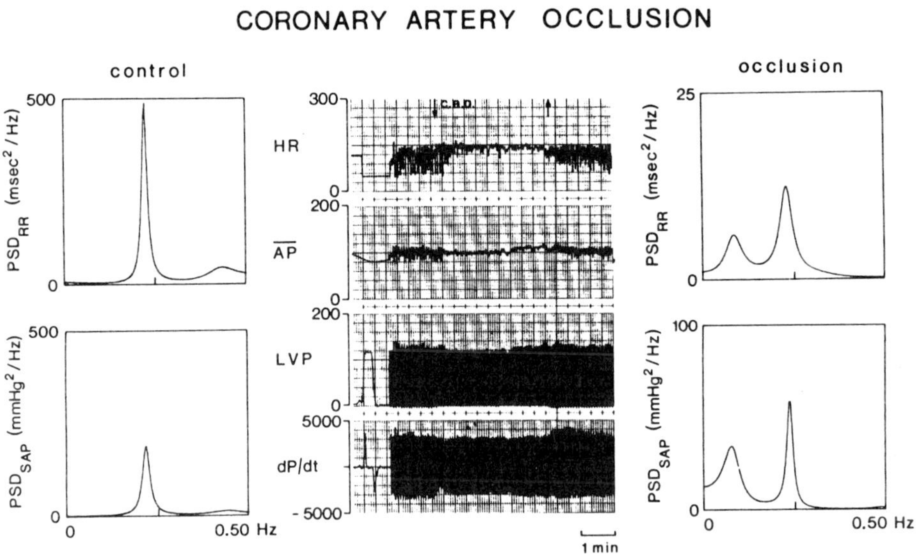

FIGURE 3. Effects of a transient, nonhypotensive, coronary artery occlusion (arrows, B) on autospectra of RR and systemic arterial pressure (SAP) variabilities (PSD) in a conscious dog. Note that during occlusion (C) there is a reduction in total variance from control (A), together with appearance of a large low-frequency component in both RR and SAP PSD, as a reflection of increase in sympathetic efferent activity. (From Rimoldi el al.[18])

THE INTERPRETATION OF RHYTHMS

Nowadays, there is an almost unanimous agreement on the relationship between vagal activity and the HF component of HRV and hence on the fact that HF might constitute a valuable marker of vagal modulation. However, while some experimental findings, such as the abolition of HF by atropine, are most commonly interpreted as a sign of a "pure" vagal nature of HF component, other views, including our own, consider this subtractive reasoning simply incapable of detecting the complex nature of HF component.[13] An HF component is clearly present in vagal and sympathetic nerve efferent discharge[1] and the hypothesis appears almost deprived of biological good sense that the HF sympathetic rhythm is practically abolished by the low-pass filter properties of sympathetic transmission.[20] It is hard to find even one single good proof for the ineffectual nature of an established biological phenomenon: thus the much more difficult hypothesis to prove, i.e. that of a continuous interaction at central and peripheral levels of sympathetic and vagal modulations in generating the rhythmic patterns, is much more appealing. As a crude observation one might simply recall that Introna et al have reported[21] that in patients undergoing spinal anaesthesia, when the spread of spinal block reached the highest thoracic segments (above T3), a remarkable abatement of variance and of both spectral components occurred. In this case, the subtractive strategy at least pointed to the complexity of the problem.

Much more debated has been the interpretation of the LF component of HRV. De Boer and coworkers[22] attributed this component to the baroreflex: in their model the arterial pressure oscillations induced by respiration stimulate the baroreceptors and through the interaction of the fast vagal response and the slower sympathetic transduction controlling arterial smooth muscle an LF component is generated in arterial pressure as a resonance phenomenon, which, in turn, stimulates baroreceptors and thereby the efferent control of RR. Sleight and coworkers[23] directly addressed this issue by mechanically stimulating at 0.1 Hz the carotid sinus region and observed an increase in LF component.

We obviously recognise the possibility that baroreflex mechanisms might interfere with LF amplitude but always questioned[1] their exclusiveness in LF pattern generation. In terms of neurophysiological thinking, an increase in the rhythmicity of a neural substratum can be induced by increasing a rhythmic input to it or by reducing some tonic inhibitory activity restraining autochthonous rhythmicity. An increased input from baroreceptors was clearly obtained by Sleight and coworkers[23] with their mechanical stimulation (but the limit of this model was, indeed, that it transformed a closed-loop system into an open input-output relation).

Usually, the tachycardia response during standing or during hypotension is attributed to a baroreflex: this attribution is obviously correct but what seems to be forgotten is that, being the baroreflex a negative-feedback mechanism, the sympathetic excitation occurs as a release phenomenon caused by a reduction of baroreceptive restraint (for instance, when recording sympathetic efferent activity in an experimental animal, the drastic reduction in baroreceptor firing obtained with bilateral carotid artery occlusion is accompanied, as a rule, by a profound change in sympathetic discharge that looses its cardiac rhythm and becomes a continuous firing). Conversely, in conditions such as mental stress or physical exercise, during which there is a simultaneous presence of tachycardia and hypertension the baroreceptor firing should increase. However also in these circumstances the LF component is increased.[1] In addition, there are conditions such as experimental regional myocardial ischaemia during which a LF increase can occur in the absence of arterial pressure changes[18] (Figure 3). Incidentally, in resting conscious dogs a LF component is usually present in arterial pressure variability but usually absent in RR variability, in spite of the high baroreflex gain. To further complicate this issue, the baroreflex gain is known to be decreased during standing, mental stress and exercise, conditions all characterised by an increased sympathetic activity: should we conclude that it is a reduction of baroreflex gain that increases sympathetic activity or vice versa?

Accordingly, we explicitly wrote that "LF component should not be considered a specific reflection of a baroreflex compensatory response – as hypothesised by Appel et al[20] – but rather a general marker of sympathetic excitation, regardless of its mechanism".[1] In brief, we now think that any *sympathetic excitation,* in the absence of strong confounders (as during physical exercise), *occurs with an LF pattern* due to its complex central-peripheral organisation. Moreover we also think that LF and HF patterns are intrinsically coupled.

The consequences of these different views in the interpretation of other phenomena can be quite relevant. For instance it was reported by Inoue and coworkers[24] that quadriplegic patients had no LF component in RR variability spectrum, a finding that was correctly interpreted as the result of an interruption of the neural pathways conveying the rhythmicity from the brain to the spinal cord. However it was subsequently demonstrated that some quadriplegic patients can have an LF component in RR and arterial pressure variabilities.[25,26] In a study by our group[25] this observation was interpreted as an emerging spinal rhythmicity influencing both sinus node pacemaker activity and the vascular bed: and indeed the LF component appeared to increase with time. Incidentally, some patients presented a LF component in RR interval, in the absence of a similar component in arterial pressure variability (a further argument against a necessary role for baroreflex mechanisms). On the other hand, in the study by Koh and coworkers,[26] the LF component was interpreted as vagally mediated in response to baroreceptor stimulation.

What is astonishing, on the whole, is how much baroreceptors are still considered as the absolute sovereigns controlling all cardiovascular reflex changes and rhythms. Incidentally, we have proven that the so-called simple baroreflex includes, in closed-loop conditions, multiple reflexes some of which of spinal origin.[27]

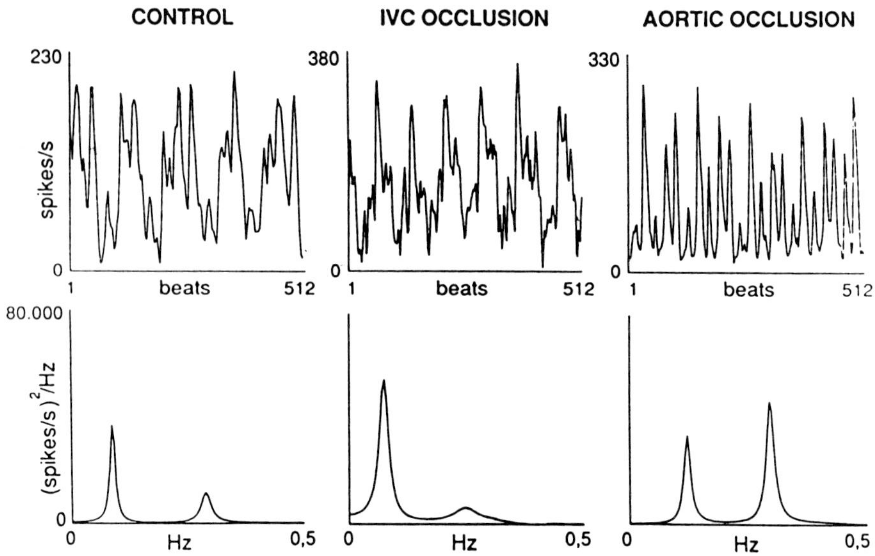

FIGURE 4. Spectral analysis of SND during control, inferior vena cava (IVC) and aortic occlusion. The time series of neural discharge is presented in the top panels, the autospectra in the lower panels. Two distinct spectral components at LF and HF are evident in control conditions. A markedly predominant LF component is present during IVC occlusion, while two well-defined components of similar magnitude are evident during aortic constriction. Data are from a decerebrate cat. (From Montano et al.[28])

THE PATTERN ORGANISATION

Experiments on acute decerebrate cats, artificially ventilated, in which ECG, arterial pressure and thoracic sympathetic preganglionic activity were recorded, furnished the first evidence of a central reciprocal relation between LF and HF rhythms.[28] For instance, it was observed (Figure 4) that a sympathetic excitation (obtained with a reduction in arterial pressure during inferior vena cava occlusion) modified the relation of LF and HF components present in sympathetic discharge towards further LF predominance. Conversely, during sympathetic inhibition (obtained with baroreceptor stimulation during aortic occlusion) the relation was clearly shifted towards HF predominance. The LF and HF components of HRV underwent similar changes.

On the other hand both LF and HF appear to be widely distributed among medullary neurons involved in the regulation of cardiovascular system[29] (Figure 5). It should be pointed out that these last experiments were carried out in sino-aortic denervated cats and that in some instances a LF component was present in medullary neurons but absent in arterial pressure variability spectrum. Hence, once again, a primary baroreceptive genesis was at least unlikely.

This is the background for our hypothesis[13] that two main rhythms, one marker of excitation (LF) and linked to sympathetic activation, and one marker of quiet (HF) and linked to vagal predominance, would be organised, in physiological conditions, in a reciprocal manner. Accordingly, a state of excitation would be accompanied by the central prevalence of LF rhythm, conveyed to the periphery by the simultaneous increase in sympathetic nerve activity, while a state of quiet would be accompanied by a central prevalence of HF rhythm, conveyed to the periphery by the simultaneous increase in parasympathetic activity. This reciprocity can be detected from the discharge of central neurons: interestingly, it was found not only that LF and HF components were present in the discharge variability of neurons belonging to various thalamic nuclei but that LF predominated during wakefulness and HF during sleep.[30] This reciprocity can be further detected in the peripheral neural outflows and in the cardiovascular target functions, according to a closed-loop organization.[1]

If this were the case it should be easy to understand that we are dealing with a neural code whose existence could not be foreseen and whose information content seems increasingly astonishing.

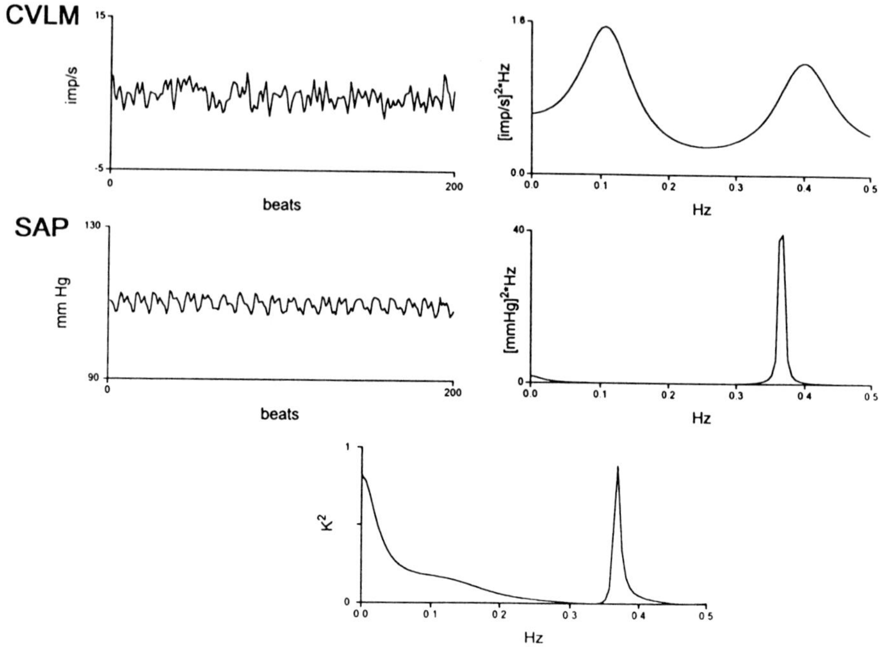

FIGURE 5: Spectral analysis of the discharge variability of caudal ventrolateral medullary neuron (CVLM) and of systolic arterial pressure (SAP) variability. Format is the same as in Figure 1. In this case the LF and HF component are clearly evident in the CVLM spectrum, while a HF component only is present in the sap spectrum. Data are from a urethane-anaesthetised cat. (From Montano et al[29]).

REFERENCES

1. Malliani A, Pagani M, Lombardi F, Cerutti S. Cardiovascular neural regulation explored in the frequency domain. *Research Advances Series, Circulation* 1991; **84**:482–492.
2. Schwartz PJ, Pagani M, Lombardi F, Malliani A, Brown AM. A cardiocardiac sympathovagal reflex in the cat. *Circ Res* 1973; **32**:215–220.
3. Kollai M, Koizumi R: Reciprocal and non reciprocal action of the vagal and sympathetic nerves innervating the heart. *J Autonom Nerv Syst* 1979; **1**:33–52.

4. Tougas G, Kamath M, Watteel G, Fitzpatrick D, Fallen EL, Hunt RH, Upton AR. Modulation of neurocardiac function by oesophageal stimulation in humans. *Clin Sci* 1997; **92**:167–174.
5. Pagani M, Lombardi F, Guzzetti S, Rimoldi O, Furlan R, Pizzinelli P, Sandrone G, Malfatto G, Dell'Orto S, Piccaluga E, Turiel M, Baselli G, Cerutti S, Malliani A. Power spectral analysis of heart rate and arterial pressure variabilities as a marker of sympathovagal interaction in man and conscious dog. *Circ Res* 1986; **59**:178–193.
6. Malik M, Camm AJ. Heart Rate Variability. Futura, Armonk 1995.
7. Task Force of the European Society of Cardiology and the North American Society of Pacing and Electrophysiology. Heart Rate Variability. Standards of Measurements, Physiological Interpretation, and Clinical Use. *Circulation* 1996; **93**:1043–1065.
8. Montano N, Gnecchi Ruscone T, Porta A, Lombardi F, Pagani M, Malliani A. Power spectrum analysis of heart rate variability to assess the changes in sympathovagal balance during graded orthostatic tilt. *Circulation* 1994; **90**:1826–1831.
9. Pagani M, Montano N, Porta A, Malliani A, Abboud FM, Birkett C, Somers VK. Relationship between spectral components of cardiovascular variabilities and direct measures of muscle sympathetic nerve activity in humans. *Circulation* 1997; **95**:1441–1448.
10. Kingwell BA, Thompson JM, Kaye DM, McPherson GA, Jennings GL, Esler MD. Heart rate spectral analysis, cardiac norepinephrine spillover, and muscle sympathetic nerve activity during human sympathetic nervous activation and failure. *Circulation* 1994; **90**:234–240.
11. Jasson S, Médigue C, Maison-Blanche P, Montano N, Meyer L, Vermeiren C, Mansier P, Coumel P, Malliani A, Swynghedauw B. Instant power spectrum analysis of heart rate variability during orthostatic tilt using a time-frequency domain method. *Circulation* 1997; **96**:3521–3526.
12. Malliani A, Pagani M, Furlan R, Guzzetti S, Lucini D, Montano N, Cerutti S, Mela GS. Individual recognition by heart rate variability of two different autonomic profiles related to posture. *Circulation* 1997; **96**:4143–4145.
13. Malliani A. Association of heart rate variability components with physiological regulatory mechanisms. In: Malik M, Camm AJ (Eds) Heart Rate Variability. Futura, Armonk 1995, 173–188.
14. Furlan R, Guzzetti S, Crivellaro W, Dassi S, Tinelli M, Baselli G, Cerutti S, Lombardi F, Pagani M, Malliani A. Continuous 24-hour assessment of the neural regulation of systemic arterial pressure and R-R variabilities in ambulant subjects. *Circulation* 1990, **81**:537–547.
15. Guzzetti S, Dassi S, Pecis, Casati R, Masu AM, Longoni P, Tinelli M, Pagani M, Malliani A. Altered pattern of circadian neural control of heart rate period in mild hypertension. *J Hypertens* 1991; **9**:831–838
16. Bernardi L, Ricordi L, Lazzari P, Solda P, Calciati A, Ferrari MR, Vandea I, Finardi G, Frantino P. Impaired circadian modulation of sympatho-vagal activity in diabetes. *Circulation* 1992; **86**:1443–1452.

17. Lombardi F, Sandrone G, Mortara A, La Rovere MT, Colombo E, Guzzetti S, Malliani A. Circadian variation of spectral indices of heart rate variability after myocardial infarction. *Am Heart J* 1992; **123:**1521–1529.
18. Rimoldi O, Pierini S, Ferrari A, Cerutti S, Pagani M, Malliani A. Analysis of short-term oscillations of R-R and arterial pressure in conscious dogs. *Am J Physiol* 1990; **258:**H967–H976.
19. Malliani A, Schwartz PJ, Zanchetti A. A sympathetic reflex elicited by experimental coronary occlusion. *Am J Physiol* 1969; **217:**703–709.
20. Appel ML, Berger RD, Saul JP, Smith JM, Cohen RJ. Beat to beat variability in cardiovascular variables: noise or music? *J Am Coll Cardiol* 1989; **14:**1139–1148.
21. Introna R, Yodlowski E, Pruett J, Montano N, Porta A, Crumrine R. Sympathovagal effects of spinal anesthesia assessed by heart rate variability analysis. *Anesth Analg* 1995; **80:**315–321.
22. DeBoer RW, Karemaker JM, van Montfrans GA. Determination of baroreflex sensitivity by spectral analysis of spontaneous blood-pressure and heart-rate fluctuations in man. In: Lown B, Malliani A, Prosdocimi M (Eds) Neural Mechanisms and Cardiovascular Disease. Liviana and Springer, Padova and Berlin 1986, 303–315.
23. Sleight P, La Rovere MT, Mortara A, Pinna G, Maestri R, Leuzzi S, Bianchini B, Tavazzi L, Bernardi L. Physiology and pathophysiology of heart rate and blood pressure variability in humans: is power spectral analysis largely an index of baroreflex gain? *Clin Sci* 1995; **88:**103–109.
24. Inoue K, Miyake S, Kumashiro M, Ogata H, Yoshimura O. Power spectral analysis of heart rate variability in traumatic quadriplegic humans. *Am J Physiol* 1990; **258:**H1722–H1726.
25. Guzzetti S, Cogliati C, Broggi C, Carozzi C, Caldiroli D, Lombardi F, Malliani A. Influences of neural mechanisms on heart period and arterial pressure variabilities in quadriplegic patients. *Am J Physiol* 1994; **266:**H1112–H1120.
26. Koh J, Brown TE, Beightol LA, Ha CY, Eckberg DL. Human autonomic rhythms: vagal cardiac mechanisms in tetraplegic subjects. *J Physiol* 1994; **474:**486–495.
27. Gnecchi Ruscone T, Lombardi F, Malfatto G and Malliani A. Attenuation of baroreceptive mechanisms by cardiovascular sympathetic afferent fibres. *Am J Physiol* 1987; **253:**H787–H791.
28. Montano N, Lombardi F, Gnecchi Ruscone T, Contini M, Finocchiaro ML, Baselli G, Porta A, Cerutti S, Malliani A. Spectral analysis of sympathetic discharge, R-R interval and systolic arterial pressure in decerebrate cats. *J Autonom Nerv System* 1992; **40:**21–32.
29. Montano N, Gnecchi Ruscone T, Porta A, Lombardi F, Malliani A, Barman SM. Presence of vasomotor and respiratory rhythms in the discharge of single medullary neurons involved in the regulation of cardiovascular system. *J Autonom Nerv Syst* 1996; **57:**116–122.

30. Montano N, Gnecchi-Ruscone T, Massimini M, Cogliati C, Porta A, Mariotti M, Malliani A. Detection of vasomotor and respiratory rhythms in the discharge of single thalamic neurons in conscious cats. Society for Neuroscience Abstracts, Annual Meeting, November 16–21, Washington D.C., 1996.
31. Malliani A, Montano N, Pagani M. Physiological background of heart rate variability. *Card Electrophys Rev* 1997; **1**:345–348.

CHAPTER EIGHT

LONG-TERM MEASUREMENT OF HEART RATE VARIABILITY

Antti E Hedman and Marek Malik

Long-term heart rate variability (HRV) refers to variations in heart rate that are measured over longer period of time (generally hours, most frequently over nominal 24 hours) compared to short-term HRV, that refers to heart rate fluctuations in electrocardiographic (ECG) recordings obtained under stationary or specific dynamic conditions (most frequently, recordings of 5 minutes are used). Commonly long-term HRV is measured from ambulatory 24 hours ECG recordings. Since the values of variance of cardiac cycles increase with the duration of recording, making measurements taken from differently long ECG records incomparable, standards have been proposed suggesting to use 24 hour ECGs with at least 18 hours of analysable data[1] (e.g. with the sum of all sinus rhythm RR intervals ≥ 18 hours). Variability of heart cycles assessed over such a period of time includes wide range of fluctuations from circadian variations in heart rate to respiratory sinus arrhythmia.

As a common rule, interpretation of HRV measurements is linked to the regulations of heart rate by the autonomic nervous system. However, it is obvious that many different factors contribute to variation in heart rate in response to environmental changes during daily life. While physiologic interpretation of some short-term measurement of HRV is widely accepted and hypotheses exist related to the interpretation of other short-term measurements,[2] physiological cor-

Malik M (Ed) Clinical guide to cardiac autonomic tests, 195–238.

relates of long-term HRV are far from understood. Despite that, depressed long-term HRV have a fully established and widely accepted clinical role, for instance as a risk stratifier of mortality and arrhythmia among survivors of acute myocardial infarction.[1,3–6] Long-term HRV has been a focus of an increasing attention as a risk stratification in various clinical conditions, and several different methods to assess HRV have been developed in studies aimed at physiology of HRV, risk stratification, improved predictive accuracy, as well as improved clinical practicality of HRV assessment. These methods include simple statistical and non-statistical time-domain measures, frequency domain measures based on spectral analysis, and methods describing heart rate dynamics in terms on non-linear mathematics.

This chapter is summarising various long-term measures of HRV as reported in recent literature and discusses some practical issues of their applications together with clinical implications of long-term HRV assessment. We have also included some novel measures of non-linear heart rate dynamics, which have been shown to be potentially promising in post-infarction risk stratification,[7,8] although their true clinical applicability remains unproven.

PHYSIOLOGICAL BACKGROUND

In the normal heart, instantaneous heart rate depends on the diastolic depolarisation rate of the sino-artial node, which is regulated by both limbs of the autonomic nervous system. It has now been firmly established that short-term HRV is mainly due to the autonomic modulations of heart rate,[1,2] although mechanical effects, such as distension of the atrial wall and the stretch of the sino-atrial node, have also been shown to contribute, positively or negatively, to heart rate fluctuations.[9–12]

Physiological background of long-term HRV is much more obscure, although an intact autonomic innervation is also required for the major part of long-term HRV. Autonomic denervation of the heart, for example in heart transplant recipients, leads to a significant reduction of long-term HRV at all frequencies[9,10,13] and diabetic patients with autonomic neuropathy show depressed long-term HRV.[14,15] It has also been demonstrated that parasympathetic blockade with atropine decreases HRV at all frequencies,[16–18] including the frequencies contributing to long-term fluctuations. Thus, the autonomic nervous system is a major contributor to long-term HRV. At the same time, controversial reports have also to be acknowledged. For instance, the reports of effects of β-blockade on HRV including the long-term HRV do not permit a clear conclusion to be made. Chronic β-blockade has been reported both to increase,[19,20] to decrease,[21,22] or not to affect HRV.[17]

Thus, intact autonomic innervation to the heart enables many factors to influence long-term HRV. Autonomic controls systems form a complex, integrated control network.[23,24] For example cardiovascular and respiratory control functions are highly integrated and connections exist at many level in the central nervous system.[25,26] This complex control network governs the end organs through efferent neural connections according to sensory information received from peripheral receptors, and according to information from the higher centres regulating emotions, behaviour and locomotion. Indeed, it has been repeatedly speculated that long-term HRV reflects mainly the responsiveness of heart rate regulation to the variable inputs coming from the surrounding environment[27–29] and that various measures of HRV are increased by training.[30–34]

Consequently, it is not surprising that many factors of very different nature have shown to contribute to HRV and have been suggested to form the physiologic background of variations with specific frequencies. These factors include thermoregulation processes,[35] haemodynamic feedback delays of blood pressure oscillators,[36–38] vascular automaticity,[39] humoral factors such as renin-agiotensin system,[40,41] and alterations in afferent chemoreceptor input.[42] Pathophysiological conditions also contribute to long-term heart rate modulations. These include haemorrhage, aortic constriction, acidosis, vasoactive agents, asphyxia, and others.[37,43,44] It has even been suggested that the heart rate control network itself can generate oscillations at various frequencies. For instance, very low frequency variations seen in heart rate have been attributed to central oscillators.[45]

Hence variability in daily activities, variable behaviour, emotions and reposes to changing requirements of the surrounding environment all modulate cardiac autonomic status and consequently contribute to long-term HRV. It is known that day and night variation in heart rate,[46] reflecting the state of arousal, is a major contributor to overall HRV during 24-hour period. Lack of vagal dominance during sleep[47] has been proposed as one of the mechanisms responsible for depressed long-term HRV in cardiac patients.[48] Furthermore, close link between locomotion and heart rate variability has also been demonstrated,[49] in addition to known effects of changes in posture or physical activity on heart rate. Indeed, selected 24-hour measures of HRV have shown to be higher in freely moving subjects as compared to immobilised ones.[50] Further, changes in breathing pattern, such as periodic breathing in patients with congestive heart failure, may also influence long-term HRV.[42,51] Nevertheless, despite the very different nature of all these mechanisms, their principal effects on heart rate, and consequently on HRV, are via the autonomic outflow that adjusts heart rate (and thus cardiac output) in response to varying haemodynamic conditions and requirements.

Although autonomic and non-autonomic regulators of heart rate are difficult to distinguish and separate, other mechanisms, mainly humoral factors, are probably also able to modulate heart rate and to contribute to HRV independent of cardiac autonomic nerves. Heart transplant patient have been shown to exhibit very low frequency fluctuations (<0.03 Hz) in heart rate,[10,52] which have been suggested to reflect hormonal modulation. Furthermore, increased sympathetic activation measures as increased plasma noradrenaline levels has shown to be associated with decreased long-term HRV.[53,54] However, detailed physiological mechanisms by which these additional mechanisms are integrated into the autonomic based regulation of heart rate are not known.

It has been clearly shown that, similar to short-term components of HRV, the measures of long-term HRV decrease with advanced age.[55,56] Gender differences with women showing lower long-term HRV than men, have been reported in some studies[55] but not found by others.[57]

In summary, global HRV expressing heart rate modulations over a period of total 24 hours is probably reflecting mainly cardiovascular responsiveness to different factors ranging from reflex responses maintaining body homeostasis, to environment triggered reflexes and will driven behaviour and locomotion. The relative contributions of individual factors and mechanisms to long-term HRV are not completely understood.

MEASUREMENT OF LONG-TERM HRV

Although other arrangements are possible,[58] HRV including long-term HRV, is conventionally measured in RR interval series derived from ECG recordings. There are several methods by which long-term HRV may be quantified. Following consensus agreement,[1] these methods include the so-called time-domain methods, frequency-domain methods and, more recently, descriptions of heart rate dynamics based on the technology of non-linear mathematics.

Recording Conditions

Duration of the Recording

The practice of assessing long-term HRV from 24-hour ECG recordings is well established.[3,4,6,7,59–65] Different long-term length of analysed RR interval time series (1 – 6 hours) has also been used in some studies.[8,66–69] However, global long-term HRV, e.g. expressed as the standard deviation of all normal-to-normal RR intervals or as the total power of power spectrum, correlates with the length of RR interval time series.[36,70] Consequently, physiologically and clinically relevant comparisons of measurements of long-term HRV are only possible when the duration of the underlying ECG record is standardised. Since, as discussed in the previous section, long-term HRV is substantially dependent on regulation reflexes and reactions triggered by the environment, the circumstances of the ECG recording period should also be standardised to allow comparisons of HRV measurements. Thus, for instance, HRV values based on mid-term ECGs recorded during the night are not analogous to HRV values derived from ECG recordings of the same duration made during the day.

For these reasons, standards have been proposed[1] recommending the long-term HRV measurements to be based on ECG recordings containing at least 18 hours of analysable data and including both the evening and morning hours. In clinical practice, parts of a long-term ECG are frequently not technically suitable for HRV processing (e.g. due to myopotential noise, arrhythmia, etc.) and a recording of full 24-hour is needed to ensure that 18 hours of analysable data are obtained. Thus the consensus recommendation[1] should be interpreted as a guideline to record 24-hour long-term ECG and to accept it for HRV analysis if no more than 6 hours, which are dispersed in the whole recording, have to be omitted due to technical reasons and/or presence of pathological rhythms. The term *nominal 24-hour recording* is used to describe a long-term ECG recording of these qualities.

Quality of the Recording

HRV measures are computed by an analysis of intervals between cardiac depolarisations of normal sinus rhythm cycles that are termed *the normal-to-normal intervals* (NN intervals). A number of long-term HRV indices are sensitive to errors and outliers in the RR interval series[71–73] which are not only caused by an incorrect distinction between normal sinus rhythm QRS complexes and atrial and ventricular ectopics but also, and perhaps more frequently, by false detection of tall T waves and myopotential patterns as well as by QRS undersensing due to variable ECG voltage in ambulatory recordings. Consequently, significant attention should be given to the quality of the NN interval series used in HRV measurements.

In practise, fully automatic computer reading of long term ECGs cannot be relied on and recordings must be visually reviewed and manually edited in order to remove premature beats and artefacts before the analysis. Whilst most modern Holter systems allow to remove artefacts and mislabelled premature beats, addition of QRS complexes that were omitted by the initial computer reading is frequently problematic. In many systems, the user can only add an element to the series of QRS fiducial points but cannot ensure that originally omitted QRS complex will be treated, in terms of fiducial point localisation, as the correctly recognised QRS complexes.[74]

The process of editing long-term ECG data is frequently cumbersome and sometimes, in cases of poor recordings with unfavourable signal to noise ratio, completely unmanageable. Some authors have therefore advocated and employed automatic filters designed to correct QRS mislabelling and remove the artefacts. Such a filtering typically identifies interbeat intervals deviating more than a specific threshold, e.g. 20–30 %, from the previous interval or exceeding a predetermined lower and upper limit.[64,75–78] Unfortunately, more detailed technical studies show that in some cases, these automatic filters behave completely unpredictably and that the quality of the NN interval data might actually be worsened rather than improved.[71]

In the time domain methods, the RR intervals surrounding premature beats or recording artefacts are usually omitted. When spectral and other methods are used, which require the time synchrony of the NN interval sequence to be maintained, the omitted or deleted intervals are replaced by an artificial value calculated from the surrounding NN intervals.[73] Several different algorithms for calculating the replacement value have been developed and used.[4,73,79,80] However, systematic experience does not exist that would permit a comprehensive comparison of the different replacement methods. With some methods, omission instead of replacement of a deleted interval is also acceptable.[73] This is especially the case when addressing long-term modulations rather than short-term modulations in long-term records. Methods of this kind have been shown to remain stable and consistent if at least 80 – 85% of the original true NN intervals are included in the analysis.[81] Usually, only segments of data containing more than 80 – 90% of ectopy and noise free data are considered acceptable for analysis with these methods.[8,75,78,81,83,84] Still, obtaining 80 – 90% of the original NN interval data without ectopics and noise may prove rather difficult and very time-consuming with some clinical recordings.

STATISTICAL TIME DOMAIN MEASURES OF LONG-TERM HRV

The term *time-domain measures* reflects the fact that the majority of these method express variability of the NN interval series in units of time (e.g. in milliseconds).

Simple time-domain measures of HRV are based on statistical methods for the quantification of NN interval data variations around the mean heart rate. Whilst such approaches are sufficient to express total HRV, they do not provide any insight into the characteristics of the variations measured.

NN Interval Variance

The simplest of these HRV measures is the standard deviation of all NN intervals (SDNN) which represents global HRV and, because of the mathematical properties of the total variance, is equal to the square root of the total power provided by the power spectral analysis. In practical terms, correspondence between SDNN and the total spectral power depends on the number and clustering of ectopic beats and artefacts that are, as a rule, deleted when computing SDNN and interpolated when computing power spectrum. In standard clinical data, this correspondence is very high.[4]

Other statistical measures include SDANN which is defined as standard deviation of the 5 minute mean NN interval over the entire recording, and which expresses long term variations with frequencies below 0.0033 Hz. Although in may clinical applications, SDNN and SDANN methods are of similar practical value (e.g. for identification of patients with reduced long-term HRV), they do not replace each other. Their numerical correlation is high but frequent discrepancies exist. (Figure 1)

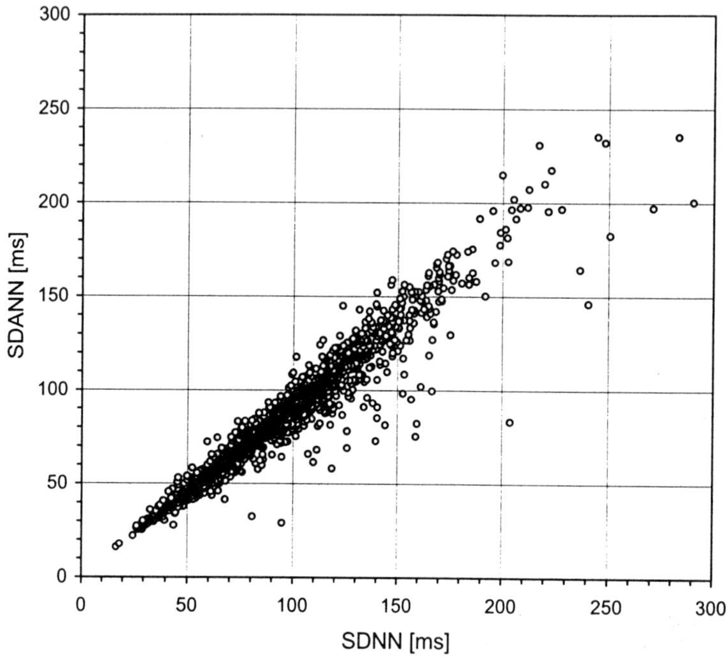

FIGURE 1. Scatter diagram of SDNN and SDANN values in a large set of nominal 24-hour Holter recordings obtained in survivors of acute myocardial infarction. Note that in some cases, SDANN is actually larger than SDNN. This happens in recordings with frequent premature beats and short periods of noise. Different 5 minute section of such a recording have a different percentage of valid data which creates a discrepancy between the representation and "weight" of different NN intervals when computing SDNN and SDANN.

The counterpart of the SDANN method is the so-called SDNN index, which is the mean of standard deviations of the 5-minute periods over the entire recording. While the SDANN has practical meaning and clinical relevance, SDNN index is a dubious measure that corresponds to averaged short-term variations. Since the averaging process of short-term variations masks their physiologic correlates, the assessment of SDNN index should better be avoided.[1]

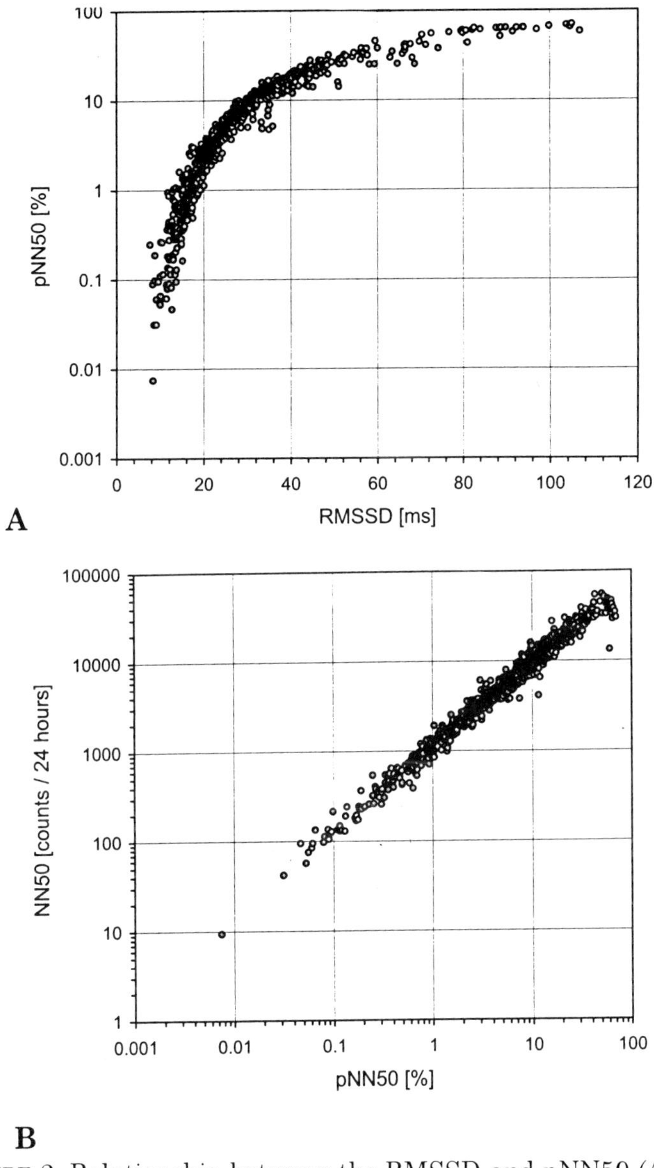

FIGURE 2. Relationship between the RMSSD and pNN50 (A), and pNN50 and NN50 (B) measures of HRV assessed from 857 nominal 24-hour Holter tapes recorded in survivors of acute myocardial infarction prior to hospital discharge. The NN50 measure used in panel (B) was normalised in respect to the length of the recording (Data of St. George's Post-infarction Research Survey Programme – Redrawn from the ESC/NASPE Task Force report[1]).

Successive NN Interval Differences

Whilst the SDNN and SDANN methods are based on the analysis of the sequence of NN intervals in which the order of the intervals is irrelevant, other time-domain methods measure the variability based on the differences between successive NN intervals. The core of these methods is the RMSSD measure, which is the root square of the mean of squares of the differences between successive NN intervals. (Since the mean of differences between successive NN intervals can be reasonably approximated by zero, RMSSD approximates the standard deviation of the successive differences.)

Other methods analysing the pairs of successive NN intervals include NN50, the number of interval differences of successive intervals greater than 50 ms, and pNN50, the proportion derived by dividing NN50 by the total number of NN intervals. All of these measurements provide estimates of short-term components of HRV that corresponds to high frequency variations in heart rate. Consequently, their values are highly correlated (Figure 2) and since the statistical properties of NN50 and pNN50 values in large clinical populations are inferior to those of RMSSD values (the distributions of NN50 and pNN50 are highly skewed), it has been proposed to use the RMSSD measure as the method of first choice[1] for the assessment of short-term variations within long-term HRV.

GEOMETRICAL TIME-DOMAIN MEASURES OF LONG-TERM HRV

The difficulties and workload related to ensuring the quality of NN intervals sequence required by the statistical methods led to the development of another category of methods for long-term HRV assessment that are much less influenced by noise and QRS recognition artefact polluting the NN data sequence. The so-called geometrical methods resulted from these needs.

In geometric methods, series of NN interval data are converted into a geometric pattern and HRV is measured from the geometrical or graphical properties of the pattern. As the noise and artefact related outliers in the NN sequence can usually be easily identified in the graphical pattern, descriptors and measures have been proposed that are, with reasonably accurate data, independent of the noise and artefact. Generally, three different approaches have been proposed.[85]

Basic Measurement of a Pattern

HRV may be assessed from basic measurements of a geometrical pattern, e.g. from the NN interval distribution histogram. Most frequently used measure of this kind is the HRV triangular index which is the integral of the density distribution, i.e. the number of all NN intervals, divided by the height of the density distribution, i.e. by the modal number of NN intervals.[1,85,86] (Figure 3) The measure is dependent on the bins used to construct the distribution histogram that

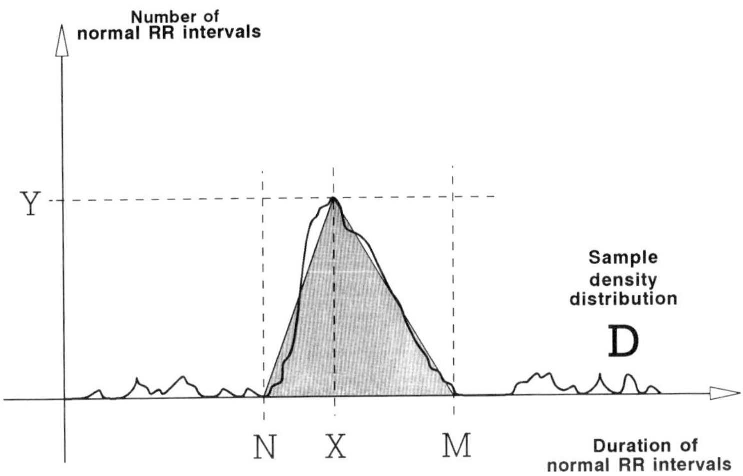

FIGURE 3. HRV triangular index and the triangular interpolation are measured from the NN interval histogram: The sample density distribution *D* assigning the number of equally long NN intervals is constructed. The most frequent NN interval length *X* is established by determining the maximum of the sample density distribution. The HRV triangular index is the value obtained by dividing the area integral *D* by the maximum *Y*.

For the triangular interpolation, the values *N* and *M* are established on the time axis based on approximation of the frequency histogram by a linear triangular function that has the same maximum point and is the nearest to the histogram as measured by the integral of the square differences.

is on the precision of the discrete scale used to measure the duration of individual NN intervals. Most experience exists with bins of 7.8125 ms that correspond to the standard 128 Hz sampling of long-term ECGs used by many manufacturers. If other sampling rate and other scale of measurement are used, it is preferable to recompute the histogram with a simulated 128 Hz sampling.

Mathematical Interpolation

HRV may also be assessed from a mathematically defined shape that interpolates the geometrical pattern constructed from the NN interval series. This approach has been applied to the NN interval histogram, histogram of successive differences between NN intervals, and return maps. An example of such a method is the triangular interpolation of NN interval histogram, which is defined as the baseline width of a triangle, with the minimum square departure from the NN interval distribution.[85,86] (Figure 3)

The methods of mathematical interpolation were proposed as a refinement of the methods based on the direct measurement of the geometrical pattern. Generally, however, the methods are not more stable and only in some exceptions, improved clinical performance in risk stratification studies has been reported.[87] For instance, differences between the triangular interpolation of NN interval histogram and the HRV triangular index are not too important.[85]

Pattern Classification

Classifications of the patterns of RR interval dynamics have also been used to describe HRV. The approach has mainly been applied to images created by plotting the length of each RR interval against that of the preceding RR interval. The result is known as scatterplot, Poincaré plot, Lorenz plot, next amplitude plot, and return map.[78,88,89]

The plots can be classified into different categories according to their shapes. A "fan shape" pattern is considered to represent normal heart rate dynamics, whereas "torpedo shape", "complex shape" or "narrow shape" patterns are considered to reflect abnormal heart rate dynamics in various conditions.[53,78,90] (Figure 4) However, such a classification is highly subjective[87] and has never been tested in a truly blinded prospective fashion. Thus, the practical utility of visual judgement of Poincaré plots is probably restricted only to the assessment of the quality of NN interval sequence. The plots make outliers and artefacts usually very visible. (Figure 5)

More recently, quantitative methods to assess HRV from these plots have been introduced (that is methods that, strictly speaking, belong to the previous categories). Basically, the form of the plot is quantified by manual measurement of the dimensions[77] or by fitting

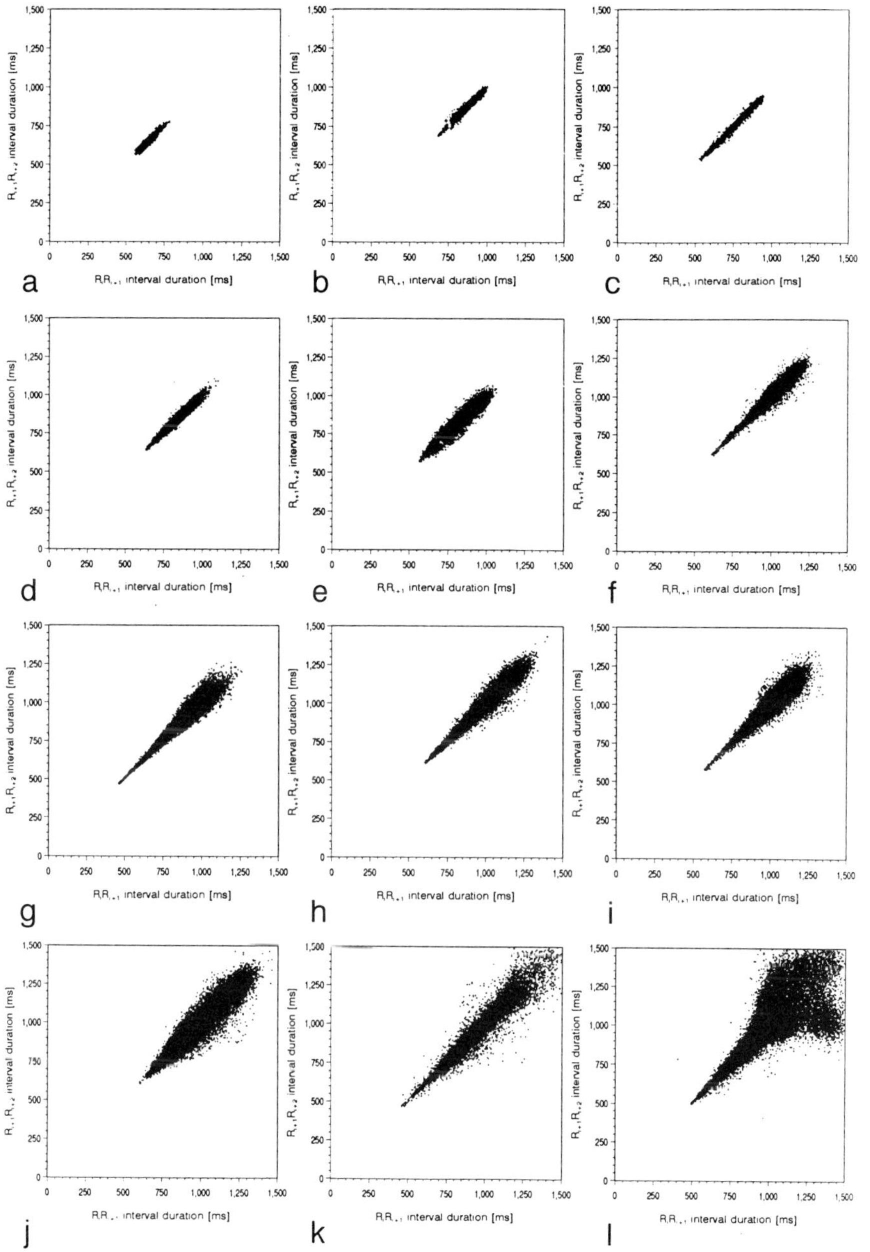

FIGURE 4. Examples of Poincaré plots arranged from minimum (a) to maximum (l) global HRV. Note that the differences in individual steps are not substantial. Consequently, setting precise borders for individual categories is frequently not possible.

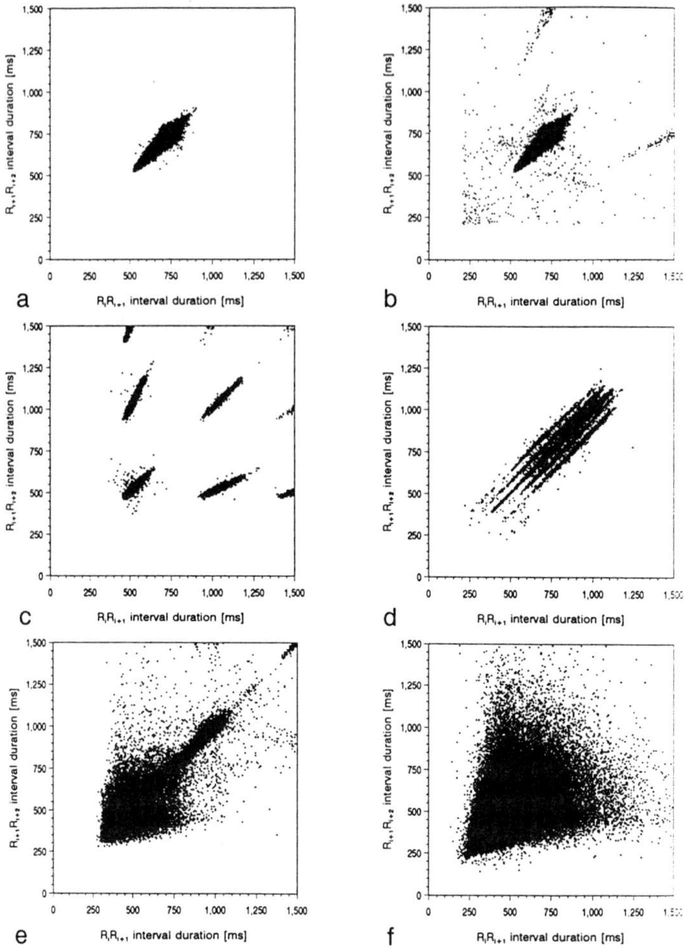

FIGURE 5. Examples of Poincaré plots used to assess the quality of the NN interval sequence in 24-hour recordings. Panel (a) shows a map of a properly edited RR interval sequence; panel (b) shows the same sequence with outliers due to artefact and Holter misrecognition; panel (c) shows a map generated by an automatic system which fails to recognise some QRS complexes and provides measurement of RR intervals which are actually composed of two or more cardiac cycles; panel (d) is a map generated by a Holter system which analysed a magnetic tape without phase-locked time track – the measured RR interval are distorted by sudden irregularities in tape revolution; panel (e) is a map of a Holter tape with episodes of paroxysmal atrial fibrillation; and panel (f) is a map of chronic atrial fibrillation.

an optimum ellipse in a two dimensional vector analysis.[78] However, no systematic experience exists with these approaches and it is likely that they will be superseded by more advanced and more physiologically relevant technologies of non-linear dynamics.

Frequency Domain Measures of Long-Term HRV

Although originally developed to handle data of technical rather than biological nature, spectral analysis has been repeatedly used for the HRV analysis.[91] The HRV frequency domain methods are either non-parametric, most frequently based on fast Fourier transformation (FFT),[92,93] or parametric, employing mainly autoregressive modelling.[94–96] In both cases, the technology assumes that the stochastic fluctuations in RR interval durations are more or less stationary during the whole analysed period and contain inherent periodicities, which are determined by the analysis.[97]

In principle, the mathematical assumptions required by spectral analysis technologies cannot be justified in long-term recordings since the environmental demands are not periodic and result in non-harmonic modulations of heart rate. This fact alone makes the physiological speculations about the nature of individual long-term spectral components difficult, if not directly inappropriate. Consequently, medical relevance of long-term spectral analysis and significance of individual components are merely based on statistical observations made in large clinical populations rather than on understanding of the inherit meaning of long-term spectral components. Since the existence of long-term periodic and harmonic modulations of heart rate is questionable, it is not obvious whether long-term spectral components of HRV have inherit meaning. Similarly, it is questionable whether long-term spectral analysis provides any practical advantages over the simpler time-domain analysis.

Data Analysed

The spectrum of the HRV signal is generally calculated either from the RR interval tachogram (RR durations versus number of progressive beat) or by interpolating the discrete event series of individual cardiac beats, thus obtaining a continuous signal as a function of time, or, less frequently, by calculating the so-called spectrum of the counts, that is of unitary pulses as a function of time corresponding to each recognised QRS complex.[98]

Such a choice may have implications on the morphology, the measurement units of the spectra and, the measurement of the relevant spectral parameters. In order to standardise the methods used, the use of RR interval tachogram with the parametric method, or the use of the regularly sampled interpolation of discrete event series with the non-parametric method has been suggested.[1] When resampling the discrete event series, the sampling frequency of interpolation has to be sufficiently high so that the Nyquist frequency of the spectrum is not within the frequency range of interest.

Analysis of Short-Term Subsections

Spectral analysis of a long-term NN interval sequence is computationally demanding and in some studies, averages of spectral analyses short subsections of the long-term record have been reported. Unfortunately, as the stationarity of heart rate modulations cannot be maintained for more than several minutes, averaging of subsequent spectra leads to results that do not correspond to any physiologic models that were derived for the interpretation of spectral analyses of short-term recordings.[99–102] Analyses of this kind are therefore misleading and should be avoided because any meaningful information obtained in this way is only a less stable and less well defined surrogate of a time-domain method.

More acceptable is the analysis of regularly or irregularly placed short-term sample windows of a long-term ECG recording with reporting the spectral analyses of individual windows separately. For instance, an analysis of the most stationary 5 minutes in each hour or each half-hour of the whole record can portray the circadian variations of individual short-term HRV components.[103] (Figure 6)

Long-Term Analysis

Consequently, if a long-term spectrum is to be obtained, it is only appropriate to perform the spectral analysis of the entire recording processed in one pass.

As a number of samples involved in such computations is large, computational necessities make it impossible to use autoregressive modelling and the non-parametric FFT analysis has to be employed. At least $2^{17} = 131{,}072$ samples are needed for a FFT of 24-hour data with resampling $\geq$ 1Hz, but more correctly $2^{18} = 262{,}144$ to $2^{20} = 1{,}048{,}576$ samples are used. While some years ago, these numbers of points required to run the FFT analysis on a special computer because of memory requirements, the computation is presently fully feasible on standard personal computers. (The complexity of autoregressive modelling concerns not only the memory requirements but

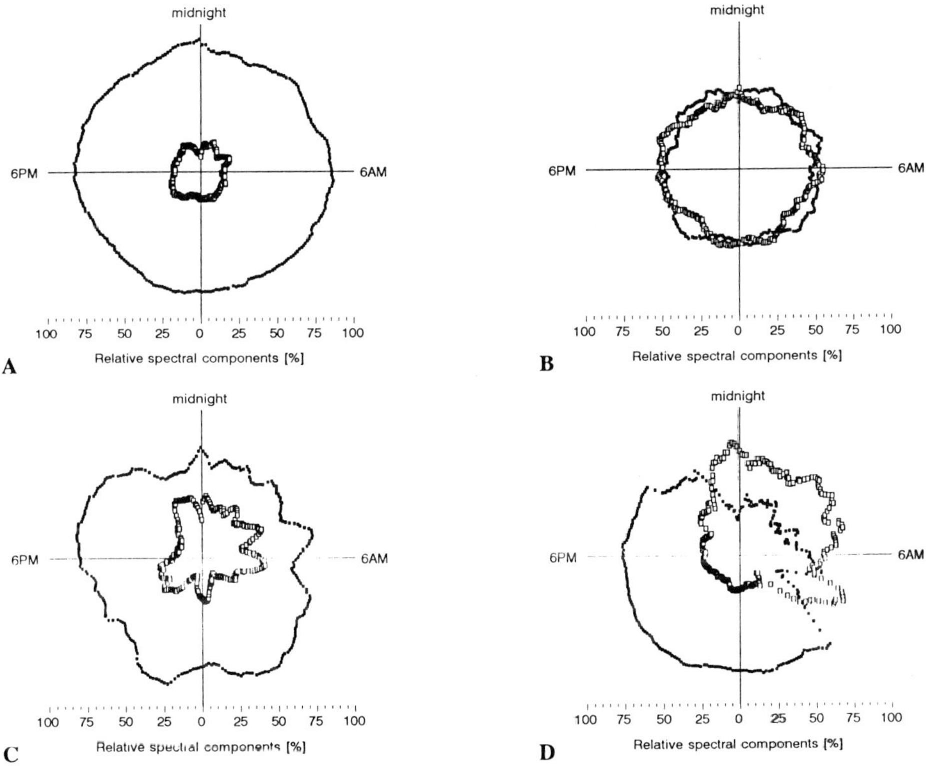

FIGURE 6. Clock plots of normalised short-term spectral components obtained in four patients surviving acute myocardial infarction. Instead of presenting the course of the time of day along a linear axis, the time is presented in angles from the upwards vertical direction. In the graphs shown here, a full circle of 360° corresponds to 24-hour interval. The lines composed of small full squares mark the trends of quasi-normalised LF (that is LF/(LF+HF)) and the lines composed of large open squares the trends of quasi-normalised HF (that is HF/(LF+HF)) components. The spectral analyses were performed using FFT applied to non-overlapping 5-minute segments of the 24-hour recording (From Malik[103]).

also computational time – even with present technology, very special matrix processors would be needed to computer an autoregressive model for the entire 24-hour NN interval series.)

Application of FFT to long-term NN interval series results in a spectrogram that permits four components to be distinguished. Similar to short-term recordings, the high (HF) and low (LF) frequency components are defined as the spectral power within the frequency bands of 0.15 – 0.4 Hz and 0.04 – 0.15 Hz, respectively. In addition to these, the very low (VLF) and the ultra low (ULF) frequency component are defined as the spectral power within 0.0033 – 0.04 Hz and below 0.0033 Hz, respectively.[1] In real recordings, the VLF and ULF components account for > 95% of the total power,[60] but their physiological correlates are not known. It is actually likely that direct physiological correlates of VLF and ULF do not exist and that these components are merely a sum of non-harmonic variations related to the environment. Because of this, distinct spectral peaks corresponding to these components are not seen and probably may only appear by chance. Indeed, even the distinction between VLF and ULF is purely arbitrary, corresponding to the wavelength of 5 minutes that distinguishes SDNN and SDANN.

Physiological interpretation of HF and LF components derived from long-term spectral analyses must not be mixed with the interpretation of HF and LF components found in short-term HRV.[99–102] Since the parasympathetic and sympathetic modulations of heart rate cannot be stable and stationary through the whole long-term recording, the standard physiological models of the vagal and sympathetic modulations are not applicable. The long-term HF and LF components correspond approximately to the averages of HF and LF components from subsequent short-term sections of the whole recording[58,104] and similar criticism as discussed in the previous sections applies. It should be remembered that the components of HRV provide measurement of the degree of autonomic modulations rather than of the level of autonomic tone[105] and that averages of modulations do not represent an averaged level of tone.

The 1/f Spectral Property

The power of 24-hour HRV spectrum spreads across the frequencies in a inversely proportional (1/f) manner[70,106] and forms a negative gradient, occasionally with peaks corresponding to LF and HF components. The slope of the ULF – VLF part of the spectrum can be assessed on a log-log scale by linear fit to the spectrogram. (Figure 7) Within a frequency band 10^{-4} – 10^{-2} Hz (that is, 2.7 hours – 1.7 minutes), this slope is approximately −1 in normal healthy subjects.[7,106,107] The variance at some given frequency bands, e.g. 10^{-2} to 10^{-3} Hz, corresponds that at other frequency bands, e.g. 10^{-3} to 10^{-4} Hz.[7] This phenomenon is called the invariance or self similarity and

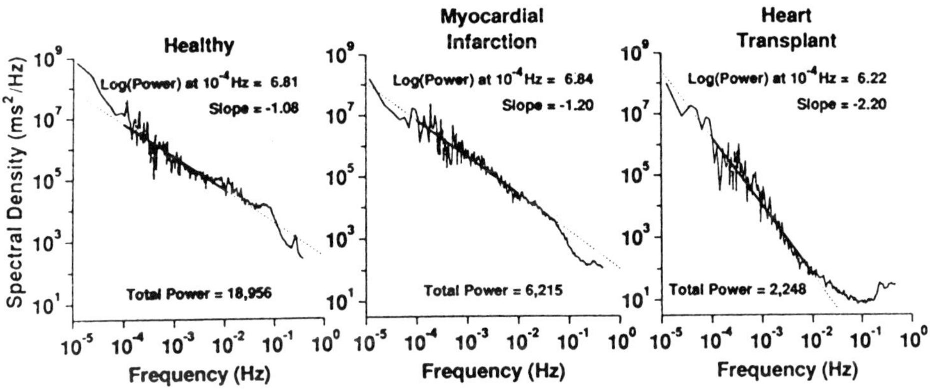

FIGURE 7. Examples of log-log graphs of 24-hour NN interval spectra for a healthy subject, a post myocardial infarction patient and a heart transplant patient. The computed power law regression lines are superimposed on the data in the segments between 10^{-4} to 10^{-2} Hz. The slope is steeper in the patient with myocardial infarction as compared to the healthy subject, and is steepest in the heart transplant patient (From Bigger et al[7]).

is characteristics for a broad band frequency spectrum in which in no single frequency component is dominant or characteristic.[108] It has been suggested that this phenomenon reflects the underlying fractal heart rate dynamics[106] and proposals have been made to use the power law behaviour of long-term HRV spectra to correct SDNN for differences in the duration of original ECG recordings.[7] However, although mathematically plausible, no practical experience exists with such speculative corrections.

The correlation between the slope the long-term spectrum and conventional power spectral components was found to be very low.[7]

NORMAL VALUES OF STANDARD LONG-TERM HRV MEASURES

Only a small number of studies addressed the normal values of long-term HRV measures. More experience exists with distributions of HRV values in clinically well defined populations, e.g. among survivors of acute myocardial infarction.

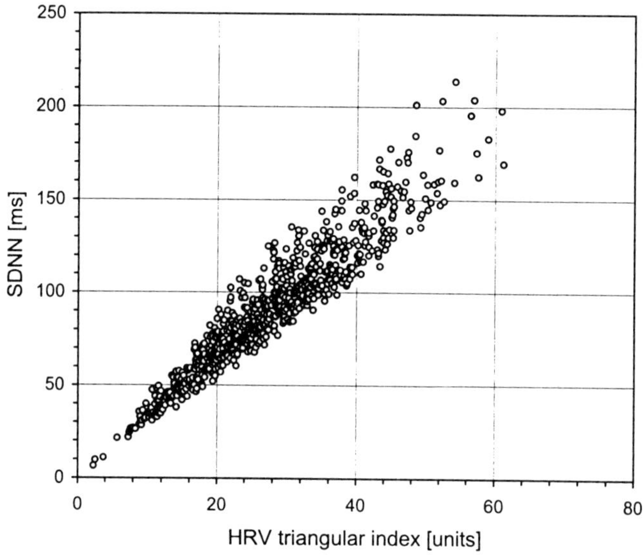

A

FIGURE 8. Scatter diagrams of SDNN and HRV trianguler index (1/128 s bins) values in sets of nominal 24-hour Holter recordings obtained from three different databases of electrocardiograms made in survivors of acute myocardial infarction. Panel (a) corresponds to a set with reasonably edited NN interval data whilst in sets of panels (b – page opposite) and (c – page opposite), the editing of the data was casual, and only very superficial, respectively. Note that as the quality of Holter editing decreases, artificially high values of SDNN are being obtained.

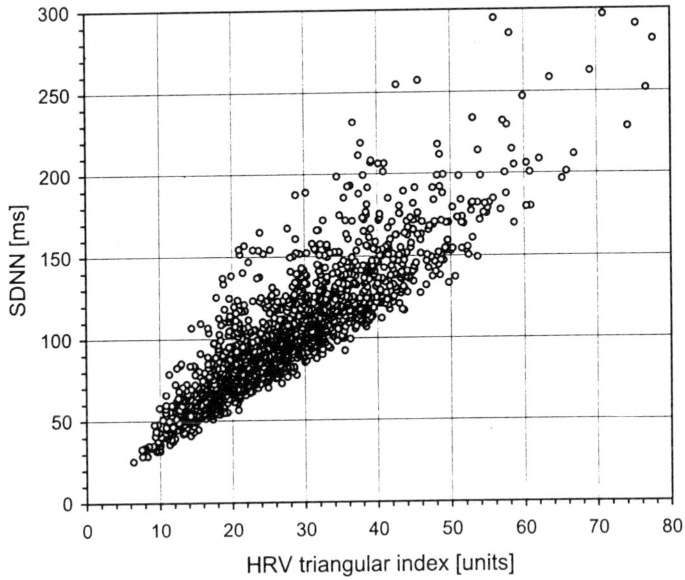

B

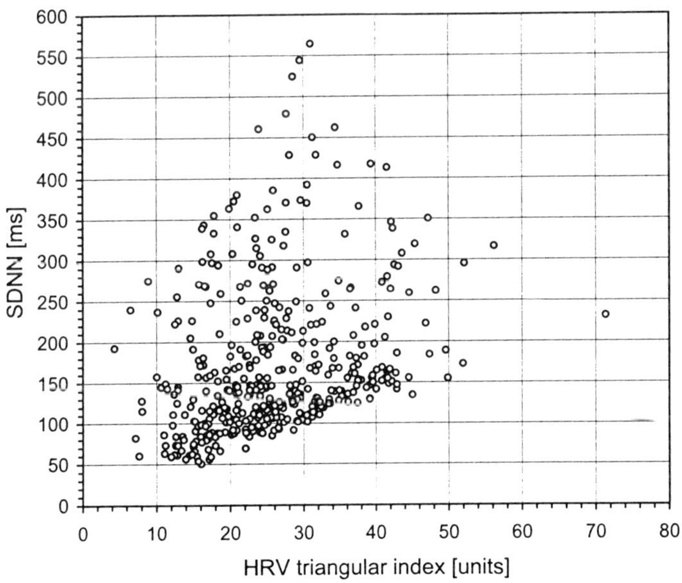

C

FIGURE 8 (Continued). See the legend opposite.

Time Domain Measures

In normal healthy middle-age population, 24-hour SDNN was found to be around 141±39 ms[57] and in a population aged 20 to 69 years, SDNN ranged from 187±40 ms to 145±51 ms.[55] Systematically, SDNN was shown to decline with age,[55,57] but sex differences with lower SDNN in women were observed only by some[55,109] whilst unconfirmed by others.[57]

In normal middle aged population, SDANN values of 127±35 ms have been reported[57] and in a population aged 20 – 69 years, values from 129±27 ms to 163±34 ms were found.[55]

In properly edited NN interval data of recordings made in cardiac patients, HRV triangular index correlates very highly and linearly with SDNN, 1 unit of HRV triangular index corresponding to approximately 2.5 ms of SDNN. (Figure 8) However, no large data exist on healthy subjects in whom the computation of HRV triangular index may be affected by a bi-modal shape of the NN interval histogram.

Values of SDNN ≤ 50 ms and HRV triangular index ≤ 20 units (128 Hz bins) have been proposed define patients with seriously depressed long-term HRV.[1] It is not known whether and how this suggestion should be modified dependent on clinical circumstances (e.g. patients on vs. patients off β-blockers).

Frequency Domain Measures

Normal values of total power, ULF and VLF components of nominal 24-hour spectra in healthy middle aged population were found to be 21,222±11,663 ms^2, 18,420±10,639 ms^2, and 1,782±965 ms^2, respectively.[57]

RHYTHM PATTERN ANALYSIS

Both the time- and frequency-domain methods share limitations imposed by the irregularity of the NN interval series since different profiles analysed by these techniques may give identical results.[110] Trends of decreasing or increasing cycle length are not symmetric[111,112] as heart rate accelerations are frequently followed by a faster decrease. In spectral analyses, this reduces the fundamental frequency peak and enlarges its basis.

Difficulties of this kind led to the idea of measuring blocks of NN intervals determined by properties of the rhythm and investigating the relationship of such blocks without considering the internal variability. The so-called "peak-valley" procedures were proposed, based either on the detection of the summit and the nadir of oscillations[113,114] or on the detection of trends of heart rate.[115] The detection may be limited to short-term changes[113] or extended to longer variations, that is second and third order peaks and troughs[114] or stepwise increase of a sequence of consecutive increasing or decreasing cycles surrounded by opposite trends.[115] The various oscillations can be characterised on the basis of the heart rate acceleration or slowing, the wavelength and/or the amplitude. In a majority of short- to mid-term recordings, the results are correlated with frequency components of HRV.[116] However, the correlation diminishes as the wavelength of the oscillations and the recording length increase. Complex demodulation uses the techniques of interpolation and detrending[117] and provides the time resolution necessary to detect short-term heart rate changes, as well as to describe the amplitude and phase of particular frequency components as functions of time.

While some clinical reports[115] have demonstrated the usefulness of these "pattern" techniques for the assessment of HRV in specific clinical populations, the technology has not been in a widespread use. Similar to complex processing of Poincaré plots, it is likely that this approach will be superseded by the possibilities of dynamic and fractal HRV measurements.

DYNAMIC AND FRACTAL MEASURES OF HRV

As discussed in the section on the physiological background of long-term HRV, modulations of heart rate cannot be assumed regular and periodic over a longer period of time. The long-term regulation of heart rate contains both short-time periodic (e.g. respiratory) modulations, the frequency of which does not remain stable, and entirely non-periodic (e.g. environment driven) fluctuations.[66] However, it has been suggested that the non-periodic fluctuations are not completely random but stem from dynamics of chaotic nature and exhibit short range correlation governed by deterministic laws.[118] Consequently, methods based on non-linear mathematics and chaos theories were proposed to describe non-periodic heart rate dynamics. It is believed that these methods, especially when used in conjunction with conventional measures of HRV, may uncover subtitle abnormalities or alterations in long-term data that are otherwise not apparent.[119]

Several methods to describe non-linear heart rate dynamics have been proposed. These include some of the simple geometrical methods applied to Poincaré plots.[78] The power law behaviour of 24-hour spectrum is also suggestive of a non-linear and chaotic nature of heart rate dynamics.[7,88] Other methods based on non-linear mathematics quantify either the fractal properties or the regularity of the analysed NN (or RR) interval time series.

Fractal Measures

A fractal is a object composed of sub-units (and sub-sub-units) that resemble the large scale structure. In heart rate dynamics, fractal (scale-invariant) processes generate irregular fluctuations in multiple time scales[108] and can be quantified by statistical analyses. Although fractals are irregular, not all irregular time series are fractal. Self-similarity is a key fractal feature, which, seen in time rather than spatial properties, extends over many time scales. Fractal properties, that is the degree of self-similarity or correlation between different time scales, can be quantified by various mathematical methods including correlation dimension,[69] fractal dimension,[120] apparent correlation dimension,[121] and coarse-graining spectral analysis.[122]

Entropy Quantifiers

Regularity of RR interval data series is quantified by methods derived from natural information mathematics.[88,119,123] The approximate entropy is a measure of regularity that quantifies the unpredictability of fluctuations in time series such as RR intervals. It reflects the likelihood that "similar" pattern of observations will not be followed by additional similar observations.[8] Approximate entropy is computed as the natural logarithm of relative prevalence of repetitive patterns of length m compared to those of $m+1$ length. If a similar patterns in the NN (or RR) interval time series is found, approximate entropy estimates the probability that the interval following each of the pattern will differ. To measure approximate entropy of a time series, the pattern length and the criterion of similarity have to be specified. Values of 2 (pattern length) and 20 % (similarity) were show to be acceptable in most cases and have been commonly used.[8,75,119,124] Completely random signals result in high values while a predictable, regularly shaped signal leads to a low value. A significant number of data points are needed for the analysis and the computation time demands are serious. Compromises were therefore proposed balancing the length of the analysed segment analysed and the time required for the analysis.[120]

Practical Applications

The fractal and entropy measures, as well as other non-linear dynamics analysis, do not quantify the actual magnitude of HRV. Rather, they describe the complexity of the underlying heart rate regulations.

Normal fractal dynamic and rhythm complexity has been reported to be lost with many pathological processes, in heart failure patients, survivors of acute myocardial infarction, etc.[69,120,75] In normal subjects, complexity can be decreases by bed rest, is lower during sleep[121] and decreases with advanced age.[110,120] The complexity of non-periodic heart rate dynamics has been shown to vary during day and night,[121] to be affected by cholinergic blockade,[125] but not by adrenergic blockade or changes in breathing pattern.[16,126,127]

Despite all these initial reports, no major breakthrough has yet been achieved by the application of non-linear analysis technology to bio-medical data including HRV analysis. It is possible that integral complexity measures are not adequate to analyse biological systems, e.g. too insensitive to detect the non-linear perturbations of RR interval that would be of physiological or practical importance. Some believe that more encouraging results are being obtained using differential, rather than integral complexity measures, e.g. the scaling index method.[128]

Since only infrequent studies have been conducted to investigate large patient populations using these methods, the non-linear methods represent only a potentially promising tools for HRV assessment. Substantial advances in both technology and the interpretation of the results are needed before these methods can be recommended for clinical studies and clinical practice.

CLINICAL IMPLICATIONS

The measurement of long-term HRV, especially by time-domain methods, satisfies several pre-requisites for clinical applications. The global measures of overall long-term HRV are stable with a satisfactory intra-subject reproducibility.[63,129–132] Compared to short-term HRV that is particularly suited to address cardiac reflexes, long-term HRV is likely to provide a better reflection of the overall responsiveness of the cardiovascular system.

Assessment of HRV has been utilised in a large variety of clinical studies investigating broad range of pathologies and pathophysiological conditions. So far, a clear practical benefit of HRV measurement has been demonstrated only in two areas, namely in risk stratification after acute myocardial infarction and in the early diagnosis of diabetic neuropathy. However, more recent studies indicate that the assessment of HRV has also its practical role under other clinical circumstances.

Although the assessment of diabetic neuropathy can benefit from long-term HRV measurement, short-term tests are more frequently used and are fully capable of an early diagnosis of the neuropathy.[133–139] The clinical conditions which we review are those having a clear advantage from long-term HRV assessment.

POST INFARCTION RISK

HRV is depressed following an acute myocardial infarction, reflecting an increased tonic sympathetic activity saturating the ability to modulate heart rate.[140–143] The predominance of sympathetic activity enhances cardiac electrical instability.[144]

A recovery of HRV measures occurs over a period of two weeks to three months,[59,140] but, on average, the values remain permanently at a depressed level.[59] The HRV based risk assessment is usually made before hospital discharge. To make the process of risk stratification more practical, patients may also be selected for long-term HRV evaluation by short-term HRV measurements.[60,145] In other words, full-scale risk stratification involving nominal 24-hour HRV monitoring can be restricted to patients with reduced short-term HRV values. Nevertheless, long-term HRV assessment is important because the predictive power of short-term assessment is substantially lower.[60,146,147]

The mechanism by which HRV is reduced after infarction is not clear. In addition to systemic sympathetic stimulation, speculations have proposed that the changes in the cardiac geometry due to necrotic and noncontracting segments may abnormally increase the firing of sympathetic afferent fibres by mechanical distortion of the sensory endings. This may attenuate the activity of vagal fibres directed to the sinus node.

The compelling evidence that depressed 24-hour HRV is a powerful risk predictor post infarction mortality has been repeatedly shown in large follow-up databases.[3,5,61,62,82] The predictive power of depressed HRV appears to be independent of demographics, New York Heart Association functional class, left ventricular ejection fraction and ventricular ectopic activity. Depressed HRV is not only capable of predicting the total cardiac mortality after infarction but also identified patients at risk of sustained symptomatic ventricular tachycardia. Indeed, it has been demonstrated that depressed HRV is a better predictor of sudden death and symptomatic ventricular tachycardia than left ventricular ejection fraction, whereas both HRV and ejection fraction perform equally well in predicting all-cause mortality in the post-infarction patients.[6] This suggests that HRV is a stronger predictor of arrhythmic events (sudden cardiac death and ventricular tachycardia) than non-arrhythmic mortality.

More recent studies confirmed that depressed HRV retains its independent prognostic significance after myocardial infarction also in patients subjected to the modern clinical management during the acute phase of myocardial infarction.[82,148]

In addition to depressed HRV, heart rate dynamics seems also to be altered in post infarction patients. Higher regularity and a loss of normal fractal characteristics were reported[81] and the slope of 24 hours FFT spectrum was shown to be steeper than in healthy subjects.[7] The slope was also steeper in post-infarction patients with low left ventricular ejection fraction compared to those with normal ejection fraction.[69]

It is not clear whether the approximate entropy (regularity measure of heart rate dynamics) has any post-infarction risk stratification meaning. It was seen both increased[75,83] and decreased[69] after myocardial infarction. Similarly, in patients with poorer prognosis the complexity was reported decreased together with higher regularity in the RR interval time series,[81] while in another study, the approximate entropy was of no prognostic significance.[8]

Predictive accuracy of depressed HRV is rather modest but in combination with other risk predictors, efficient stratification of patients at high post infarction risk is possible.[149] (Figure 9) Indeed, combinations of different risk predictors seem to be able to distinguish patients prone to arrhythmic and non-arrhythmic death.[150] Arrhythmic death is predominantly associated with depressed HRV and runs of ventricular tachycardia, whereas low ejection fraction, ventricular ectopic beats, and depressed HRV elicited non-arrhythmic death. This corresponds to the present understanding of the pathophysiology of risk prediction. Reduced ejection fraction reflects a li-

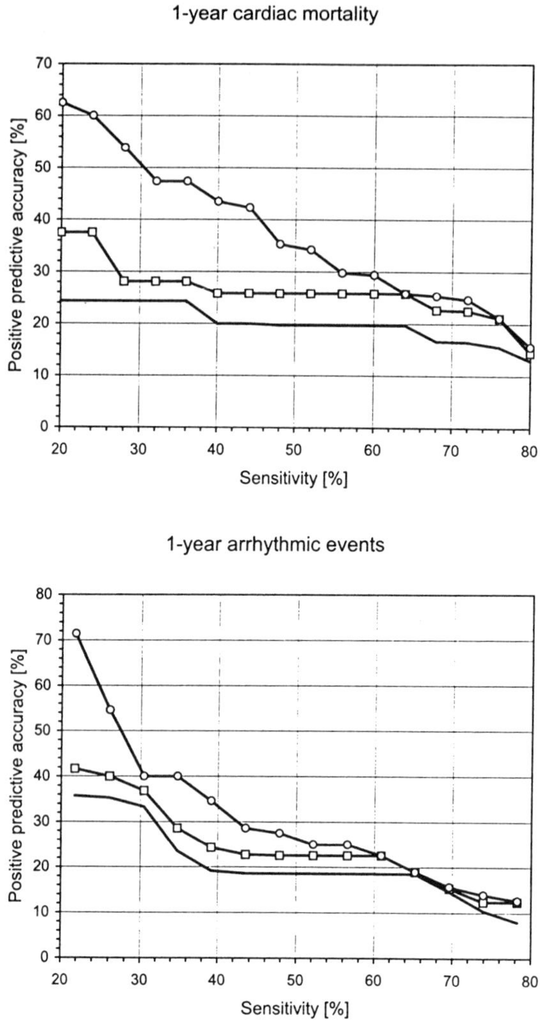

FIGURE 9. Comparison of positive predictive characteristics of HRV (solid lines) and of combinations of HRV with left ventricular ejection fraction (solid lines with open squares) and of HRV with left ventricular ejection fraction and ectopic counts on 24-hour ECGs (solid lines with open circles) used for identification of patients at risk of 1-year cardiac mortality (top panel) and 1-year arrhythmic events (sudden death and/or symptomatic sustained ventricular tachycardia – bottom panel) after acute myocardial infarction (Data of St. George's Post-infarction Research Survey Programme – Redrawn from the ESC/NASPE Task Force report[1]).

ability to cardiac failure and is likely to identify patients at risk of non-sudden death. On the other hand, reduced HRV implies a lack of an autonomic protection against ventricular arrhythmias and is therefore more powerful in identifying those patients at high risk of sudden death or arrhythmic complication.

VENTRICULAR ARRHYTMIAS

Decreased HRV has been reported to be associated with malignant ventricular arrhythmias both with and without coronary artery disease.[78,83,151–154] To some extent, however, the individual reports are inconsistent. Global long-term HRV was seen lower in patients with the propensity to ventricular tachycardia as compared to those without the propensity or to normal controls.[78] In another study, long-term global HRV (SDNN and variability assessed from Poincaré plots) were reduced in post-infarction patients with vulnerability to ventricular fibrillation but not in patients with ventricular tachycardia.[155]

A loss of fractal stability of heart rate dynamics and an increased value of the approximate entropy was seen in patients with vulnerability to ventricular tachycardia[83] indicating a more irregular interbeat dynamics despite reduced global HRV (SDNN and long-term HRV in Poincaré plots).

RISK OF CARDIOVASCULAR MORBIDITY

Depressed long-term HRV was associated with four fold increase in risk for sudden death in 6693 patients with heart disease, but without history of myocardial infarction.[156]

In a large epidemiological study,[68] the Framingham Heart Study Group showed that in a population which was apparently free of coronary heart disease or congestive heart failure, depressed HRV was associated with an increased incidence of subsequent cardiac events, namely angina, myocardial infarction, coronary heart disease death and congestive heart failure. The mechanism for such an association is unknown.

In a separate epidemiological study, the Framingham Heart Study Group reported that reduced HRV is associated with an increased risk for all-cause mortality.[157] These observations are in accord with other smaller studies showing that depressed HRV is able to predict sudden cardiac death.[158–160]

HEART FAILURE

Earlier studies of patients with chronic heart failure documented autonomic dysfunction similar to that seen after acute infarction, that is a withdrawal of parasympathetic tone together with a chronic activation of the adrenergic system and impaired arterial baroreflex sensitivity.[161,162]

Mechanisms underlying depressed HRV in patients with congestive heart failure are not yet fully established. Increased sympathetic activation, assessed by, for instance, plasma noradrenaline levels, might be one of the explanations.[53,54] On the other hand, periodic patterns of deep breathing, seen in these patients, have been suggested to increase long-term HRV.[42,51] The patients may also show very low frequency fluctuations in blood pressure associated with increased hypoxic chemosensitivity which lead to similar modulations of heart rate.[42]

Although disputed in some smaller studies[163,164] heart failure patients with worsening New York Heart Association functional class seem to have a significant progressive decrease in HRV in both time and frequency domains.[165,166] Studies investigating this development of HRV with heart failure progression also found a strong correlation between different measures of HRV and depression in left ventricular ejection fraction.[166,167] Similarly, an association between peak oxygen consumption and HRV has been observed.[166] This all suggests that in heart failure, reduction of HRV is mainly driven by systemic sympathetic overdrive rather then by primarily vagal withdrawal.

While depressed HRV is a predictor of arrhythmic risk in post-infarction patients, such prediction is less clear in patients with chronic heart failure. Controversial reports suggest that in chronic heart failure patients, there are diverse mechanisms in the pathogenesis of ventricular arrhythmias and impaired autonomic tone is probably only one of multiple trigger mechanisms. A larger study is still needed to clarify the role of reduced HRV in identifying chronic heart failure patients at high risk of ventricular arrhythmias.

Several studies showed that HRV has an independent value in predicting total survival of patients with chronic heart failure.[168,169] For instance, it has been proposed that patients with 24-hour SDNN < 100 ms and a peak oxygen consumption < 14 ml/kg/min have a particularly poor prognosis with the 1-year mortality of 37 %.[76] Thus HRV has an important prognostic implication in patients with moderate to severe congestive heart failure. Together with other variables such as left ventricular ejection fraction and peak oxygen consumption, it may have a role in identifying patients for consideration of cardiac transplantation.

Abnormal HRV complexity in 24-hours Poincaré plot analysis had an independent prognostic value in patients with mild to moderate congestive heart failure and identified a patient group with a 6 – 7 fold increased risk of sudden cardiac death and all cause mortality, although none of the conventional time and frequency domain measures of HRV were related to survival.[53] Patients with congestive heart failure show a loss of the physiological heart rate dynamics, which, in combination with traditional HRV measures, may improve the predictive power of their risk stratification.[8,170]

A rather recent study analysed time domain measures of HRV in patients with familial dilated cardiomyopathy, their relatives with left ventricular enlargement which is a possible early stage of the disease, and healthy controls.[171] Overall 24-hour HRV was reduced in cardiomyopathy patients compared to the controls or relatives and was much lower in patients who developed progressive heart failure compared to those who remained clinically stable. Thus depressed HRV might be an early marker of clinical deterioration of familial dilated cardiomyopathy although it does not seem to be useful in an early diagnosis and in family screening of the disease.

CARDIAC TRANSPLANTATION

Earlier studies of HRV in heart transplant recipients suggested that development of random 'white noise' patterns of HRV may serve as warning signs of transplant rejection.[13] However, these conclusions have been more recently disputed. Specifically, a new study[172] of patients with and without graft arteriosclerosis, episodes of rejection, or earlier cytomegalovirus infection did not find any difference in the power spectral measures of HRV. On the contrary, an increase of HRV with post-transplantation time has been noted. The changes of HRV after transplantation are now being intensively studied[173–177] and the HRV increase is considered as a sign of reinnervation. Further studies are needed as in some reports, the authors were able to find only sympathetic reinnervation[173] while in others studies, gradual parasympathetic reinnervation[174] of the sinus node was detected.

REFERENCES

1. Task Force of The European Society of Cardiology and The North American Society of Pacing and Electrophysiology. Heart rate variability. Standards of measurement, physiological interpretation, and clinical use. *Circulation* 1996; **93:**1043–1065.
2. Berntson GG, Bigger JT Jr, Eckberg DL, Grossman P, Kaufmann PG, Malik M, Nagaraja HN, Porges SW, Saul JP, Stone PH, van der Molen MW. Heart rate variability: Origins, methods, and interpretative caveats. *Psychophysiology* 1997; **34:**623–648.
3. Kleiger RE, Miller JP, Bigger JT, Moss AJ, and the Multicenter Post-infarction Research Group. Decreased hear rate variability and its association with increased mortality after acute myocardial infarction. *Am J Cardiol* 1987; **59:**256–262.
4. Bigger JT, Kleiger RE, Fleiss JL, Rolnitzky LM, Steiman RC, Miller JP, and the Multicenter Post-infarction Research Group. Components of heart rate variability measured during healing of acute myocardial infarction. *Am J Cardiol* 1988; **61:**208–215.
5. Cripps TR, Malik M, Farrell TG, Camm AJ. Prognostic value of reduced heart rate variability after myocardial infarction: clinical evaluation of a new analysis method. *Br Heart J* 1991; **65:**14–19.
6. Odemuyiwa O, Malik M, Farrell T, Bashir Y, Poloniecki, Camm J. Comparison of the predictive characteristics of heart rate variability index and left ventricular ejection fraction for all-cause mortality, arrhythmic events and sudden death after acute myocardial infarction. *Am J Cardiol* 1991; **68:**434–439.
7. Bigger JT, Steinman RC, Rolnitzky LM, Fleiss JL, Albrecht P, Cohen RJ. Power law behavior of RR-interval variability in healthy middle-aged persons, patients with recent acute myocardial infarction, and patients with heart transplants. *Circulation* 1996; **93:**2142–2151.
8. Ho KKL, Moody GB, Peng C-K, Mietus JE, Larson MG, Levy D, Goldberger AL. Predicting survival in heart failure case and control subjects by use of fully automated methods for deriving nonlinear and conventional indices of heart rate dynamics. *Circulation* 1997; **96:**842–848.
9. Bernardi L, Salvucci F, Suardi R, Soladá PL, Calcati A, Perlini S, Falcone C, Ricciardi L. Evidence for an intrinsic mechanism regulating heart rate variability in transplanted and the intact heart during submaximal dynamic exercise? *Cardiovasc Res* 1990; **24:**969–981.
10. Bernardi L, Leuzzi S, Valle F, Martinelli L, Passino C, Rinaldi M, Spadacini MV, Fianrdi G, Sleight P. Effect of different interventions on heart rate variability after heart transplantation: non-autonomic vs autonomic factors. *Clin Sci* 1996; **91**(Suppl 22):22–24.
11. Perlini S, Solda PL, Piepoli M, Sala-Gallini G, Calcati A, Finardi G, Bernardi L. Determinants of respiratory sinus arrhythmia in the vagotomized rabbit heart. *Am J Physiol* 1995; **269:**H909–H915.

12. Horner SM, Murphy CF, Coen B, Dick DJ, Harrison FG, Vespalcova Z, Lab MJ. Contribution to heart rate variability by mechanoelectric feedback. Stretch of the sinoatrial node reduces heart rate variability. *Circulation* 1996; **94**:1762–1767.
13. Sands KEF, Appel ML, Lilly LS, Schoen FJ, Mudge GH, Cohen RJ. Power spectral analysis of heart rate variability in human cardiac transplant recipients. *Circulation* 1989; **79**:76–82.
14. Ewing DJ, Neilson JMM, Travis P. New method for assessing cardiac parasympathetic activity using 24 hour electrocardiograms. *Br Heart J* 1984; **52**:396–402.
15. Bernardi L, Ricordi L, Lazzari P, Soladá P, Calciati A, Ferrari MP, Vandela I, Finardi G, Frantino P. Impaired circadian modulation sympathovagal activity in diabetes. A possible explanation for altered temporal onset of diabetic autonomic neuropathy. *Circulation* 1992; **86**:1443–1452.
16. Zwiener U, Richter A, Schumann NP, Glaser S, Witte H. Heart rate fluctuations in rabbits during different behavioral states. *Biomed Biochim Acta* 1990; **49**:59–68.
17. Hayano J, Sakakibara Y, Yamada A, Yamada M, Mukai S, Fujinami T, Yokoyama K, Watanabe Y, Takata K. Accuracy of assessment of cardiac vagal tone by heart rate variability in normal subjects. *Am J Cardiol* 1991; **67**:199–204.
18. Hedman AE, Hartikainen JEK, Tahvanainen KUO, Hakumäki MOK. Power spectral analysis of heart rate and blood pressure variability in anaesthetized dogs. *Acta Physiol Scand* 1992; **146**:155–164.
19. Tuininga YP, Crijns HJGM, Brouwer J, van den Berg MP, Man in't Veld AJ, Mulder L, Lie KI. Evaluation of importance of central effects of atenolol and metoprolol measured by heart rate variability during mental performance tasks, physical exercise, and daily life in stable postinfarct patients. *Circulation* 1995; **92**:3415–3423.
20. Cook Jr, Bigger JT Jr, Kleiger RE, Fleiss JL, Steinman RL, Rolnizky LM. Effect of atenolol and diltiatzem on heart period variability in normal persons. *Am J Coll Cardiol* 1991; **17**:480–484.
21. Coumel P, Hermida JS, Wennerblööm B, Leenhard A, Madson-Blanche P, Cauchemez B. Heart rate variability and heart failure, and the effect of beta-blockade. A non-spectral analysis of heart rate variability in the frequency domain and in the time domain. *Eur Heart J* 1991; **12**:412–422.
22. Elghozi JL, Japundzic N, Grichois ML, Zitou P. Nervous mechanisms of spontaneous oscillations of systolic blood pressure and heart rate. *Arch Mal Coeur Vais* 1990; **83**:1065–1068.
23. Loewy AD. Central autonomic pathways. In: Loewy, AD, Spyer KM (Eds) Central regulation of autonomic functions. Oxford University Press, New York 1990, 88–103.
24. Koepchen HP, Abel HH, Klussendorf D, Lazar H. Respiratory and cardiovascular rhythmicity. In: von Euler C, Lagergrantz H (Eds) Neurobiology of the Control of Breathing. Raven Press, New York 1986, 303–312.

25. Richter DW, Spyer KM. Cardiorespiratory control. In: Loewy, AD, Spyer KM (Eds) Central regulation of autonomic functions. Oxford University Press, New York 1990, 189–207.
26. Spyer KM. Central nervous mechanisms contributing to cardiovascular control. *J Physiol* 1994; **474.1**:1–19.
27. Freitas J, Lago P, Puig J, Carvalho MJ, Costa O, de Freitas AF. Circadian heart rate variability rhythm in shift workers. *J Electrocardiol* 1997; **30**:39–44.
28. Freeman R, Weiss ST, Roberts M, Zbikowski SM, Sparrow D. The relationship between heart rate variability and measures of body habitus. *Clin Auton Res* 1995; **5**:261–266.
29. Lee C, Katsuura T, Harada H, Kikuchi Y. Effects of handgrip work and heat load on heart rate variability. (Japanese) *Ann Physiol Anthropol* 1994; **13**:233–243.
30. Sacknoff DM, Gleim GW, Stachenfeld N, Coplan NL. Effect of athletic training on heart rate variability. *Am Heart J* 1994; **127**:1275–1278.
31. Adamopoulos S, Ponikowski P, Cerquetani E, Piepoli M, Rosano G, Sleight P, Coats AJ. Circadian pattern of heart rate variability in chronic heart failure patients. Effects of physical training. *Eur Heart J* 1995; **16**:1380–1386.
32. Boutcher SH, Stein P. Association between heart rate variability and training response in sedentary middle-aged men. *Eur J Appl Physiol Occupat Physiol* 1995; **70**:75–80.
33. Malfatto G, Facchini M, Bragato R, Branzi G, Sala L, Leonetti G. Short and long term effects of exercise training on the tonic autonomic modulation of heart rate variability after myocardial infarction. *Eur Heart J* 1996; **17**:532–538.
34. Fujimoto S, Uemura S, Tomoda Y, Yamamoto H, Matsukura Y, Hashimoto T, Dohi K. Effects of physical training on autonomic nerve activity in patients with acute myocardial infarction. (Japanese) *J Cardiol* 1997; **29**:85–93.
35. Kitney RI. An analysis of the nonlinear behaviour of the human thermal vasomotor control system. *J Theor Biol* 1975; **52**:231–248.
36. Saul JP, Arai Y, Berger RD, Lilly LS, Colucci WS, Cohen RJ. Assessment of autonomic regulation in chronic congestive heart failure by heart rate spectral analysis. *Am J Cardiol* 1988; **61**:1292–1299.
37. Madwed JB, Albrecht P, Mark RG, Cohen RJ. Low frequency oscillations in arterial pressure and heart rate: A simple computer model. *Am J Physiol* 1989; **256**:H1573–H1579.
38. Di Rienzo M, Castiglioni P, Parati G, Mancia G, Pedotti A. Effects of sino-aortic denervation on spectral characteristics of blood pressure and pulse interval variability: A wide-band approach. *Med Biol Eng Comp* 1996; **34**:133–141.
39. Siegel G. Principles of vascular rhythmogenesis. *Prog Appl Microcirc* 1983; **3**:40–62.
40. Akselrod S, Gordon D, Uber FA, Shannon DC, Barger AC, Cohen RJ. Power spectral analysis of heart rate fluctuation: A quantitative probe of beat-to-beat cardiovascular control. *Science* 1981; **213**:220–222.

41. Bonaduce D, Marciano F, Peretta M, Migaux ML, Morgano G, Bianchi V, Salemme L, Valva G, Condorelli M. Effects of converting enzyme inhibition on heart period variability in patients with acute myocardial infarction. *Circulation* 1994; **90**:108–113.
42. Ponikowski P, Chua TP, Amadi AA, Piepoli M, Harrington D, Volterrani M, Colombo R, Mazzuero, Giordano A, Coats AJS. Detection and significance of a discrete very low frequency rhythm in RR interval variability in chronic congestive heart failure. *Am J Cardiol* 1996; **77**:1320–1326.
43. Koepchen HP. History of studies and concepts of blood pressure waves. In: Miyakawa K, Koepchen HP, Polosa C (Eds) Mechanisms of blood pressure waves. Springer, Tokyo 1984, 3–23.
44. Madwed JB, Cohen RJ. Heart rate response to hemorrhage-induced 0.05–Hz oscillations in arterial pressure in conscious dogs. *Am J Physiol* 1991; **260**:H1248–H1253.
45. Preiss G, Iscoe S, Polosa C. Analysis of a periodic breathing pattern associated with Mayer waves. *Am J Physiol* 1975; **228**:786–774.
46. Furlan R, Cuzzetti S, Crivellario W, Dassi S, Tinelli M, Baselli G, Cerutti S, Lombardi F, Pagani M, Malliani A. Continuous 24-hour assessment of the neural regulation of systemic arterial pressure and RR variabilities in ambulant subjects. *Circulation* 1990; **81**:537–547.
47. Bonnet MH, Arand DL. Heart rate variability: sleep stage, time of night, and arousal influences. *Electroencephalography Clin Neurophysiol* 1997; **102**:390–396.
48. Vanoli E, Adamson PB, Ba-Lin, Pinna GD, Lazzara R, Orr WC. Heart rate variability during specific sleep stages. A comparison of healthy subjects with patients after myocardial infarction . *Circulation* 1995, **91**:1918–1922.
49. Niizeki K, Kawahara K, Miyamoto Y. Interaction among cardiac, respiratory and locomotor rhythms during cardiolocomotor synchronization. *J Appl Physiol* 1993; **75**:1815–1821.
50. Osterhues HH, Hanzel SR, Kochs M, Hombach V. Influence of physical activity on 24-hour measurements of heart rate variability with coronary artery disease. *Am J Coll Cardiol* 1997; **80**:1437–1437.
51. Mortara A, Sleight P, Pinna GD, Maestri R, Prpa A, La Rovera MT, Cobelli F, Tavazzi L. Abnormal awake respiratory patterns are common in chronic heart failure and may prevent evaluation of autonomic tone by measures of heart rate variability. *Circulation* 1997; **96**:246–252.
52. Arai Y, Saul JP, Albrecht P, Hartley LH, Lilly LS, Cohen RJ, Colucci WS. Modulation of cardiac autonomic activity during and immediately after exercise. *Am J Physiol* 1989; **256**:H132–H141.
53. Brouwer J, van Veldhuisen DJ, Man in't Veld AJ, Haaksma J, Dijk WA, Visser KR, Boomsma F, Dumselman PHJM, Lie KI, for the Deutch Ibapamine Multicenter Trial Study Group. Prognostic value of heart rate variability during long-term follow-up in patients with mild to moderate heart failure. *Am J Coll Cardiol* 1996; **28**:1183–1189.
54. Woo Ma, Stevenson WG, Moser DK, Middlekauff HR. Complex heart rate variability and serum norepinephrine levels in patients with advanced heart failure. *J Am Coll Cardiol* 1994; **23**:565–569.

55. Jensen-Urstad K, Storck N, Bouvier F, Ericson M, Lindblad E, Jensen-Urstad M. Heart rate variability in healthy subjects in related to age and gender. *Acta Physiol Scand* 1997; **160**:235–241.
56. Yeragani VK, Sobolewski E, Kay J, Jampala VC, Igel G. Effect of age on long-term heart rate variability. *Cardiovasc Res* 1997; **35**:35–42.
57. Bigger JT, Fleiss JL, Steinman RC, Rolnitzky LM, Schneider WJ, Stein PK. RR variability in healthy, middle-aged persons compared with patients with chronic coronary heart disease or recent acute myocardial infarction. *Circulation* 1995; **91**:1936–1943.
58. Berger RD, Akselrod S, Gordon D, Cohen RJ. An efficient algorithm for spectral analysis of heart rate variability. *IEEE Trans Biomed Eng* 1986; **33**:900–904.
59. Bigger JT, Fleiss Jl, Rolnitzky LM, Steinman RC, Schneider WJ. Time course of recovery of heart period variability after myocardial infarction. *J Am Coll Cardiol* 1991; **18**:1643–1649.
60. Bigger JT, Fleiss JL, Steinman RC, Rolnizky LM, Kleiger RE, Rottman JN. Correlation among time and frequency domain measures of heart period variability two weeks after acute myocardial infarction. *Am J Cardiol* 1992; **69**:891–898.
61. Bigger JT, Fleiss JL, Steinman RC, Rolnizky LM, Kleiger RE, Rottman JN. Frequency domain measures of heart period variability and mortality after myocardial infarction. *Circulation* 1992; **85**:164–171.
62. Farrell TG, Bashir Y, Cripps T, Malik M, Poloniecki J, Bennett ED, Ward DE, Camm AJ. Risk stratification for arrhythmic events in postinfarction patients based on heart rate variability, ambulatory electrocardiographic variables and the signal-averaged electrocardiogram. *J Am Coll Cardiol* 1991; **18**:687–697.
63. Kleiger RE, Bigger JT, Bosner MS, Chung MK, Cook JR, Rolnitzky L, Steinman R, Fleiss JL. Stability over time of variables measuring heart rate variability in normal subjects. *Am J Cardiol* 1991; **68**:626–630.
64. Vaishnav S, Stevenson R, Marchant B, Lagi K, Rajadayalan K, Timmis AD. Relation between heart rate variability early after acute myocardial infarction and long-term mortality. *Am J Cardiol* 1994; **73**:653–657.
65. Huang J, Sopher SM, Leatham E, Redwood S, Camm AJ, Kaski JC. Heart rate variability depression in patients with unstable angina. *Am Heart J* 1995; **130**:772–779.
66. Yamamoto Y, Hughson RL. On the fractal nature of heart rate variability in humas: effect of data length and β-adrenergic blockade. *Am J Physiol* 1994; **266**:R40–R49.
67. Tsuji H, Venditti Jr FJ, Manders ES, Evans JC, Larson MG, Feldman C, Levy D. Determinants of heart rate variability. *J Am Coll Cardiol* 1996; **28**:1539–1546.
68. Tsuji H, Larson MG, Venditti Jr FJ, Manders ES, Evans JC, Feldman CL, Levy D. Impact of reduced heart rate variability on risk for cardiac events. The Framingham heart study. *Circulation* 1996; **94**:2850–2855.

69. Lombardi F, Sandrone G, Mortara A, Torzillo D, La Rovere MT, Signorini MG, Cerutti S, Malliani A. Linear and nonlinear dynamics of heart rate variability after acute myocardial infarction with normal and reduced left ventricular ejection fraction. *Am J Cardiol* 1996; **77**:1283–1288.
70. Saul JP, Albrecht P, Berger RD, Cohen RJ. Analysis of long term heart rate variability: methods, 1/f scaling and implications. In: Computers in Cardiology 1987. IEEE Computer Society press, Washington 1988, 419–422.
71. Malik M, Xia R, Odemuyiwa O, Staunton A, Poloniecki J, Camm AJ. Influence of the recognition artefact in the automatic analysis of long-term electrocardiograms on time-domain measurement of heart rate variability. *Med Biol Eng Comput* 1993; **31**:539–544.
72. Xia R, Odemuyiwa O, Gill J, Malik M, Camm AJ. Influence of recognition errors of computerised analysis of 24-hour electrocardiograms on the measurement of spectral components of heart rate variability. *Int J Bio-Med Comput* 1993; **32**:223–235.
73. Lippman N, Stein KM, Lerman BB. Comparison of methods for removal of ectopy in measurement of heart rate variability. *Am J Physiol* 1994; **267**:H411–H418.
74. Friesen GM, Jannett TC, Jadalloh MA, Yates SL, Quint SR, Nogle HT. A comparison of the noise sensitivity of nine QRS detection algorithms. *IEEE Trans Biomed Eng* 1990; **37**:85–98.
75. Mäkikallio TH, Seppänen T, Niemelä M, Airaksinen KEJ, Tulppo M, Huikuri HV. Abnormalities in beat to beat complexity of heart rate dynamics in patients with a previous myocardial infarction. *Am J Coll Cardiol* 1996; **28**:1005–1011.
76. Ponikowski P, Anker S, Chua TP, Szelemej, Piepolo M, Adamopoulos S, Webb-Peploe K, Harrington D, Banasiak W, Wrabec K, Coats AJS. Depressed heart rate variability as an independent predictor of death in chronic congestive heart failure secondary to ischemic or idiopathic dilated cardiomyopathy. *Am J Cardiol* 1997; **79**:1645–1650.
77. Copie X, Le Heyzey J-Y, Iliou M-C, Khouri K, Lavergne T, Pousset F, Guize L. Correlation between time-domain measures of heart rate variability and scatterplots in post infarction patients. *Pacing Clin Electrophysiol* 1996; **19**:342–247.
78. Huikuri HV, Seppänen T, Koistinen MJ, Airaksinen KEJ, Ikäheimo MJ, Castellanos A, Myerburg RJ. Abnormalities in beat-to-beat dynamics of heart rate before the spontaneous onset of life-threatening ventricular tachyarrhythmias in patients with prior myocardial infarction. *Circulation* 1996; **93**:1836–1844.
79. Sugihara G, May RM. Nonlinear forecasting as a way of distinguishing chaos from measurement error in time series. Nature 1990; **344**:734–741.
80. Kamath MV, Fallen EL. Correction of the heart rate variability signal for ectopics and missing beats. In: Malik M, Camm AJ (Eds) Heart rate variability. Futura, Armonk 1995, 75–85.
81. Mäkikallio TH, Ristimäe T, Airaksinen KEJ, Peng C-K, Goldberger AL, Huikuri HV. Heart rate dynamics in patients with stable angian pectoris and utility of fractal and complexity measures. *Am J Cardiol* 1998; **81**:27–31.

82. Zuanetti G, Neilson JMM, Latini R, Santoro E, Maggioni AP, Ewing DJ on Behalf of GISSI-2 Investigators. Prognostic significance of heart rate variability in post-myocardial infarction patients in the fibrinolytic era. *Circulation* 1996; **94**:432–436.
83. Mäkikallio TH, Seppänen T, Airaksinen KEJ, Koistinen J, Tulppo MP, Peng C-K, Goldberger AL, Huikuri HV. Dynamic analysis of heart rate may predict subsequent ventricular tachycardia after myocardial infarction. *Am J Cardiol* 1997; **80**:779–782.
84. Valkama JO, Huikuri HV, Koistinen MS, Ylä-Mäyry S, Airaksinen KEJ, Myeburg RJ. Relation between heart rate variability and spontaneous and induced ventricular arrhythmias in patients with coronary artery disease. *J Am Coll Cardiol* 1995; **25**:437–447.
85. Malik M. Geometrical methods for heart rate variability assessment. In: Malik M, Camm AJ (Eds) Heart rate variability. Futura, Armonk 1995, 47–61.
86. Malik M, Farrell T, Cripps T, Camm AJ. Heart rate variability in relation to prognosis after myocardial infarction: selection of optimal processing techniques. *Eur Heart J* 1989; **10**:1060–1074.
87. Hnatkova K, Copie X, Staunton A, Malik M. Numeric processing of Lorenz plots of R-R intervals from long-term ECGs. Comparison with time-domain measures of heart rate variability for risk stratification after myocardial infarction. *J Electrocardiol* 1995; **28** Suppl:74–80.
88. Denton TA, Diamond GA, Helfant RH, Khan S, Karagueuzan H. Fascinating rhythm: A primer on chaos theory and its application to cardiology. *Am Heart J* 1990; **120**:1419–1440.
89. Gilham J. Application of scatterplots to Holter ECG data. *J Electrocardiol* 1993; **26**:31-33.
90. Woo MA, Stevenson WG, Moser DK, Middlekauff HR. Complex heart rate variability and serum norepinephrine levels in patients with advanced heart failure. *J Am Coll Cardiol* 1994; **23**:565–569.
91. Kay S, Marple SL. Spectrum analysis – a modern perspective. *Proc IEEE* 1981; **69**:1380–1419.
92. Challis RE, Kitney RI. Biomedical signal processing: The frequency transform and their inter-relationships. *Med Biol Eng Comput* 1991; **29**:1–17.
93. Challis RE, Kitney RI. Biomedical signal processing: The power spectrum and coherence function. *Med Biol Eng Comput* 1991; **29**:225–241.
94. Baselli G, Cerutti S, Livraghi M, Meneghini C. Causal relationship between heart rate and arterial blood pressure variability signals. *Med Biol Eng Comput* 1988; **26**:374–378.
95. Pagani M, Lombardi F, Guzzetti S, Rimoldi O, Furlan R, Pizzinelli P, Sandrone G, Malfatto G, DellOrto S, Piccaluga E, Turel M, Baselli G, Cerutti A, Malliani A. Power spectral analysis of heart rate and arterial pressure variabilities as a marker of sympatho-vagal interaction in man and conscious dog. *Circ Res* 1986; **59**:178–193.

96. Bianchi AM, Mainardi LT, Petrucci E, Signorini MG, Mainardi M, Cerutti S. Time-variant power spectrum analysis for the detection of transient episodes in HRV signal. *IEEE Trans Biomed Eng* 1993; **40**:136–144.
97. Kamath MV, Fallen EL. Power spectral analysis of heart rate variability: a noninvasive signature of cardiac autonomic function. *Critical Rev Biomed Eng* 1993; **21**:245–311.
98. De Boer RW, Karemaker JM, Strackee J. Comparing spectra of a series of point events, particularly for heart-rate variability spectra. *IEEE Trans Biomed Eng* 1984; **31**:384–387.
99. Hirsh JA, Bishop B. Respiratory sinus arrhythmia in humans, how breathing pattern modulates heart rate. *Am J Physiol* 1981; **241**:H620–H629.
100. Pomeranz M, Macaulay RJB, Caudill MA, Kutz I, Adam D, Gordon D, Kilborn KM, Barger AC, Shannon DC, Cohen RJ, Benson M. Assessment of autonomic function in humans by heart rate spectral analysis. *Am J Physiol* 1985; **248**:H151–H153.
101. Malliani A, Pagani M, Lombardi F, Cerutti S. Cardiovascular neural regulation explored in the frequency domain. *Circulation* 1991; **84**:1482–1492.
102. Montano N, Ruscone TG, Porta A, Lombardi F, Pagani M, Malliani A. Power spectrum analysis of heart rate variability to assess the changes in sympathovagal balance during graded orthostatic tilt. *Circulation* 1994; **90**:1826–1831.
103. Malik M. Graphical representation of circadian patterns of heart rate variability components. *Pacing Clin Electrophys* 1995; **18**:1575–1580.
104. Rottman JN, Steinman RC, Albrecht P, Bigger JT, Rolnitzky LM, Fleiss JL. Efficient estimation of the heart period power spectrum suitable for physiologic or pharmacologic studies. *Am J Cardiol* 1990; **66**:1522–1524.
105. Malik M, Camm AJ. Components of heart rate variability – What they really mean and what we really measure. *Am J Cardiol* 1993; **72**:821–822.
106. Yamamoto Y, Huhgson RL. Extracting fractal components from time series. *Physica D* 1993; **68**:250–264.
107. Kobayashi N, Musha T. 1/f fluctuation of heart beat period. *IEEE Trans Biomed Eng* 1992; **29**:456–457.
108. Goldberger AL. Non-linear dynamics for clinicians: chaos theory, fractals, and complexity at the bedside. *Lancet* 1996; **347**:1312–1314.
109. Huikuri HV, Pikkujämsä SM, Airaksinen KEJ, Ikäheimo MJ, Rantala AO, Kauma H, Lilja M, Kesöniemi A. Sex-related differences in autonomic modulation of heart rate in middle-aged subjects. *Circulation* 1996; **94**:122–125.
110. Kaplan DT. The analysis of variability. *J Cardiovasc Electrophysiol* 1994; **5**:16–19.
111. Katona PG, Jih F. Respiratory sinus arrhythmia : a non invasive measure of parasympathetic cardiac control. *J Appl Physiol* 1975; **39**:801–805.
112. Eckberg DL. Human sinus arrhythmia as an index of vagal cardiac outflow. *J Appl Physiol* 1983; **54**:961–966.

113. Fouad FM, Tarazi RC, Ferrario CM, Fighaly S, Alicandri C. Assessment of parasympathetic control of heart rate by a noninvasive method. *Am J Physiol* 1984; **246:**H838–H842.
114. Schechtman VL, Kluge KA, Harper RM. Time-domain system for assessing variation in heart rate. *Med Biol Eng Comput* 1988; **26:**367–373.
115. Coumel Ph, Hermida JS, Wennerblöm B, Leenhardt A, Maison-Blanche P, Cauchemez B. Heart rate variability in myocardial hypertrophy and heart failure, and the effects of beta-blocking therapy. A non-spectral analysis of heart rate oscillations. *Eur Heart J* 1991; **12:**412–422.
116. Grossman P, Van Beek J, Wientjes C. A comparison of three quantification methods for estimation of respiratory sinus arrhythmia. *Psychophysiology* 1990; **27:**702–714.
117. Shin SJ, Tapp WN, Reisman SS, Natelson BH. Assessment of autonomic regulation of heart rate variability by the method of complex demodulation. *IEEE Trans Biomed Eng* 1989; **36:**274–283.
118. Babloyantz A, Destexhe A. Is the normal heart a periodic oscillator? *Biol Cybern* 1988; **58:**203–211.
119. Pincus SM, Goldberger AL. Physiologic time-series analysis: what does regularity quantify? *Am J Physiol* 1994; **226:**H1643–H1656.
120. Signorini MG, Cerutti S, Guzzetti S, Parola R. Non-linear dynamics of cardiovasculal variability signals. *Meth Inform Med* 1994; 33:81–84.
121. van Leeuwen P, Bettermann H, an der Heiden U, Kümmel HC. Circadian aspects of apparent correlation dimension in human heart rate dynamics. *Am J Physiol* 1995;269:H130–H134.
122. Yamamoto Y, Hughson RL. Coarse-graining spectral analysis: new method for studying heart rate variability. *J Appl Physiol* 1991; **71:**1143–1150.
123. Kaplan DT, Goldberger AL. Chaos in cardiology. *J Cardiovasc Electrophysiol* 1991; **2:**342–354.
124. Pincus SM. Approximate entropy as a measure of system complexity. *Proc Natl Acad Sci USA* 1991; **88:**2297–2301.
125. Yamamoto Y, Nakamura Y, Sato H, Yamamoto M, Kato K, Hughson RL. On the fractal nature of heart rate variability in humans: effect of vagal blockade. *Am J Physiol* 1995; **269:**H830–H837.
126. Kanters J, Højgaard MV, Agner E, Holstein-Rathlou N-H. Influence of forced respiration on nonlinear dynamics in heart rate variability. *Am J Physiol* 1997; **272:**R1149–R1154.
127. Yamamoto Y, Fortrait J-O, Hughson RL. On the fractal nature of heart rate variability in humans: effect of respiratory sinus arrhythmia. *Am J Physiol* 1995; **269:**H480–H486.
128. Schmidt G, Monfill GE. Nonlinear methods for heart rate variability assessment. In: Malik M, Camm AJ (Eds) Heart rate variability. Futura, Armonk 1995, 87–98.
129. Huikuri HV, Kessler KM, Terracell E, Castellanos A, Linnaluoto MK, Myerburg RJ. Reproducibility and circadian rhythm of heart rate variability in healthy subjects. *Am J Cardiol* 1990; **65:**391–393.

130. Kautzner J, Hnatkova K, Staunton A, Camm AJ, Malik M. Day-to-day reproducibility of time-domain measures of heart rate variability in survivors of acute myocardial infarction. *Am J Cardiol* 1995; **76**:309–312.

131. Kamalesh M, Burger AJ, Kumar S, Nesto R. Reproducibility of time and frequency domain analysis of heart rate variability in patients with chronic stable angina. *Pacing Clin Electrophys* 1995; **18**:1991–1994.

132. van Hoogenhuyze D, Weinstein N, Martin GJ, Weiss JS, Schaad JW, Sahyouni XN, Fintel D, Remme WJ, Singer DH. Reproducibility and relation to mean heart rate of heart rate variability in normal subjects and in patients with congestive heart failure secondary to coronary artery disease. *Am J Cardiol* 1991; **68**:1668–1676.

133. Pagani M, Malfatto G, Pierini S, Casati R, Masu AM, Poli M, Guzzetti S, Lombardi F, Cerutti S, Malliani A. Spectral analysis of heart rate variability in the assessment of autonomic diabetic neuropathy. *J Auton Nerv System* 1988; **23**:143–153.

134. Smith S. Reduced sinus arrhythmia in diabetic autonomic neuropathy: diagnostic value of an age related normal range. *Br Med J* 1982; **285**:1599–1601.

135. O'Brien IA, O'Hare P, Corrall RJM. Heart rate variability in healthy subjects: effect of age and the derivation of normal ranges for tests of autonomic function. *Br Heart J* 1986; **55**:348–354.

136. Ewing DJ, Neilson JMM, Shapiro JA, Reid W. Twenty four hour heart rate variability: effects of posture, sleep and time of day in healthy controls and comparison with bedside tests of autonomic function in diabetic patients. *Br Heart J* 1991; **65**:239–244.

137. Malpas SC, Maling TJB. Heart rate variability and cardiac autonomic function in diabetes. *Diabetes* 1990; **39**:1177–1181.

138. Bianchi A, Bontempi B, Cerutti S, Gianogli P, Comi G, Natali Sora MG. Spectral analysis of heart rate variability signal and respiration in diabetic subjects. *Med Biol Eng Comput* 1990; **28**:205–211.

139. Bellavere F, Balzani I, De Masi G, Carraro M, Carenza P, Cobelli C, Thomaseth K. Power spectral analysis of heart rate variation improves assessment of diabetic cardiac autonomic neuropathy. *Diabetes* 1992; **41**:633–640.

140. Casolo GC, Stroder P, Signorini C, Calzolari F, Zucchini M, Balli E, Sulla A, Lazzerini S. Heart rate variability during the acute phase of myocardial infarction. *Circulation* 1992; **85**:2073–2079.

141. Schwartz PJ, Vanoli E, Stramba-Badiale M, De Ferrari GM, Billman GE, Foreman RD. Autonomic mechanisms and sudden death. New insights from the analysis of baroreceptor reflexes in conscious dogs with and without a myocardial infarction. *Circulation* 1988; **78**:969–979.

142. Malliani A, Schwartz PJ, Zanchetti A. A sympathetic reflex elicited by experimental coronary occlusion. *Am J Physiol* 1969; **217**:703–709.

143. Brown AM, Malliani A. Spinal sympathetic reflexes initiated by coronary receptors. *J Physiol* 1971; **212**:685–705.

144. Lown B, Verrier RL. Neural activity and ventricular fibrillation. *N Engl J Med* 1976; **294**:1165–1170.

145. Faber TS, Staunton A, Hnatkova K, Camm AJ, Malik M. Stepwise strategy of using short- and long-term heart rate variability for risk stratification after myocardial infarction. Pacing Clin Electrophys 1996; **19**:1845–1851.
146. Malik M, Camm AJ. Significance of long-term components of heart rate variability for the future prognosis after acute myocardial infarction. *Cardiovasc Res* 1990; **24**:793–803.
147. Malik M, Farrell T, Camm AJ. Circadian rhythm of heart rate variability after acute myocardial infarction and its influence on the prognostic value of heart rate variability. *Am J Cardiol* 1990; **66**:1049–1054.
148. Touboul P, Andre-Fouët X, Leizorovicz A, Itti R, Lopez M, Sayegh Y, Milon H, Kirkorian G, for the Groupe d'etude du pronostic de l'infarctus du myocarde (GREPI). Risk stratification after myocardial infarction. A reappraisal in the era of thrombolysis. *Eur Heart J* 1997; **18**:90–107.
149. Redvood S, Odemuyiwa O, Hnatkova K, Staunton A, Poloniecki J, Camm AJ, Malik M. Selection of dichotomy limits for multifarctorial prediction of arrhythmic events and mortality in survivors of acute myocardial infarction. *Eur Heart J* 1997; **18**:1278–1287.
150. Hartikainen JEK, Malik M, Staunton A, Poloniecki J, Camm AJ. Distinction between arrhythmic and nonarrhythmic death after acute myocardial infarction based on heart rate variability, signal-averaged electrocardiogram, ventricular arrhythmias and left ventricular ejection fraction. *J Am Coll Cardiol* 1996; **28**:296–304.
151. Farrell TG, Paul V, Cripps TR, Malik M, Bennett ED, Ward D, Camm AJ. Baroreflex sensitivity and electrophysiological correlates in patients after acute myocardial infarction. *Circulation* 1991; **83**:945–952.
152. Fei L, Anderson MA, Katritsis D, Sneddon J, Statters DJ, Malik M, Camm AJ. Depressed heart rate variability in survivors of sudden cardiac death not associated with coronary artery disease. *Br Heart J* 1994; **71**:16–21.
153. Fei L, Statters DJ, Hnatkova K, Poloniecki J, Malik M, Camm AJ. Change of autonomic influence on the heart immediately before the onset of spontaneous idiopathic ventricular tachycardia. *J Am Coll Cardiol* 1994; **24**:1515–1522.
154. Filipiecki A, Trusz-Gluza M, Szydlo K, Giec L. Value of heart rate variability parameters for prediction of serious arrhythmic events in patients with malignant ventricular arrhythmias. *Pacing Clin Electrophys* 1996; **19**:1852–1856.
155. Perkiömäki JS, Huikuri HV, Koistinen JM, mäkikallio T, Castellanos A, Myerburg RJ. Heart rate variability and dispersion of QT interval in patients with vulnerability to ventricular tachycardia and ventricular fibrillation after previous myocardial infarction. *J Am Coll Cardiol* 1997; **80**:779–783.
156. Algra A, Tijssen JGP, Roelandt JRTC, Pool J, Lubsen J. Heart rate variability from 24-hours electrocardiography and 2-year risk for sudden death. *Circulation* 1993; **88**:180–185.
157. Tsuji H, Venditti FJ, Manders ES, Evans JC, Larson MG, Feldman CL, Levy D. Reduced heart rate variability and mortality risk in an elderly cohort: the Framingham Heart Study. *Circulation* 1994; **90**:878–883.

158. Huikuri HV, Linnaluoto MK, Seppanen T, Airaksinen KEJ, Kessler KM, Takkunen JT, Myerburg RJ. Circadian rhythm of heart rate variability in survivors of cardiac arrest. *Am J Cardiol* 1992; **70**:610–615.
159. Myers GA, Martin GJ, Magid NM, Barnett PS, Schaad JW, Weiss JS, Lesch M, Singer DH. Power spectral analysis of heart rate variability in sudden cardiac death: comparison to other methods. *IEEE Trans Biomed Eng* 1986; **33**:1149–1156.
160. Martin GJ, Magid NM, Myers G, Barnett PS, Schaad JW, Weiss JS, Lesch M, Singer DH. Heart rate variability and sudden death secondary to coronary artery disease during ambulatory ECG monitoring. *Am J Cardiol* 1986; **60**:86–89.
161. Eckberg DL, Drabinsky M, Braunwald E. Defective cardiac parasympathetic control in patients with heart disease. *N Engl J Med* 1971; **285**:877–883.
162. Hasking GJ, Esler MD, Jennings GL, Burton D, Johns JA, Korner PI. Norepinephrine spillover to plasma in congestive heart failure: Evidence of increased cardiorenal and total sympathetic nerve activity. *Circulation* 1986; **73**:615–621.
163. Rich MW, Saini JS, Kleiger RE, Carney RM, te Velde A, Freeland KE. Correlation of heart rate variability with clinical and angiographic variables and late mortality after coronary angiography. *Am J Cardiol* 1988; **62**:714–717.
164. Kienzle M, Ferguson DW, Birkett CL, Myers GA, Berg WJ, Marino DJ. Clinical, haemodynamic and sympathetic neural correlates of heart rate variability in congestive heart failure. *Am J Cardiol* 1992; **69**:761–767.
165. Casolo GC, Stroder P, Sulla A, Chelucci A, Freni A, Zerauscheck M. Heart rate variability and functional severity of congestive heart failure secondary to coronary artery disease. *Eur Heart J* 1995; **16**:360–367.
166. Szabo BM, van Veldhuisen DJ, Brouwer J, Haaksma J, Lie KL. Relation between severity of disease and impairment of heart rate variability parameters in patients with chronic congestive heart failure secondary to coronary artery disease. *Am J Cardiol* 1995; **76**:713–716.
167. Nolan J, Flapan AD, Capewell S, MacDonald TM, Neilson JMM, Ewing DJ. Decreased cardiac parasympathetic activity in chronic congestive heart failure and its relation to left ventricular function. *Br Heart J* 1992; **67**:482–485.
168. Szabo BM, van Veldhuisen DJ, van der Veer N, Brouwer J, De Graeff PA, Crijns HGM. Prognostic value of heart rate variability in chronic congestive heart failure secondary to idiopathic or ischaemic dilated cardiomyopathy. *Am J Cardiol* 1997; **79**:978–980.
169. Jiang W, Hathaway WR, McNulty S, Larsen RL, Hansley K, Zhang Y, O'Connor CM. Ability of heart rate variability to predict prognosis in patients with advanced congestive heart failure. *Am J Cardiol* 1997; **80**:808–811.
170. Butler GC, Ando S, Floras SJ. Fractal component of heart rate variability and systolic pressure in congestive heart failure. *Clin Sci* 1997; **92**:543–550.

171. Yi G, Goldman JH, Keeling PJ, Reardon M, McKenna WJ, Malik M. Heart rate variability in idiopathic dilated cardiomyopathy: relation to disease severity and prognosis. *Heart* 1997; **77**:108–114.
172. Koskinen P, Virolainen J, Koskinen PK, Hayry P, Kupari M. Evolution of heart rate variability in cardiac transplant recipients: a clinical study. *J Int Med* 1996; **239**:443–449.
173. Halpert I, Goldberg AD, Levine AB, Levine TB, Kornberg R, Kelly C, Lesch M. Reinnervation of the transplanted human heart as evidenced from heart rate variability studies. *Am J Cardiol* 1996; **77**:180–183.
174. Ramaekers D, Ector H, Vanhaecke J, van Cleemput J, van de Werf F. Heart rate variability after cardiac transplantation in humans. *Pacing Clin Electrophys* 1996; **19**:2112–2119.
175. Bernardi L, Valle F, Leuzzi S, Rinaldi M, Marchesi E, Falcone C, Martinelli L, Vigano M, Finardi G, Radaelli A. Non-respiratory components of heart rate variability in heart transplant recipients: evidence of autonomic reinnervation? *Clin Sci* 1994; **86**:537–545.
176. Frey AW, Dambacher M, Uberfuhr P, Reichart B, Ziegler S, Roskamm H, Schwaiger M. Clinical relevance of heart rate variability changes after heart transplantation. *Clin Sci* 1996; **91** Suppl:146–150.
177. Radaelli A, Valle F, Falcone C, Calciati A, Leuzzi S, Martinelli L, Goggi C, Vigano M, Finardi G, Bernardi L. Determinants of heart rate variability in heart transplanted subjects during physical exercise. *Eur Heart J* 1996; **17**:462–471.

CHAPTER NINE

NEURAL RECORDINGS IN THE ASSESSMENT OF CARDIAC AUTONOMIC BALANCE

Nicola Montano, Virend K Somers,
Alberto Porta and Massimo Pagani

Considerable evidence has so far accumulated to substantiate the hypothesis that a high level of resting beat by beat heart rate variability, and a brisk capacity to respond to moderate stimuli, such as standing up, is a sign of health. Conversely, disease states are associated to a progressive loss of resting variability, and responsiveness to stimuli.

Extremes of stress, both in health and disease, demand maximal engagement and interaction of different, possibly opposite, cardiovascular control mechanisms, such as inhibitory and excitatory reflexes. Examples of these conditions are heavy physical exercise or severe heart failure, which are both characterised by elevated sympathetic and reduced vagal activity.

In the last two decades, spectral analysis of beat to beat cardiovascular variability has emerged as a convenient tool to study the dynamic interaction of sympathetic and vagal control mechanisms, also in real life conditions and with non-invasive measurements.[1]

In our laboratory we followed the general hypothesis that power spectrum analysis of the short term fluctuations affecting RR interval and arterial pressure might provide indices of neural regulation, and, in particular, of the balance between sympathetic and parasympa-

Malik M (Ed) Clinical guide to cardiac autonomic tests, 239–255.

thetic cardiovascular modulations. Over these years a large body of evidence has been collected supporting this hypothesis, however direct comparisons between cardiovascular and neural rhythms was scanty and controversial.

Accordingly we felt that it was crucial to address directly the investigation of the rhythmical content of the activity of neural pathways involved in cardiovascular control. In particular we sought to evaluate the presence of coherent oscillatory patterns between sympathetic outflows and cardiovascular target functions, considering various states of the sympatho-vagal balance.

A BRIEF HISTORY OF SYMPATHETIC RECORDINGS

Adrian and coworkers[2] were the first to record from sympathetic nerves an efferent impulse activity displaying a dual rhythm in phase with arterial pulse and with respiration. Several decades then elapsed before Fernandez de Molina and Perl[3] observed, in spinal cats, rhythmic increases in sympathetic efferent discharge simultaneous to those slow arterial pressure oscillations, which are also known as Mayer waves.[4]

In a quite recent study[5] performed in decerebrate, artificially ventilated cats, we observed the presence of two main rhythmicities, one (low frequency, LF) centered around 0.1 Hz, corresponding to Mayer's waves, and the other one (high frequency, HF) synchronous with ventilation rate.

Both these oscillatory components resulted to be highly coherent with the similar spectral components detectable in the variability of RR interval and systolic arterial pressure and underwent parallel changes during hemodynamic manoeuvres increasing or decreasing sympathetic efferent discharge (i.e., increase of LF and decrease of HF in the variability signals following a drops in arterial pressure and, conversely, decrease of LF and increase of HF during raises in arterial pressure).

As to the impulse activity of vagal efferent fibres, the existence of cardiac and respiratory rhythms was also reported before the description of an additional rhythm in phase with slow wave arterial oscillations.[6]

There has been much debate over the years as to the origin of these LF neural rhythms: indeed they could depend on peripheral modulatory reflexes with negative feedback characteristics, on autochthonous oscillatory properties of the central nervous system or on both mechanisms.[7] Similarly cardiovascular oscillations could be due to neural rhythms,[8] cause them[9] or both influences could be interacting.[10,11]

However the main limitation of the above studies can be the use of anaesthetised or decerebrate animal models. Accordingly it might be inappropriate to extrapolate these findings to human physiology.

The advent, in recent years, of direct intraneural microneurographic recordings of efferent sympathetic nerve to both muscle blood vessels (MSNA) and skin blood vessels (SSNA) has increased the breadth of our understanding of the mechanisms governing neural control of blood vessels in humans. The definition of a new autoregressive spectral algorithm capable of analysing the amplitude and the oscillatory power of the nerve signal gave us the unique opportunity to investigate the oscillatory characteristics of the sympathetic outflow and the mechanisms modulating it.

MEASUREMENT OF MSNA

The microneurographic technique has been widely described in numerous publications.[12–14] It involves the recording of multiunit sympathetic nerve discharges from a peripheral nerve, usually the peroneal nerve. In brief, tungsten microelectrodes are inserted percutaneously into sympathetic fascicles in the peroneal nerve. One electrode is used for direct intraneural recordings and the other electrode, inserted about 3 cm away, serves as the reference electrode. The electrodes are connected to a high impedance preamplifier and the nerve signal, through a band pass filter, is routed through an amplitude discriminator to a storage oscilloscope. For recording and analysis the filtered neural signal is fed through a resistance capacitance integrating network to a tape recorder or, after appropriate A/D conversion, to a PC. With usual approaches MSNA is analysed in terms of sympathetic nerve bursts, which are synchronous with blood pressure pulses. However, while this method is well suited to describe large changes in overall activity it might be inadequate to follow subtle firing characteristics such as spontaneous beat by beat oscillations, which requires specific techniques to be visualised.

SPECTRAL ANALYSIS OF CARDIOVASCULAR SIGNALS AND MSNA USING AUTOREGRESSIVE TECHNIQUES

The methodology and the software for data acquisition and spectral analysis of RR interval, blood pressure and respiration has been extensively described.[15,16] We will briefly summarise some major points. A derivative-threshold algorithm provides the continuous series of RR intervals (tachogram) from the ECG signal. From the continuous arterial pressure signal, beat-by-beat systolic (systogram) and diastolic (diastogram) values are calculated, and the signal of respiratory activity is sampled once per cardiac beat.

Regarding MSNA, traditional analysis is based on visual computation of the rate (bursts/min) and amplitude of the bursts. Conversely, a neural time series better suited for spectral analysis is provided through the integration of the continuous MSNA signal according to the following:

$$\mathrm{MSNA}(i) = 1/t\,(i) \int \mathrm{MSNA}(t)\,dt$$

where each integral was performed over the time window delimited by the i-th cardiac cycle of period t(i) (Figure 1). This series of variability measures of MSNA is synchronous with the other variability signals, such as tachogram and systogram on a beat-by-beat basis as requested by the method.[17]

The time series obtained are then analysed by means of autoregressive techniques, which can automatically provide the number, centre frequency and power of each relevant component. Power is expressed both in absolute as well as in normalised units (nu), which are obtained by dividing the power of each spectral component by the total variance from which the VLF component has been subtracted, and multiplying this value by 100.[16]

The method of assessing the LF and HF components is a crucial and controversial point in spectral methodology. We showed[18] that spectral analysis of HRV, using nu or LF/HF ratio appeared to be capable of providing a non-invasive quantitative evaluation of graded changes in the state of the sympathovagal balance. Regarding the normalisation procedure of MSNA spectral components, it minimises the drawback of this approach, which does not provide an absolute calibration of nerve activity.

Because all variability signals are synchronised, cross-spectral analysis can be performed to compute a squared coherence function (K^2), which is a measure of the statistical link between two variability

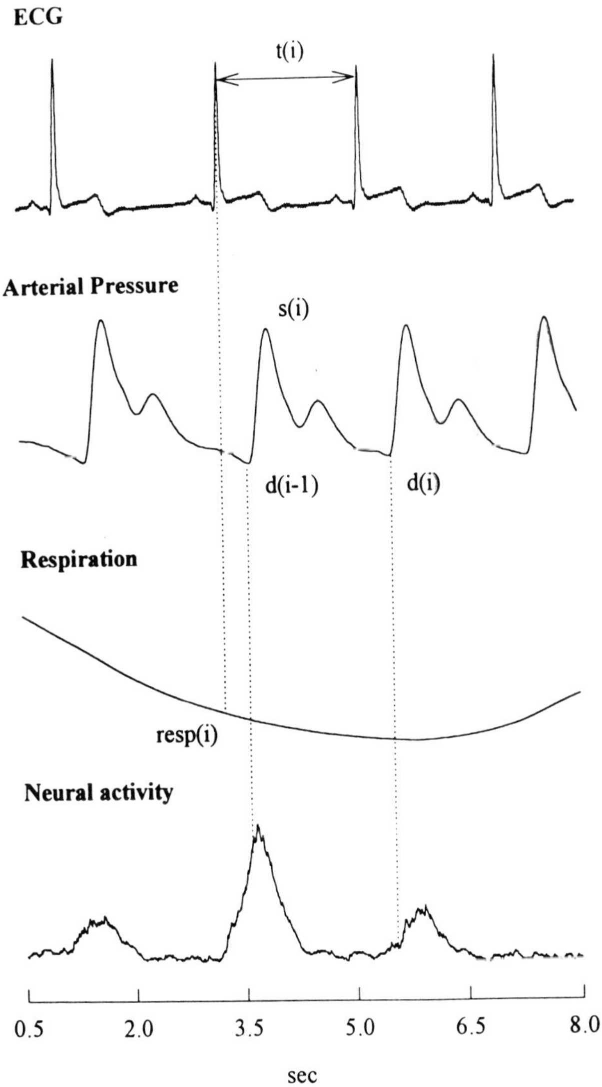

FIGURE 1. Traces from top to bottom display an example of ECG, arterial pressure, respiration and integrated muscle sympathetic nerve activity (MSNA) as recorded in subjects. t(i) represents the RR interval and s(i), d(i) and resp(i) represent the systolic, diastolic and respiratory values at the corresponding t(i), respectively. The interval between two consecutive diastolic values [d(i–1)] (or between two R waves whether arterial pressure is not present) is the time window for numerical integration to obtain MSNA(i) (From Pagani et al[17]).

series at any given frequency and is expressed as a number between 1 and 0.

STUDIES IN HEALTHY SUBJECTS

Previous studies in human healthy volunteers reported the presence of both LF and HF components in spectral profile of MSNA.[19,20]

Our preliminary work[17] has involved designing, testing and perfecting an autoregressive algorithm for simultaneous power spectral analysis of key variables including intra-arterial pressure, RR interval, respiration and MSNA. Our study has revealed that MSNA display clear cut LF and HF oscillations that are highly coherent with the corresponding blood pressure and RR interval spectral components.

We intended to determine the effects of sympathetic excitation (inducing graded hypotension) and of sympathetic inhibition (inducing graded hypertension) of the traditional indices of MSNA (bursts and amplitude) and on spectral components of MSNA (Figure 2)

During sympathetic activation induced by i.v. nitroprusside we observed an increase in predominance of the LF oscillation in the variability of MSNA, RR and arterial pressure (Figure 3). Conversely, sympathetic inhibition and vagal activation induced by i.v. phenilephrine was assessed with a predominance of the HF components in all variability signals (Figure 3).

Over the limited range of pressure changes that were explored, the absolute values of MSNA spectral components correlated with measures of nerve activity such as burst frequency and amplitude. The strong positive correlation between SAP and the normalised LF component of MSNA variability indicate that reflexly induced increases in average levels of MSNA are accompanied by a shift (i.e. concentration) of the spectral power of MSNA variability toward the LF region. The HF component of MSNA, expressed in nu, was instead negatively correlated with SAP, indicating that increases in sympathetic nerve traffic were associated with a decrease in the normalised power of the respiratory modulation of MSNA.

The simultaneous evaluation of spectral measures of MSNA and RR interval variability allowed us to directly address the issue of which spectral measure of RR interval variability better reflects changes in autonomic state. Our results showed that the use of nu or the LF/HF ratio provides the strongest correlation with attendant changes in MSNA (assessed by its amplitude or by normalised power

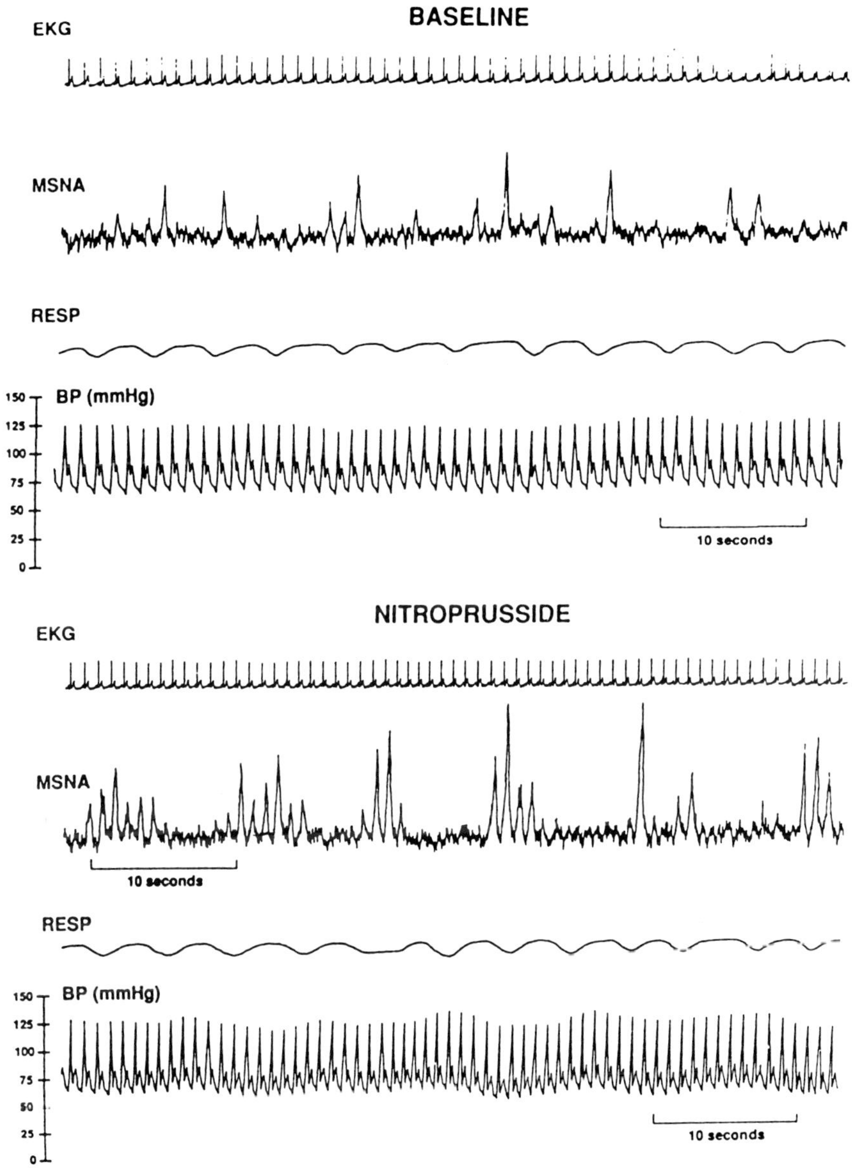

FIGURE 2. ECG, MSNA, respiration (Resp), and blood pressure (BP) in a single subject during baseline measurement (top) and during infusion of 1.8 µg/kg/min of nitroprusside (bottom). During sympathetic activation induced by nitroprusside, a 10-second period (~ 0.1 Hz) is clearly detectable in MSNA as well as in BP (From Pagani et al[17]).

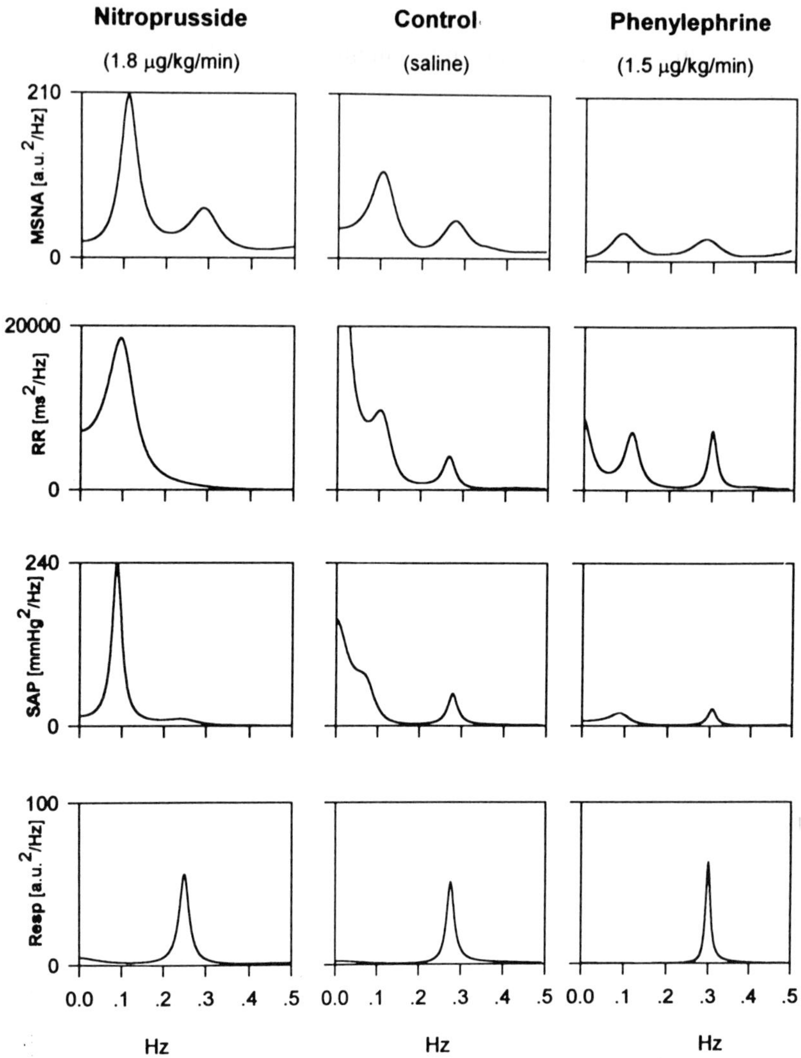

FIGURE 3. Power spectra of MSNA, RR interval, systolic arterial pressure (SAP) and respiration variabilities in a single subject during infusions of saline (Control), nitroprusside, and phenylephrine. During sympathetic activation induced by nitroprusside (left), the LF component of neural and cardiovascular variability signals predominates relative to the HF component. Conversely, during sympathetic inhibition and vagal activation, induced by phenylephrine (right), there is an increase of the HF component relative to the LF component. a.u. indicates arbitrary units (From Pagani et al[17]).

of spectral components rather than bursts/min) (Figure 4). The synchronous changes in the LF and HF of both MSNA and RR interval variabilities during different levels of sympathetic drive are suggestive of common central mechanisms governing both sympathetic and parasympathetic cardiovascular modulation.

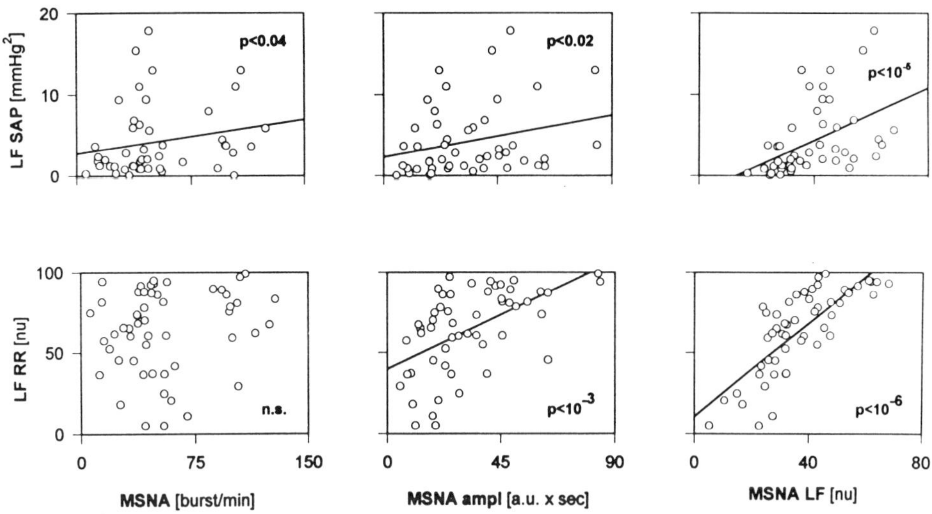

FIGURE 4. Correlation analysis using Theil's regression of the relationships between MSNA burst rate, MSNA amplitude (ampl), and the normalised (nu) LF oscillation of MSNA, with the LF of SAP (top) and the normalised LF of RR interval (bottom) (From Pagani et al[17]).

An increase in the HF of direct intraneural recording of sympathetic activity during vagal activation and a decrease in the HF of MSNA during sympathetic activation and vagal inhibition, contradict the hypothesis that the HF oscillation is purely a vagally mediated phenomenon.[21] Rather these data suggest that the two main oscillatory rhythms are markers of excitation (LF) and inhibition (HF), and that in healthy subjects, these rhythms are modulated in a reciprocal manner, that can be detected from the discharge of central neurons[22] and in peripheral neural outflows and that is reflected by target function variability, according to a closed-loop organisation. Changes in the central excitatory-inhibitory balance might induce reciprocal changes not only in average impulse activity of central vagal and sympathetic motor neurons but also in the balance between LF and HF rhythms that can be observed in peripheral targets The finding that vagal activation induced by phenilephrine is able to modulate the oscillatory characteristics of sympathetic neural discharge, and that oscillatory components of both MSNA and RR interval exhibit synchronous changes, further supports the concept of a common central organisation[23] governing both vagal and sympathetic rhythmic activity.

REPRODUCIBILITY OF CARDIOVASCULAR AND MSNA SPECTRAL MEASURES

After assessing the correlation between oscillatory characteristics of cardiovascular variables and MSNA we asked whether LF and HF components of RR interval, blood pressure and MSNA change systematically as these measurements fluctuate spontaneously when measured on different days.[24] Thus, we obtained sequential recordings of RR interval, blood pressure and MSNA at day 1, one week, one month and three months in a group of seven normal subjects.

These data underwent interim analysis to test the hypothesis that oscillation in RR and MSNA variabilities change systematically with spontaneous changes in resting MSNA, blood pressure and heart rate over repeated measurement. Data showed that, spontaneous RR interval shortenings were associated with increases in LFnu ($r=-0.4$ $p<0.01$) and decreases in HFnu of RR variability. Faster heart rates (decreases RR interval) were also linked to an increase in the LF/HF of RR (0.43 $p<0.01$) and also linked to an increase in MSNA ($r=36$ $p<0.01$); the LF/HF ratio of RR interval correlated with the LF/HF of MSNA ($r=0.3$ $p<0.05$) (Figure 5). As systolic blood pressure decreased, the LFnu of MSNA increased ($r= -0.41$ $p<0.01$), HFnu of MSNA decreased ($r= -.52$ $p<0.0001$). Remarkably, blood pressure changes did not correlate with absolute changes in MSNA, but did correlate with the spectral oscillations in the MSNA signal. .The relationship between spontaneous changes in arterial pressure, RR interval and MSNA variabilities observed over the 3 months of repeated measurements are very similar to our data obtained during induced pressure changes in a single session using vasoactive drugs. Again, they suggest that there is a tight link between spontaneous long-term fluctuations in absolute measures of arterial pressure and the oscillatory characteristics of MSNA measurements.

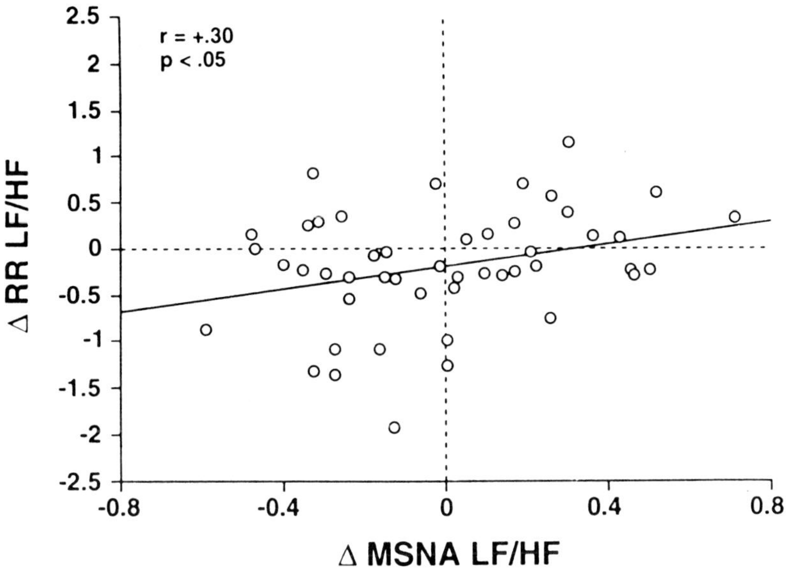

FIGURE 5. Regression analysis between changes in the low to high frequency ratio of the variability of muscle sympathetic nerve activity (ΔLF/HF MSNA) and changes in the low to high frequency ratio of the variability of RR interval (ΔLF/HF RR). LF/HF ratios (log transformations) of both RR and MSNA changed in parallel during the repeated measurements ($r=+0.30$, $p<0.05$) (From van de Borne et al[24]).

PRELIMINARY CLINICAL STUDIES

An increased sympathetic discharge, both to muscle blood vessels and to the heart is a hallmark of severe heart failure.[25,26] Moreover some Authors have reported that the LF component of RR variability was absent in heart failure[27,28] in relationship with the severity of the disease.[29] This finding was attributed to a reduced target responsiveness related to α,β-adrenergic down regulation or to saturation of the sinus node by a very high sympathetic drive that renders the sinus node less capable of maintaining a rhythmic modulation.

Therefore, we performed a study[30] in 21 CHF patients and observed that although heart failure was associated with a very high level of MSNA, there was an absence of LF component of both MSNA and RR interval variability in 14 out of 21 patients (Figure 6). Thus the lack of a detectable LF in RR and MSNA variabilities cannot be explained simply in terms of changes in end organ responsiveness but rather by abnormalities in central autonomic outflow in these patients. Further analysis showed that MSNA was higher, heart rate was faster, RR variance was lower and left ventricular ejection fraction was lower in those heart failure patients with absence of the LF oscillations in MSNA and RR (n=14), in comparison to those in whom the LF was present (n=7). At nine-month follow up, 25% of patients without LF oscillation had died, with zero mortality in heart failure patients in whom the LF oscillation in RR variability was present.

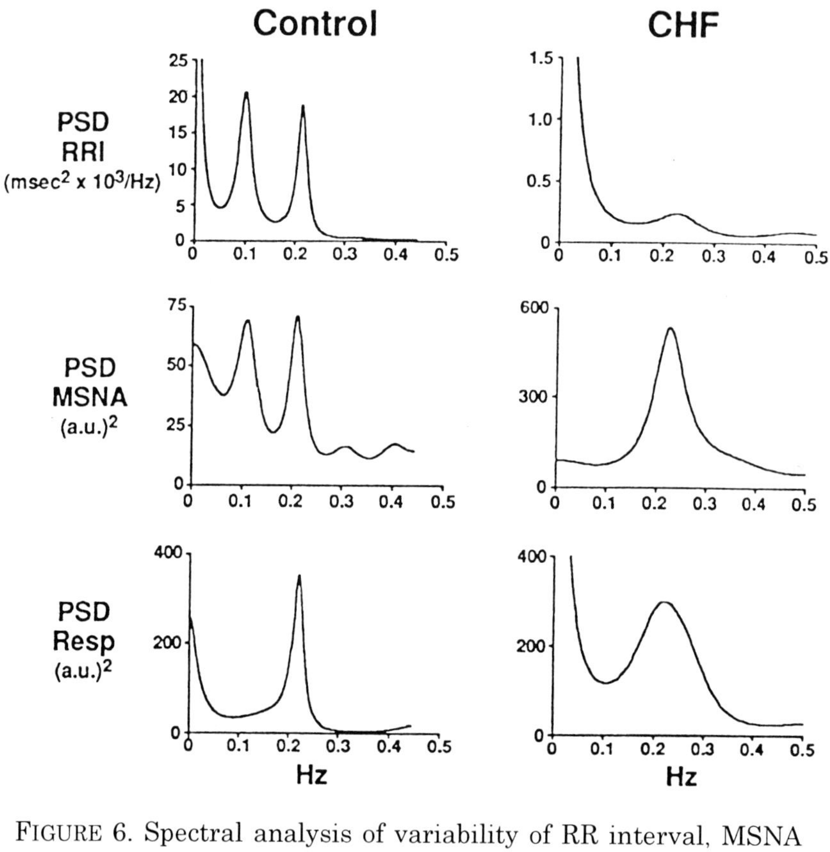

FIGURE 6. Spectral analysis of variability of RR interval, MSNA and respiration in a control subject (left) and a patient with severe chronic heart failure (CHF, right). Control subject has an LF (0.1 Hz) and a HF (0.22 Hz) components in both RR and MSNA, whereas only an HF (0.23 Hz) component is present in these measurements in the patient with heart failure (From van de Borne et al[30]).

Remarkably, similar results have been obtained by Ando and coworkers[31] who observed a reduced LF component in MSNA spectra of CHF patients compared with control subjects. Since coherence between blood pressure and MSNA in the HF range was high also in CHF patients, suggesting that the modulation of MSNA by arterial baroreflex and respiration was preserved, the Authors concluded that the higher sympathetic activity observed in CHF would be caused by factors other than the loss of regulation by baroreflex mechanisms, namely by an impaired neuroeffector transduction.[31]

Thus, despite the diverse neurohumoral, metabolic and other abnormalities that characterise heart failure, there is a preservation of the coherence between the variabilities of RR interval and sympathetic neural discharge. The common variability patterns and abnormalities present in the two autonomic neural outflows, namely the sympatho-vagal balance modulating the SA node and sympathetic neural discharge directed to the muscle vasculature, suggest that both vagal and sympathetic autonomic modulatory nuclei manifest unitary alterations in oscillatory characteristics in patients with heart failure. We therefore hypothesized[30] that the pathophysiology of heart failure may include primary central abnormalities in autonomic modulation. Increased sympathetic activation in heart failure but a decreased LF component in RR interval and MSNA and a decreased LF/HF ratio does not appear to follow the traditional paradigm used to explain the LF and HF oscillation in cardiovascular variability. If it did we would expect a relatively increased LF component in most heart failure patients, consistent with high sympathetic drive. Why should there be this dissociation between high sympathetic drive and the LF component of MSNA and RR variabilities? We believe that while the traditional paradigm may hold true for a physiologic range of autonomic drives in normal humans, it cannot be simply extrapolated to include pathological conditions such as severe heart failure, which is characterised by very high sympathetic activity.

LIMITATIONS

Some important limitations of the approach must be indicated. First, MSNA is not a direct measure of sympathetic activity directed to the heart or other internal organs. However, experimental studies with simultaneous recordings from different sympathetic nerves indicate a high degree of similarity among the various outflows that were examined. Likewise, norepinephrine spillover studies[32] from various organs confirm such a similarity also in man. Accordingly, it might be surmise that changes in average or oscillatory properties of MSNA could be taken as representative, at least as a first approximation, of similar trends occurring in other efferent outflows.

Second, there is at the moment a complete lack of direct information on efferent vagal activity in humans, or in anaesthetised animals.

Accordingly, the concept of a continuous balance between sympathetic and vagal outflows, at the heart level, although based on a large body of circumstantial evidence, should be considered as a convenient model, still requiring direct experimental proof.

CONCLUSIONS

Along the continuum of central neural activation,[33] the change between excitation and quite[23] seems to be signalled by the respective prevalence of low (0.1 Hz) or high frequency (respiratory) oscillations of critical structures.[22]

A similar dynamic balance seems to occur between sympathetic and vagal structures of cardiovascular control.[34] Analysis of peripheral nerve activity suggests that control of the circulation is coded simultaneously in two different modalities: amplitude (average strength of nerve signal, indicated as tonic activity) and frequency modulation (oscillation of the neural signal around the average, related to phasic activation). Changes from excitation to inhibition are accompanied by changes in strength of neural signal, together with the relative distribution of power between the low and high frequency oscillations.[17]

In pathophysiological conditions near equilibrium, the balance between LF and HF oscillations, best assessed with normalised units or the LF/HF ratio, reflects the dynamics of the relationship between vagal and sympathetic drives.

REFERENCES

1. Task Force of the European Society of Cardiology and the North American Society of Pacing and Electrophysiology. Heart rate variability. Standards of measurement, physiological interpretation and clinical use. *Circulation* 1996; **93:**1043–1065.
2. Adrian ED, Bronk DW., Phillips G. Discharges in mammalian sympathetic nerves. *J Physiol* 1932; **74:**115–133.
3. Fernandez de Molina A, Perl ER. Sympathetic activity and the systemic circulation in the spinal cat. *J Physiol* 1965, **181:**82–102
4. Penaz J. Mayer waves: history and methodology. *Automedica* 1978; **2:**135–141.
5. Montano N, Lombardi F, Gnecchi-Ruscone T, Contini M, Finocchiaro ML, Baselli G, Porta A, Cerutti S, Malliani A. Spectral analysis of sympathetic discharge, RR interval and systolic arterial pressure in decerebrate cats. *J Autonom Nerv Syst* 1992. **40:**21–32.
6. Polosa C. Central nervous system origin of some types of Mayer waves. In: Miyakawa K, Koepchen HP, Polosa C (Eds) Mechanism of Blood Pressure Waves. Springer, Berlin and Tokyo 1984, 277–292.
7. Gebber GL. Central oscillators responsible for sympathetic nerve discharge. *Am J Physiol* 1980, **239:**H143–H155.
8. Preiss G, Polosa C. Patterns of sympathetic neuron activity associated with Mayer waves. *Am J Physiol* 1974, **226:**724–730.
9. Appel ML, Berger RD, Saul PJ, Smith JM, Cohen RJ. Beat to beat variabilitiy in cardiovascular variables: noise or music? *J Am Coll Cardiol* 1989; **14:**1139–1148.
10. Koepchen HP. History of studies and concepts of blood pressure waves. In: Miyakawa K, Koepchen HP, Polosa C (Eds) Mechanism of Blood Pressure Waves. Springer, Berlin and Tokyo 1984, 3–23.
11. Malliani A, Pagani M, Lombardi F, and Cerutti S. Cardiovascular neural regulation explored in the frequency domain. *Circulation* 1991; **84:**482–492.
12. Hagbarth KE, Vallbo AB. Pulse and respiratory grouping of sympathetic impulses in human muscle nerves. *Acta Physiol Scand* 1968; **74:**96–108.
13. Wallin BG. Intraneural recording and autonomic function in man. In: Bannister R (Ed) Autonomic Failure. Oxford University Press, London 1983, 36–51.
14. Mark AL, Victor RG, Nerhed C, Wallin BG. Microneurographic studies of the mechanisms of sympathetic nerve responses to static exercise in humans. *Circ Res* 1985; **57:**461–469.
15. Baselli G, Cerutti S, Civardi S, Liberati D, Lombardi F, Malliani A, Pagani M. Spectral and cross-spectral analysis of heart rate and arterial blood pressure variability signals. *Comput Biomed Res* 1986; **19:**520–534.
16. Pagani M, Lombardi F, Guzzeti S, Rimoldi O, Furlan R, Pizzinelli P, Sandrone G, Malfatto G, Dell'Orto S, Piccaluga E, Turiel M, Baselli G, Cerutti S, Malliani A. Power spectral analysis of heart rate and arterial pressure variabilities as a marker of sympatho-vagal interaction in man and conscious dog. *Circ Res* 1986; **59:**178–193.

17. Pagani M, Montano N, Porta A, Malliani A, Abboud FM, Birkett C, Somers VK. The relationship between spectral components of cardiovascular variabilities and direct measures of muscle sympathetic nerve activity in humans. *Circulation* 1997; **95:**1441–1448.
18. Montano N, Gnecchi Ruscone T, Porta A, Lombardi F, Pagani M, Malliani A. Power spectrum analysis of heart rate variability to assess a changes in sympathovagal balance during graded orthostatic tilt. *Circulation* 1994; **90:**1826–1831.
19. Eckberg DL, Nerhed C, Wallin BG. Respiratory modulation of muscle sympathetic and vagal cardiac outflow in man. *J Physiol (Lond)* 1985; **365:**181–196.
20. Saul JP, Arai Y, Berger RD, Lilly LS, Colucci WS, Cohen RJ. Assessment of autonomic regulation in chronic congestive heart failure by spectral analysis. *Am J Cardiol* 1988; **61:**1292–1299.
21. De Boer RW, Karemaker JM, Strackee J. hemodynamic fluctuations and baroreflex sensitivity in humans: A beat to beat model. *Am J Physiol* 1987; **253:**H680–H689.
22. Montano NT, Ruscone G, Porta A, Lombardi F, Malliani A, Barman SM. Presence of vasomotor and respiratory rhythms in the discharge of single medullary neurons involved in the regulation of cardiovascular system. *J Auton Nerv Syst* 1996; **57:**116–122.
23. Hobson JA. Sleep and dreaming. *J Neurosci* 1990; **10:**371–382.
24. van de Borne P, Montano N, Zimmermann B, Pagani M, Somers VK. The relationship between repeated measures of hemodynamics, muscle sympathetic nerve activity and their spectral oscillations. *Circulation* 1997; **96:**4326–4332.
25. Leimbach WN, Wallin BG, Victor RG, Aylward PE, Sundlof G, Mark AL. Direct evidence from intraneural recordings for increased central sympathetic outflow in patients with heart failure. *Circulation* 1986; **73:**913–919.
26. Ferguson DW, Berg WJ, Sanders JS, Kempf JS. Clinical and hemodynamic correlates of sympathetic nerve activity in normal humans and patients with heart failure: evidence from direct microneurographic recordings. *J Am Coll Cardiol* 1990; **16:**1125–1134.
27. Saul JP, Rea RF, Eckberg DL, Berger RD, Cohen RJ. Hear rate and muscle sympathetic nerve variability during reflex changes of autonomic activity. *Am J Physiol* 1990; **258:**H713–H721.
28. Szabo BM, van Veldhuisen DJ, Brouwer J, Haaksma J, Lie KI. Relation between severity of disease and impairment of heart rate variability parameters in patients with chronic congestive heart failure secondary to coronary artery disease. *Am J Cardiol* 1995; **76:**713–716
29. Guzzetti S, Cogliati C, Turiel M, Crema C, Lombardi F, Malliani A. Sympathetic predominance followed by functional denervation in the progression of chronic heart failure. *Eur Heart J* 1995; **16:**1100–1107
30. van de Borne P, Montano N, Pagani M, Oren R, Somers VK. Absence of low-frequency variability of sympathetic nerve activity in severe heart failure. *Circulation* 1997; **95:**1449–1454.

31. Ando S, Dajani HR, Floras JS. Frequency domain characteristics of muscle sympathetic nerve activity in heart failure and healthy humans. *Am J Physiol* 1997; **273:**R205–R212.
32. Wallin BG, Esler M, Dorward P, Eisenhofer G, Ferrier C, Westerman R, Jennings G. Simultaneous measurements of cardiac noradrenaline spillover and sympathetic outflow to skeletal muscle in humans. *J Physiol* 1002; **453:**45–58.
33. Moruzzi G. Sleep and instinctive behavior. *Arch Ital Biol* 1969; **107:**175–216.
34. Malliani A. Association of heart rate variability components with physiological regulatory mechanisms. In: Malik M, Camm AJ (Eds) Heart Rate Variability. Futura, Armonk 1995, 173–188.

Chapter Ten

ASSESSMENT OF BAROREFLEX SENSITIVITY

Maria Teresa La Rovere, Gian Domenico Pinna and Andrea Mortara

By governing autonomic outflow to the heart and circulation, arterial baroreceptors play a central role in controlling short-term blood pressure responses to the continuous perturbations produced by various stimuli occurring in daily life. Although in a physiological setting baroreflex mediated changes in autonomic outflow affect heart rate, myocardial contractility and peripheral resistance, the baroreflex control of heart rate in pathological conditions is of particular interest, since it has been associated with an increased propensity for cardiac mortality and sudden cardiac death.[1]

The mechanisms by which cardiovascular diseases may affect baroreflex control of heart rate and blood pressure and the resultant clinical implications have been extensively reviewed.[2]

In this chapter we shall review the influence of arterial baroreceptors in the regulation of the autonomic outflow to the heart, we shall describe the methods currently used for assessing baroreflex sensitivity and shall discuss their application in different physiological and pathological conditions.

Malik M (Ed) Clinical guide to cardiac autonomic tests, 257–281.

PHYSIOLOGICAL BASIS FOR THE BAROREFLEX CONTROL

The sensory endings of the arterial baroreflex are located in the wall of the carotid sinus and in the wall of the aortic arch. They are known as baroreceptors, a term which would imply that nerve endings are sensitive to pressure per se. The relationship between blood pressure and baroreceptor discharge is more complex than this since pressure is sensed by the baroreceptors as mechanical deformation or strain of the vessel wall that is followed by electrical transduction in the receptors themselves. Wall deformation and intravascular pressure are not linearly related. Differences among baroreceptors and patterns of baroreceptors discharge have been reviewed.[3] Several factors can modulate baroreceptor responses at the peripheral level. For instance norepinephrine can activate carotid sinus smooth muscle and change discharge at a constant wall radius[4] and stimulation of the sympathetics to the carotid sinus increases the sensitivity of the carotid baroreceptors, whereas decreases in sympathetic nerve activity to this region have opposite effects.[5]

The afferent fibres carrying sensory information from the carotid sinus enter the brain stem with the glosso-pharyngeal nerve while the aortic afferents join the vagus nerve to travel to the nucleus of the tractus solitarius in the medulla oblongata. Here, through one or more synaptic connections, the information from baroreceptors projects to those areas controlling vagal and sympathetic outflow to the heart and peripheral circulation.

The simultaneous interaction on the medullary centres of afferent inputs from the multiple reflexogenic areas in the heart, lungs and in the vessels modulate arterial baroreflex responses. Indeed, activity from cardiac mechanoreceptors, chemoreceptors, cardiopulmonary receptors, and respiration, exercise or postural changes together with excitement and arousal may modify central responsiveness to incoming baroreceptor information and lead to reflex interactions of great complexity and diversity. Additionally, humoral substances (e.g. angiotensin) also act at a central level to modify baroreceptor responses.

An increase in systemic arterial pressure increases the firing rate by baroreceptors causing sympathetic deactivation and parasympathetic activation which decreases heart rate, vasoconstriction and contractility with the aim of buffering the blood pressure change.[6,7] In contrast, a fall in systemic arterial pressure, by decreasing baroreceptor firing rate, results in sympathetic excitation and parasympathetic withdrawal. As reflex sympathetic responses are slower than vagal ones,[8,9] the fastest baroreflex control involves changes in heart rate which are solely vagally mediated. In normal subjects and for heart rates less than 75 beats per minute, the conduction time from baroreceptos to the heart has been estimated around 475 ms.[10] At higher heart rates, the reflex response is partly mediated by the sympathetic with a greater delay in the onset of bradycardia.

The relative contribution of vagal and sympathetic mechanisms to control of the sinus node depends on their tonic levels of activity in various conditions. Under normal circumstances and when arterial pressure is normal, there is a prevailing level of arterial baroreceptor discharge so that vagal efferent activity is tonically excited and sympathetic activity is partially inhibited. During circumstances such as exercise vagal activity is low and heart rate control is mediated principally by sympathetic mechanisms.

Abnormalities in baroreceptor mechanisms occurring in a variety of pathological conditions, could compromise their tonic restraining influence on the heart and circulation leading to sympathetic excitation and parasympathetic withdrawal.

ASSESSMENT OF BAROREFLEX SENSITIVITY

Baroreflexes operate not only on a dynamic basis, to control abrupt changes in arterial pressure, but also continuously in steady state condition. As a consequence, baroreceptor activity can be investigated either by a variety of mechanical or pharmacological manipulations that cause sudden changes sensed by the receptors, or by analysing spontaneous fluctuations of arterial pressure. The conceptual differences between the two approaches are not trivial. When an external input of adequate amplitude is applied to baroreceptors (either directly or indirectly, by forcing a rapid change in arterial pressure), blood pressure changes are mainly under the control of this external stimulus and the relationship between pressure changes and RR changes can be approximated by an 'open loop' model. In contrast, the analysis of spontaneously occurring changes in arterial pressure and RR interval is made under a condition characterised by all reflexes and control mechanisms being fully active, which provides a 'closed loop' estimation of the baroreflex gain. The first approach allows a better study of the 'whole' capability of the system to react to a given stimulus whereas the second is more limited to the range of responses occurring at baseline conditions. Both have advantages and limitations.

The quantitative approaches developed for evaluating baroreflex sensitivity (BRS) include pharmacologically induced increases or decreases in arterial pressure; neck suction or pressure; and analysis of spontaneous fluctuations of arterial pressure and RR intervals.

VASOACTIVE DRUGS

The most widely used method for studying baroreflex control of heart rate is the one devised in the 1960s by the Oxford group.[11] The underlying concept is to study the reflex heart rate response to physiological activation or deactivation of the baroreceptors secondary to drug-induced changes in arterial pressure.

When systolic arterial pressure is increased by 20–40 mm Hg baroreceptors are activated to operate in the linear portion of their reflex response. Initially angiotensin was used as the pressor agent; however, because of direct effects on cardiac contractility and on the central nervous system,[12,13] it has been replaced by the use of phenylephrine, a pure alpha-adrenoreceptor stimulant. In normal subjects 25–100 μg of phenylephrine are flushed into a vein to obtain the required blood pressure increase or to obtain an obvious increase in RR interval when the pressure rise is below this threshold.[14] Given the rapidity of vagal responses (within one or two beats), it is commonly assumed that each heart period value is mainly related on a cause-effect basis to the previous systolic pressure peak and that this relation is linear.

Prolongation of successive RR intervals with respect to baseline values (i.e. pre-injection) are therefore plotted as a function of preceding systolic pressure changes and an analysis windows selected between the beginning and the end of the first significant increase in systolic arterial pressure. The slope of the regression line fitted to the selected points (expressed as ms / mm Hg) – that is, the absolute increase in pulse interval produced by a 1 mm Hg rise in systolic blood pressure – is used to quantify the sensitivity of the arterial baroreflex control of heart rate. Non linear effects such as saturation in the response are not generally taken into account.

In the evaluation of a BRS test, the goodness of the linear association between systolic arterial pressure and RR interval changes is assessed by Pearson's correlation coefficient. When the observed correlation coefficient is not statistically significant (i.e. when there is no evidence of a linear relation between pressure and RR changes) the test result should be disregarded (see below). Indeed, although a better correlation coefficient is obtained when each response is analysed by discriminating inspiratory from expiratory pulse intervals, in patients with a normal respiratory pattern the correlation coefficients obtained taking the whole sets of data do not lose their statistical significance.[11]

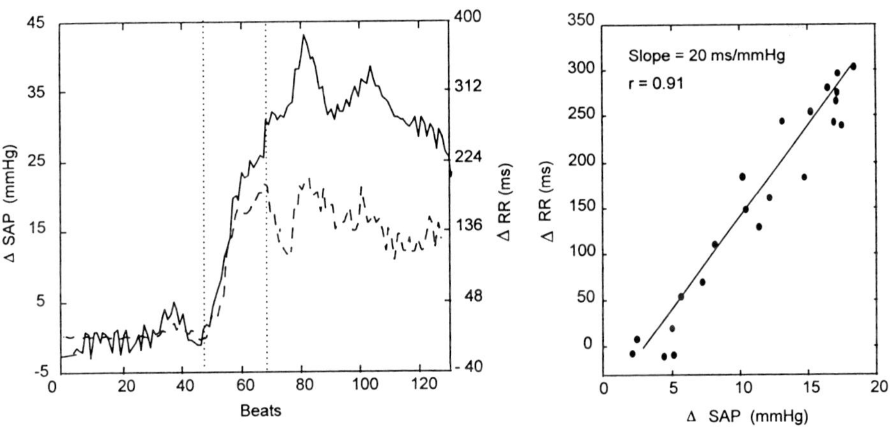

FIGURE 1. Example of a sensitive BRS. On the left beat-to-beat changes in systolic blood pressure (SAP) (dotted line) and in RR interval (full line) with respect to baseline values are reported. Analysis is limited to the first major increase in systolic arterial pressure with the attendant changes in RR interval (points included between dotted lines). These points are used for calculation of the regression line (on the right). The increase in SAP is greater than 20 mm Hg and is accompanied by a consistent increase in RR interval. Accordingly the slope of the regression line is 20 ms/mm Hg and represents a response primarily characterised by an increase in efferent vagus nerve activity to the sinus node.

Several injections are generally repeated at intervals of a few minutes and the corresponding slopes are averaged in order to reduce measurement variability between tests. This variability is mainly due to the rate of injection, steepness of arterial pressure increase and selection of the time window used for analysis. The final slope is generally obtained by at least three slopes with the higher correlation coefficients. Sometimes such variance is best handled by performing weighted averages of slopes, with those slopes displaying the least variance given the greatest weight.[15]

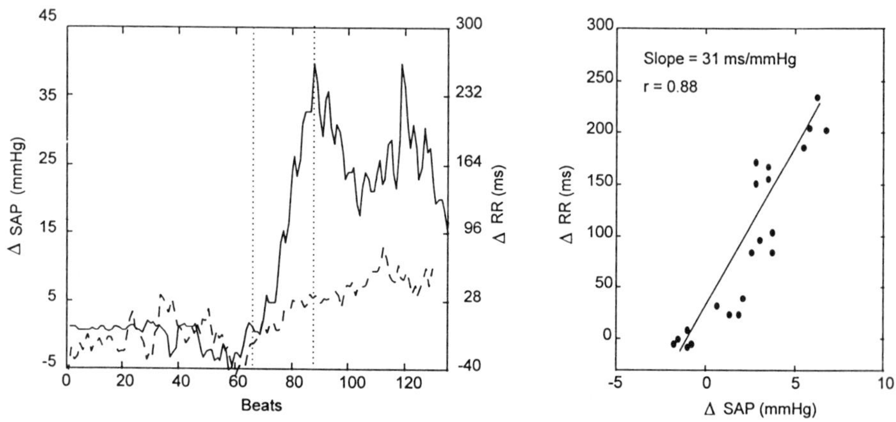

FIGURE 2. Example of a sensitive BRS. Detailed description as in Figure 1. The increase in SAP is below the threshold of 20 mm Hg, however it is accompanied by an obvious increase in RR interval.

In normal young subjects average values of baroreflex sensitivity of 14.8 ± 9.2 ms / mm Hg (SD)[14] and 16.0 ± 1.8 ms / mm Hg (SEM)[16] have been reported. When analysing a test of BRS such a slope is generally interpreted as the result of the interplay between effective vagal reflexes and tonic sympathetic activity (Figures 1 and 2). Overall, BRS decreases whenever the autonomic balance shifts toward sympathetic dominance and increases whenever the autonomic balance shifts toward parasympathetic dominance. The gain of the baroreflex modulates reciprocal changes in sympathetic and parasympathetic activity which characterise several physiological conditions. Indeed, BRS has circadian variations[17] and increases during sleep.[18] (Figure 3) On the other hand, physical exercise[19,20] and mental arousal[21] are associated with a reduction in baroreflex control of heart rate while the blood pressure control is left relatively unaffected.[22] Increasing age significantly reduces the baroreflex control of heart rate[23] and this factor is of meaningful importance when dealing with the prognostic value of baroreflex sensitivity. Significant impairment in the baroreflex control of heart rate is produced by pathological conditions including hypertension[14] and cardiac disease.[16,24] In hypertensive patients BRS averaged 3.0 ± 0.9 ms / mm Hg[11] and similar values have also been observed in patients with heart disease of different etiologies. In patients after myocardial infarction the mean value from several studies averaged 7.5 ms / mm Hg[24–26] and the consistency of these data has been recently confirmed over a large population.[27] Accordingly a depressed BRS slope is interpreted as a reduced capability of vagal reflexes which may also be associated with the presence of elevated sympathetic activity (Figure 4).

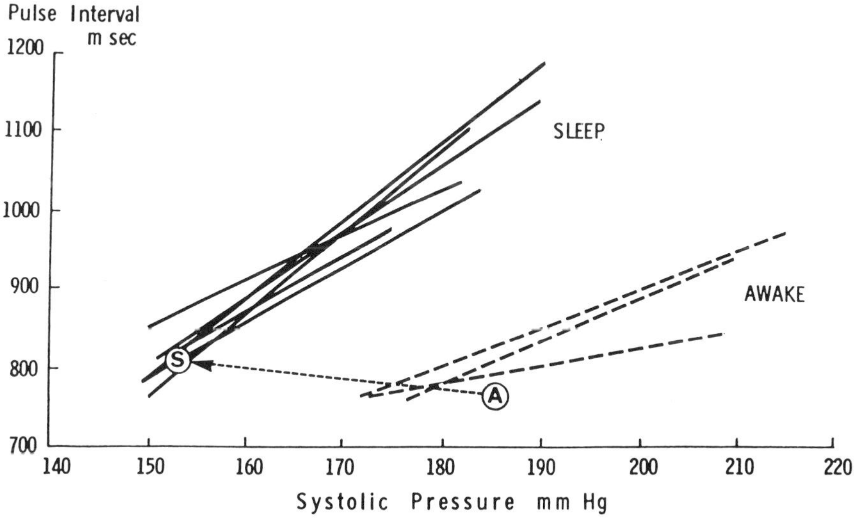

FIGURE 3. Baroreflex changes with sleep in one subject. Each line represents one injection of phenylephrine. A – average systolic blood pressure awake, both before and after sleep; S – average systolic pressure during slow wave sleep. (Reproduced with permission from Bristow et al[18]).

Several methodological aspects of the phenylephrine technique deserve some comments and considerations when dealing with the application of this test in patients.

The dosage of phenylephrine and time of injection may vary widely according to the patients under evaluation. In patients after a recent myocardial infarction[24] phenylephrine has been injected over 30 seconds, and boluses of 150–350 μg (2–4 μg/Kg) have been administered. In patients with congestive heart failure[28] even higher doses (up to 10 μg/kg) have been used to elicit baroreceptor responses and the rate of injection has been increased.

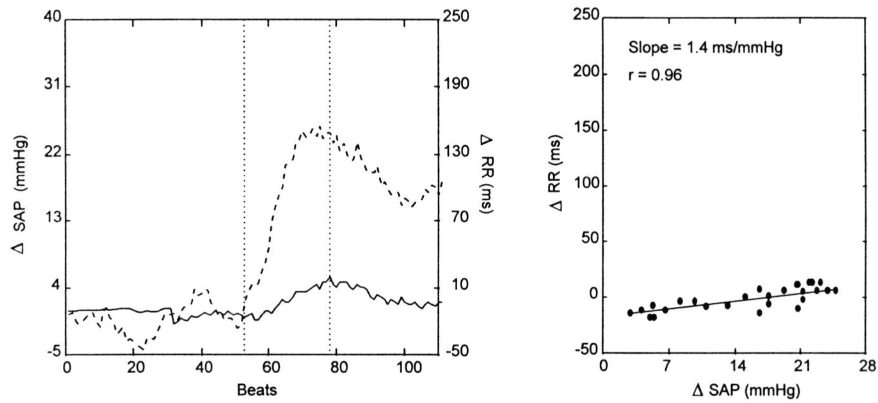

FIGURE 4. Example of a poor BRS. Detailed description as in Figure 1. The increase in SAP is accompanied by a modest increase in RR interval and the slope of the regression line is 1.4 ms/mm Hg and represents a response characterised by weak vagal reflexes.

The value of the correlation coefficient also deserves several comments. When the correlation coefficient is higher than 0.7 – 0.8, and the pairs of data included in the analysis are greater than 10, a test of significance is not required since this correlation is obviously statistically significant. For lower values a test of significance is required to test the association between systolic arterial pressure and RR interval changes. When BRS values are near zero (–0.5 – +0.5 ms / mm Hg) as it is commonly observed among patients with severely depressed left ventricular function and congestive heart failure, the correlation coefficient is obviously not significant. However, provided an adequate (> 15 mm Hg)[24] systolic arterial pressure increase has been obtained, the BRS value is retained, independently of the not significant correlation coefficient[28] (Figure 5). In fact, this slope faithfully describes the observed phenomenon, namely the virtual absence of response between pressure elevation and RR interval. Paradoxically in some cases, (as in the transplant patient of Figure 6), a BRS value near to 0 may have a statistically significant correlation coefficient. Finally, the correlation coefficient is not simply related to the scatter of data in the regression analysis as in some cases it can be improved by changing the lag between systolic pressure pulses and RR intervals. Abnormal respiratory patterns such as periodic breathing or Cheyne-Stokes respiration which are often observed in patients with heart failure[29] frequently induce such an erratic behaviour of heart rate that definitely affects the correlation coefficient. In such cases this erratic behaviour of heart rate can be markedly attenuated and a measurable BRS is obtained by the use of controlled breathing.

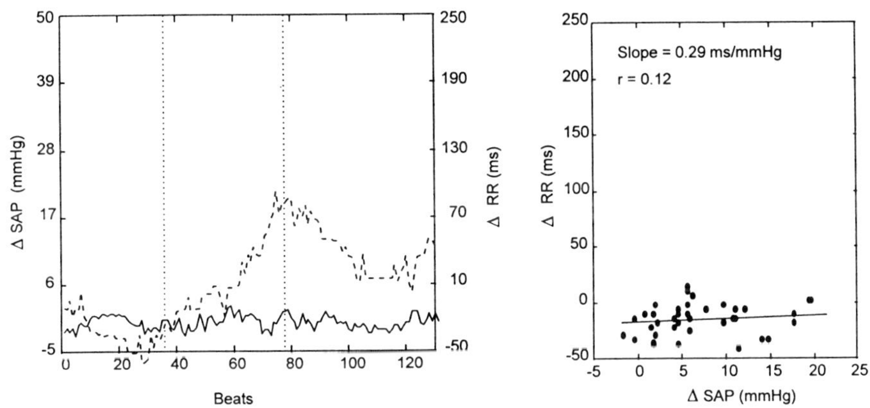

FIGURE 5. Example of a poor BRS in a patient with severe congestive heart failure. Detailed description as in Figure 1. The increase in SAP is accompanied by small, erratic changes in RR interval. The slope of the regression line is near to 0 and the correlation coefficient is not statistically significant.

As with blood pressure increases, a similar but generally less steep linear relationship also exists when the RR intervals shorten following a reduction in blood pressure induced by vasodilatory agents (i.e. nitroglycerin, sodium nitroprusside). Bolus injections of

100–200 μg of nitroglycerin are able to induce an immediate and progressive decrease in blood pressure lasting for 8–15 beats with a nadir at a value of about 20 mm Hg less than the baseline value.[30,31] The fact that the baroreflex slopes obtained by increasing arterial pressure are significantly steeper than those obtained by decreasing pressure to a similar extent, suggests that responses to rising and falling arterial pressure are asymmetrical in humans[32] or that the tachycardia is produced by a direct effect on pacemaker cells.[33]

Several theoretical objections can be raised to the use of vasodilatory agents in patients with congestive heart failure. Actually, these patients are likely to lie in a saturation and no longer linear portion of the response curve that could imply a physiological limitation to evoking further sympathetic activation. More importantly, the hemodynamic response to a vasodilator agent in congestive heart failure is an important confounder of BRS assessment because the drug induced reduction of peripheral resistances causes a beneficial increase in stroke volume and cardiac output that deactivates arterial baroreflexes, leading to bradycardia and not, as expected, to shortening of the RR interval.[34]

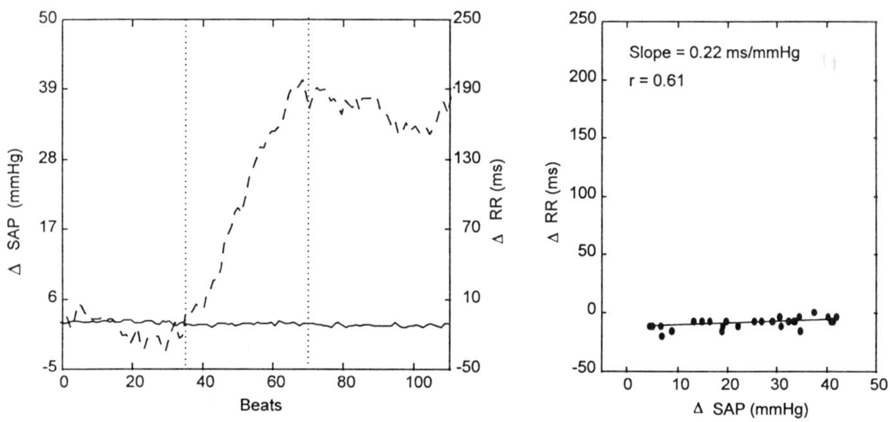

FIGURE 6. Example of a poor BRS in a patient with recent heart transplantation. Detailed description as in Figure 1. The increase in SAP is accompanied by a negligible change in RR interval. The slope of the regression line is near to 0, however the correlation coefficient is statistically significant ($p < 0.05$).

Vasoactive drugs provide a physiological perturbation of the baroreflex system since all the arterial baroreceptors areas are stimulated (activated or deactivated) in the same direction. Moreover, the subject is unaware of the stimulus, thus avoiding the influence of emotional factors that, by increasing sympathetic activity, could reduce heart rate response. This has practical importance in that, in the presence of similar physiological and laboratory conditions and in the absence of pharmacological interventions, baroreflex responses are highly reproducible in healthy subjects.[23]

While with the Oxford method a rapid increase of arterial pressure is produced and the associated dynamic changes in RR interval are analysed, with Korner's method,[35] reflex RR interval lengthening is assessed at different steady levels of blood pressure increase, which are maintained for several minutes by phenylephrine infusion. The long time required for the infusion limits the clinical use of this procedure.

The use of vasoactive drugs is not free from criticism. One is that the direct effect of vasoactive drugs on blood pressure and cardiovascular resistance prevents evaluation of baroreflex control of arterial pressure. Another points to the simultaneous activation of reflexes from the cardiopulmonary region and of changes in baroreceptor transduction due to the vasoconstrictor effect on smooth muscle fibres in the arterial wall. However, when BRS is used to draw inferences on the amount of autonomic mediators (acetylcholine, norepinephrine) released at cardiac level, a most important question for cardiac electrophysiology, this is no longer a limitation. Indeed, the fact that vasoactive drugs induce a simultaneous activation of multiple reflexogenic areas is actually an advantage for the purpose of using BRS as a measure of net autonomic balance to the heart.

Phenylephrine also produces a direct alpha-adrenergic stimulation of the sinus node.[36] However, these direct effects are not usually seen at low (< 5 μg/kg) doses of phenylepherine and have no clinical consequences. In our laboratory, more than three thousand bolus injections of phenylephrine have never precipitated clinical events. Even higher doses of phenylephrine (> 10 μg/kg) have not produced deleterious effects in patients with severe heart failure.[28] In some of these patients, particularly those with severe mitral regurgitation, a paradoxical tachycardia has been observed,[28] which is more likely to be due to a further hemodynamic impairment accompanied by an uncompensated increase in mitral regurgitation than to a direct effect on sinus node function. This is supported by the observation that BRS may be restored by hemodynamic improvement and reduction of severity of mitral regurgitation.

The use of vasoactive drugs as bolus injections obviously requires a beat-to-beat arterial pressure measurement. This is commonly done by a direct recording from the radial or brachial artery. However, the development of devices for non-invasive recording of blood pressure has facilitated more widespread use of BRS measurement.[36,37]

NECK CHAMBER TECHNIQUE

The neck chamber technique allows local activation or deactivation of carotid baroreceptors by application of measurable positive and negative pneumatic pressures to the neck region. The first neck chamber device was described in 1957 by Ernsting and Parry.[38] Among the different types subsequently developed, the one proposed by Eckberg[39] (a lead collar which encircles only the anterior and lateral aspects of the neck which can be modelled to suit the neck of each individual subject) and the one proposed by Ludbrook[40] (which completely surrounds the neck) have been more extensively used. An increase in neck chamber pressure is sensed by the baroreceptors as a decrease in arterial pressure and, thus, activates a double response that determines vagal withdrawal and sympathetic activation to the heart and the arterial vessels. Conversely, a decrease in neck chamber pressure results in reflex reduction of blood pressure and heart rate.

Neck suction is presently the preferred stimulus as it is easier to use than pressure and is well tolerated by the subject. Neck succion is critically dependent on the rate of the stimulus and its duration. Different investigators have used quite different protocol from one beat to 5 or 10 seconds. The negative pressure is applied in separate steps and ranges in magnitude from –7 to –40 mm Hg. Each step is generally repeated three times and the steps are applied in random order. The maximal lengthening in RR interval observed over the three beats following the neck suction application generally represents the reflex response and the slope of the regression of RR interval over neck pressure is taken as the carotid baroreflex sensitivity. Neck chamber baroreflex slopes after myocardial infarction are consistently lower than in normal subjects.[30] However, it needs to be pointed out that there is no way to know the actual stimulation delivered to baroreceptors for each value of the pressure applied to the neck, that is the 'equivalent arterial pressure increase'.

Therefore the slopes obtained by these two techniques are not comparable. In a group of patients studied 1–3 days after admission to the coronary care unit for an acute myocardial infarction, the average phenylephrine BRS was 7.9 ms / mm Hg, while the average neck suction BRS slope was 3.8 ms / mm Hg.[30]

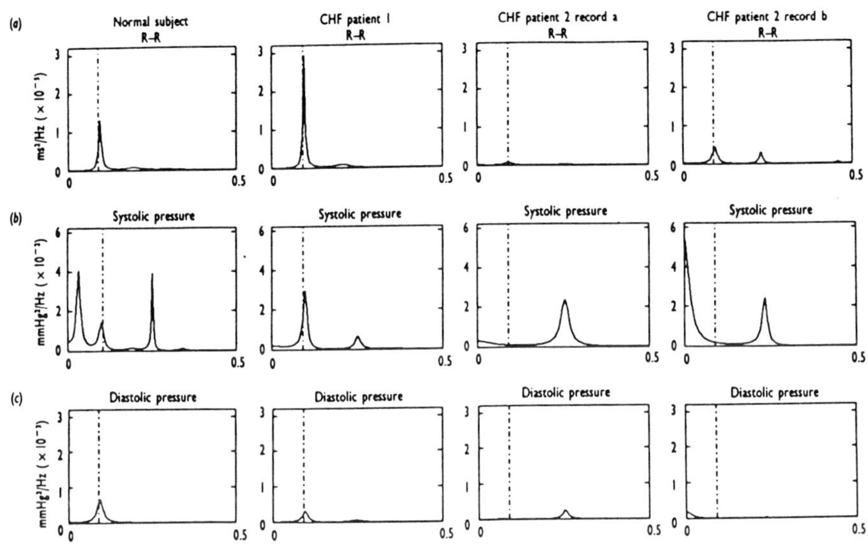

FIGURE 7. Power spectral density of (a) tachogram, (b) systogram and (c) diastogram during neck suction a 0.1 Hz in a normal subject and two patients with congestive heart failure with good (patient 1) and poor (patient 2 record a) baroreceptor reflexes. Record 2 b refers to patient 2 after compensation and clinical improvement. The strong stimulus to the baroreceptors at 0.1 Hz dominates the power spectrum in all three parameters in the normal subject and in the patient with good baroreceptor reflexes (patient 1). In patient 2 in record a very limited power at 0.1 Hz is present only in RR interval and it increases in record b after improvement in baroreflex gain (Reproduced with permission from Sleight et al[41]).

A very interesting feature of the neck suction technique is the possibility of analysing the dynamic behaviour of baroreceptors. This can be performed by connecting the lead collar to a suction device able to follow a predefined waveform derived from a function generator.[41,42] Optimal performance can be obtained by servo-controlled systems providing continuous monitoring of the pressure inside the neck and adjusting the suction accordingly. Two kinds of waveform can be used to stimulate baroreceptors. The first are pure sinusoids given at a set of discrete frequencies within the spectrum of short-term blood pressure variability (0.02÷0.25 Hz). The amplitude of these sinusoids is commonly in the range of 0÷–30 mm Hg. At each stimulating frequency the magnitude squared coherence function between the suction signal and the RR interval is first estimated, in order to assess whether the oscillation of the latter at that frequency can be considered linearly related to the corresponding oscillation of baroreceptor stimulation.[41,42] This function attains a value between 0 and 1 and is conceptually equivalent to the estimation of the correlation coefficient in the vasoactive drug technique. However, instead of using statistical testing, a coherence value of > 0.5 is commonly considered to guarantee a significant relation between neck suction and RR changes.[42] Once the association at a given frequency has been ascertained, the power of the corresponding spectral component of the RR signal is computed.[41,42] A sinusoidal stimulation of the carotid baroreceptors at a frequency > 0.2 can be followed only by vagal efferent activity while stimulation at 0.1 Hz can be followed normally by both vagal and sympathetic reflex activity. By this technique it has been found that carotid baroreceptors have an important role in the determination of the RR interval oscillations particularly in the low frequency band[41] (around 0.1 Hz) (Figure 7) and that in heart transplant patients the occurrence of sympathetic reinnervation may be easily detected by the presence of a reflex oscillation at 0.1 Hz in response to baroreceptor stimulation at the same frequency.[42] The second kind of waveform is a broad band signal which contains virtually all frequencies of the band of interest (Figure 8).

In Figure 8, note in the autospectrum of neck suction that the frequency content of the stimulating signal is actually broadened over the whole frequency band. Note also in the autospectrum of the RR interval a prominent peak related to respiratory activity. The coherence is statistically significant in all the LF band and most of the HF band, with highest values in the former. The phase spectrum indicates a rather constant delay of about 1 second between neck suction and RR interval. Finally, the transfer function magnitude

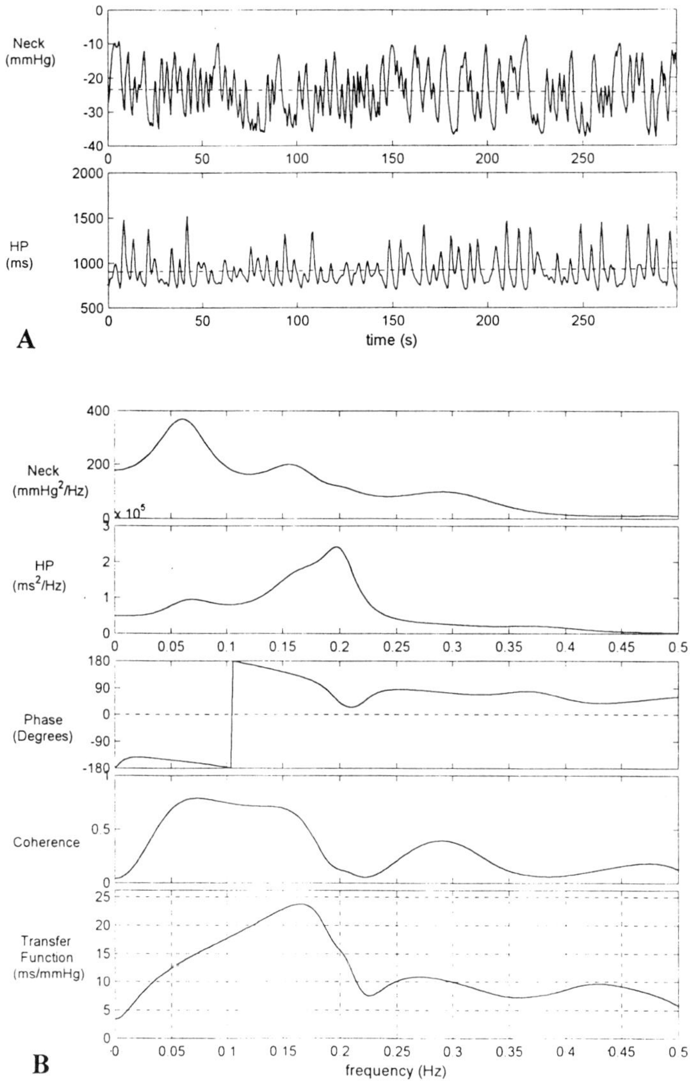

FIGURE 8. A – Time course of neck suction (top panel) and corresponding heart period (HP) (bottom panel) time series during broad-band stimulation. B – Bivariate spectral analysis of neck suction and heart period (HP) signals during broad-band stimulation. From top to bottom: neck suction autospectrum, heart period autospectrum, phase spectrum, coherence function and transfer function magnitude.

(gain) shows a clear maximum around 0.7 Hz with a smooth roll-off for lower and higher frequencies.

The potential advantage of this technique, which is currently under development in our laboratory, is to be far more efficient than the discrete frequency approach, as it provides the gain of baroreceptors at all frequencies of the spectrum at the same experimental cost in time of a few discrete stimulations. However this promising method is at present applicable only in research laboratories and for particular pathophysiological purposes.

Two distinctive features of the neck chamber technique need to be emphasised: (1) the stimulus is, to a great extent, 'pure', that is, it is confined to one baroreceptor area. This is a quite important advantage with respect to other manoeuvres, which can be used to study baroreceptor activity. For example, the changes in baroreceptor activity induced by the Valsalva manoeuvre occur through important alterations in intrathoracic pressure, venous return, pulmonary and systemic pressures, inotropism and left/right ventricular wall stress, so that separating the autonomic from the mechanical effects is extremely complex and (2) unlike the vasoactive drugs, neck stimulation due to its specific and confined action, allows observation of the response of both the heart rate and blood pressure.

There are several limitations of the neck chamber technique. The technique is complicated to use and the collection of data is time consuming. Furthermore, neck chamber devices are not comfortable to wear, and the subject, aware of the stimulus applied, may have some emotional reaction. The hypothesis that the aortic baroreflex, not directly stimulated, may interfere with the response at the carotid level, appears mainly theoretical; since the heart rate response occurs within the same or the next heart period, while the arterial pressure response is more gradual (within the next 5 to 10 heart periods from the rapid bradycardia) this may not be a limitation when immediate heart rate changes are being evaluated.

There are conflicting data on the reproducibility of human responses to baroreceptor stimulation by the neck chamber technique probably depending on different methodologies.[43,44] As with the vasoactive drugs, an asymmetry in the control of the circulation by activation or deactivation of the carotid baroreceptor has been shown.[44]

ANALYSIS OF SPONTANEOUS BAROREFLEX SENSITIVITY

Since the baroreflexes not only control abrupt changes in arterial pressure, but are continuously activated by small variations of systolic arterial pressure around the set point for that particular patient or condition, spontaneous fluctuations in arterial pressure and RR interval may also provide information about cardiovascular control. The major advantage in analysing the spontaneously occurring changes in arterial pressure and RR interval is that the length of the recording is not limited by the duration of the perturbating stimulus, therefore these methods can be applied to investigate long term control of the cardiovascular system and in conditions other than a laboratory set-up.

Two basic approaches have been proposed and validated: one based on 'time domain' and one based on 'frequency-domain' measurements.

With the time domain approach, a computer program automatically identifies those sequences of three or more consecutive heart beats characterised by a progressive increase in systolic arterial pressure and a progressive lengthening in RR interval, or characterised by a progressive reduction in systolic arterial pressure and a progressive shortening in RR interval. The threshold value for including beat-to-beat systolic blood pressure and pulse interval changes in a sequence are set at 1 mm Hg and 6 ms, respectively. Similarly to that which is done when vasoactive drugs are injected, information on the sensitivity of baroreflex control of heart rate can be obtained by computing the slope of the regression line between changes in systolic arterial pressure and the following changes in RR interval.[45] The validity of this approach has been verified by the almost complete disappearance of the sequences produced by sino-aortic denervation in animals.[46]

The 'frequency domain' approach uses the spectral analysis of simultaneously obtained sequences of RR interval and blood pressure. The idea is that each spontaneous oscillation in blood pressure elicits an oscillation of similar frequency in the RR interval by the effect of arterial baroreflex activity. Two major oscillations are usually considered: that centred around 0.1 Hz, within the low frequency (LF) band (0.04÷0.15 Hz), and that associated with respiratory activity in the high frequency (HF) band (0.15÷0.40 Hz). Hence, a rough estimate of

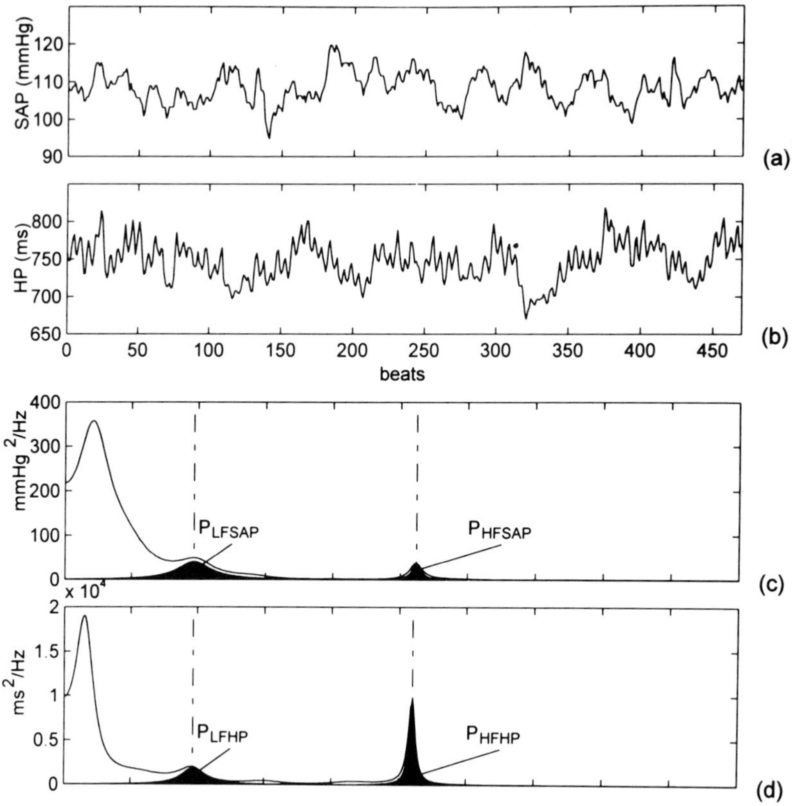

FIGURE 9. Examples of the computation of the spectral indexes of Pagani et al[47] and Robbe et al[48]. Panels a and b show representative examples of Systolic Arterial Pressure (SAP) and Heart Period (HP) time series. The corresponding autoregressive spectra are in panels c and d, respectively, the coherence function between SAP and HP in panel e (opposite), and the transfer function in panel f (opposite). The dark areas P_{LFSAP}, P_{HFSAP}, P_{LFHP}, and P_{HFHP} represent the spectral components involved in the computation of the alpha-LF and alpha-HF. The alpha-LF is computed by dividing P_{LFHP} (≈LF component of HP) by the P_{LFSAP} (≈LF component of SAP) and taking the square root. The same holds for alpha-HF considering the areas P_{HFHP} and P_{HFSAP} . In panel f, the portion of the transfer function in the frequency region 0.07 to 0.14 Hz, where the coherence is $\geq$ 0.5, is plotted in bold (Reproduced with permission from Maestri et al[52]).

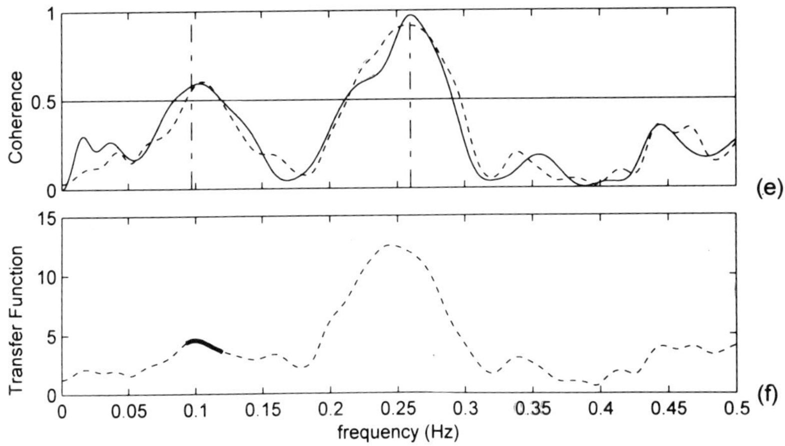

FIGURE 9 (Continued). See the legend opposite.

the baroreflex gain can be obtained by dividing the amplitude of the RR oscillation in each band by the amplitude of the corresponding oscillation in the blood pressure. Some authors indicate the two gains as 'alfa-LF' and 'alfa-HF'.[47] Under the assumption that the transfer from systolic pressure to RR interval is linear, 'alfa-LF' and 'alfa-HF' can be simply computed by the square root of the ratio between the spectral component of the RR spectrum and that of the systolic pressure spectrum in the LF and HF band, respectively. In order to be confident that the assumption of linearity holds, the magnitude squared coherence function between pressure and RR interval is computed first. A coherence > 0.5 is commonly required to obtain a reliable estimate. However, using this approach, contributions to the RR oscillations not due to the pure effect of the baroreflex might be erroneously taken into account. A more accurate method to quantify the gain of the baroreflex at each oscillatory frequency is the modulus of the so-called transfer function between systolic pressure and RR interval. This is commonly performed only in the LF band,[48] as the relationship between systolic pressure and heart period in this band best satisfies the mathematical assumption for transfer function estimation[49,50] (Figure 9).

Hence spectral technique estimate BRS exploiting specific rhythms shared by the RR interval and systolic blood pressure, whereas time domain estimates are not related to any specific rhythm and virtually can exploit all kinds of fluctuations hidden in the variability signals, provided the conditions for the changes in systolic pressure and RR interval, as previously described, are satisfied.

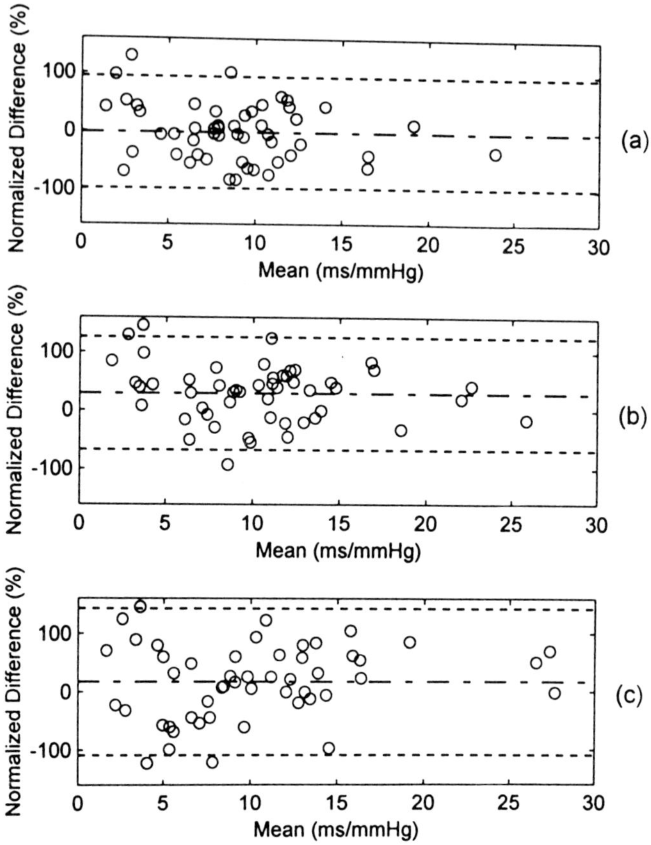

FIGURE 10. Bland-Altman plot of the normalised difference between spectral and phenylephrine measurement of baroreflex gain (Phe-BRS) against their mean value. In panel a normalised differences between the index of Robbe and the Phe-BRS. In panel b normalised differences between the alpha-LF index and the Phe-BRS. In panel c normalised differences between the alpha-HF index and the Phe-BRS. The fine dashed lines indicate the limits of agreement, the semi-dashed lines indicate the mean difference (bias) (reproduced with permission from Maestri et al[52]).

Thus depending on the particular distribution of the overall variability among the systolic pressure – RR relationship in the LF and/or HF bands or may reflect other fluctuating components (e.g. those associated with VLF band). Moreover, spectral BRS, requiring stationarity recording, is commonly obtained from supine resting recordings and hence reflects this autonomic condition. On the contrary, time domain techniques, when obtained from 24-hour recordings, can span several conditions of daily life including exercise.

Several investigators have proposed that spectral estimations of baroreflex gain are reliable alternatives to the phenylephrine test.[47,48,51] However these results have been obtained only in small groups of normal subjects or hypertensive patients and have been based on simple correlation analysis. We have compared spectral measurements and the phenylephrine measured BRS in a larger group of patients with a previous myocardial infarction and different degrees of left ventricular dysfunction and analysed the agreement between the two approaches using more sophisticated statistical techniques.[52] We found that despite a substantial linear association, the agreement between spectral measurements and phenylephrine in the estimation of baroreflex gain is weak because the difference can be as large as the BRS value being estimated (Figure 10). It is not surprising that the phenylephrine and spectral methods do not provide similar quantification of baroreflex gain in cardiac patients. Besides theoretical and methodological differences, from a clinical point of view, the phenylephrine baroreflex gain – which is the result of the complex interplay of multiple receptor areas and the hemodynamic burden produced by the after-load increase – yields information on the 'whole' capability of the system to evoke an increase in vagal activity, whereas the spectral baroreflex gain yield information on the level of autonomic modulation on a continuous basis. The differences between the two methodologies may become exaggerated particularly when dealing with a diseased heart and there could be a major clinical impact when the different measures are used for risk assessment.[53]

CONCLUSIONS

We have reviewed the methods currently used to assess baroreceptor reflexes and discussed their implications in clinical use. However the system is complex and many features of its regulation still need to be clarified. We believe that continued investigation also by means of the new techniques under development will better delineate the mechanisms involved in the control of physiological and pathological conditions.

REFERENCES

1. Schwartz PJ, La Rovere MT, Vanoli E. Autonomic nervous system and sudden cardiac death. *Circulation* 1992; **85** (Suppl I):I 77–I 91.
2. Eckberg DL, Sleight P. Human baroreflexes in heart and disease. Clarendon Press, Oxford 1992.
3. Brown AM. Receptors under pressure. An update on baroreceptors. *Circ Res* 1980; **46**:1–10.
4. Bergel DH, Brooks DE, Mac Dermott AJ, Robinson JL, Sleight P. Baroreceptor firing frequency and activation of carotid sinus vascular smooth muscle. *J Physiol (Lond)* 1978; **275**:37–38.
5. Felder RB, Heesch CM, Thames MD. Reflex modulation of carotid sinus baroreceptor activity in the dog. *Am J Physiol* 1983; **244**:H437–H443.
6. Kirchheim HR. Systemic arterial baroreceptor reflexes. *Physiol Rev* 1976; **56**:100–176.
7. Abboud FM, Thames MD. Interaction of cardiovascular reflexes in circulatory control. In: Shepherd JT, Abboud FM (Eds) The Cardiovascular System. American Physiological Society, Bethesda 1983, 675–754.
8. Coleman TG. Arterial baroreflex control of heart rate in the conscious rat. *Am J Physiol* 1980; **238**:H515–H520.
9. Thames MD, Kontos HA. Mechanisms of baroreceptor-induced changes in heart rate. *Am J Physiol* 1970; **218**:251–256.
10. Pickering TG, Davies G. Estimation of the conduction time of the baroreceptor-cardiac reflex in man. *Cardiovasc Res* 1973; **7**:213–219.
11. Smyth HS, Sleight P, Pickering GW. Reflex regulation of arterial pressure during sleep in man: a quantitative method for assessing baroreflex sensitivity. *Circ Res* 1969; **24**:109–121.
12. Koch-Weser J. Nature of inotropic action of angiotensin on ventricular myocardium. *Circ Res* 1965; **16**:230–237.
13. Scroop GC, Whelan RF. A central vasomotor action of angiotensin in man. *Clin Sci* 1966; **30**:70–90.
14. Bristow JD, Honour AJ, Pickering JW, Sleight P, Smyth HS. Diminished baroreflex sensitivity in high blood pressure. *Circulation* 1969; **39**:48–54.
15. Sleight P. Arterial Baroreceptor and Hypertension. Oxford University Press, Oxford 1980, 471.
16. Eckberg DL, Drabinsky M, Braunwald E. Defective cardiac parasympathetic control in patients with heart disease. *N Engl J Med* 1971; **285**:877–883.

17. Hossman V, Fitzgerald GA, Dollery CT. Circadian rhythm of baroreflex reactivity and adrenergic vascular response. *Cardiovasc Res* 1969; **3**:476–485.
18. Bristow JD, Honour AJ, Pickering TG, Sleight P. Cardiovascular and respiratory changes during sleep in normal and hypertensive subjects. *Cardiovasc Res* 1969; **3**:476–485.
19. Pickering TG, Gribbin B, Petersen ES, Cunningham DJ, Sleight P. Comparison of the effect of exercise and posture on the baroreflex in man. *Cardiovasc Res* 1971; **5**:582–586.
20. Bristow JD, Brown EB Jr, Cunningham DJC, Howson MG, Strange Petersen E, Pickering TG, Sleight P. Effect of bicycling on the baroreflex regulation of pulse interval. *Circ Res* 1971; **28**:582–592.
21. Sleight P, Fox P, Lopez R, Brooks DE. The effect of mental arithmetic on blood pressure variability and baroreflex sensitivity in man. *Clin Sci* 1978; **55**:381S–382S.
22. Sherrer U, Pryor SL, Bertocci LA, Victor RG. Arterial baroreceptor buffering of sympathetic activation during exercise-induced elevation of arterial pressure. *J Clin Invest* 1990; **86**:1855–1861.
23. Gribbin B, Pickering TG, Sleight P, Peto R. Effect of age and blood pressure on baroreflex sensitivity in man. *Circ Res* 1971; **29**:424–431.
24. La Rovere MT, Specchia G, Mortara A, Schwartz PJ. Baroreflex sensitivity, clinical correlates and cardiovascular mortality among patients with a first myocardial infarction. A prospective study. *Circulation* 1988; **78**:816–824.
25. Schwartz PJ, Zaza A, Pala M, Locati E, Beria G, Zanchetti A. Baroreflex sensitivity and its evolution during the first year after myocardial infarction. *J Am Coll Cardiol* 1988; **12**:629–636.
26. Farrell TG, Paul V, Cripps TR, Malik M, Bennett ED, Ward D, Camm AJ. Baroreflex sensitivity and electrophysiological correlates in patients after acute myocardial infarction. *Circulation* 1991; **83**:945–952.
27. La Rovere MT, Bigger JT Jr, Marcus FI, Mortara A, Schwartz PJ for the ATRAMI Investigators. Baroreflex sensitivity and heart rate variability in prediction of total cardiac mortality after myocardial infarction. *Lancet* 1998; **351**:478–484.
28. Mortara A, La Rovere MT, Pinna GD, Prpa A, Maestri R, Febo O, Pozzoli M, Opasich C, Cobelli F, Tavazzi L. Arterial baroreflex modulation of heart rate in chronic heart failure. Clinical and hemodynamic correlates and prognostic implications. *Circulation* 1997; **96**:3450–3458.
29. Mortara A, Sleight P, Pinna GD, Maestri R, Prpa A, La Rovere MT, Cobelli F, Tavazzi L. Abnormal awake respiratory patterns are common in chronic heart failure and may prevent evaluation of autonomic tone by measures of heart rate variability. *Circulation* 1997; **96**:246–252.
30. Osculati G, Grassi G, Giannattasio C, Seravalle G, Valagussa F, Zanchetti A, Mancia G. Early alterations of the baroreceptor control of heart rate in patients with acute myocardial infarction. *Circulation* 1990; **81**:939–948.
31. Pomidossi G, Saino A, Perondi F, Gregorini L, Alessio P, Rimini A, Omboni S, Zanchetti A, Mancia G. Impairments of the arterial baroreflex during

symptomatic and silent myocardial ischemia in humans. *J Am Coll Cardiol* 1993; **22**:1866–1872.
32. Pickering TG, Gribbin B, Sleight P. Comparison of the reflex heart rate response to rising and falling arterial pressure in man. *Cardiovasc Res* 1972; **6**:277–283.
33. Musialek P, Lei M, Brown HF, Paterson DJ, Casadei B. Nitric oxide can increase heart rate by stimulating the hyperpolarization-activated inward current If. *Circ Res* 1997; **81**:60-68.
34. Marin-Neto JA, Maciel BC, Gallo L Jr. Effect of lowering atrial pressure on arterial baroreflex control of heart rate in patients with congestive heart failure. In: Sleight P (Ed) Arterial Baroreceptor and Hypertension. Oxford University Press, Oxford 1980, 499–509.
35. Korner PI, West MJ, Shaw J, Uther JB. "Steady state" properties of the baroreceptor-heart rate reflex in essential hypertension in man. *Clin Exp Pharmacol Physiol* 1974; **1**:65–76.
36. Rosen MR, Bilexikian JP, Cohen IS, Robinson RB. Adrenergic modulation of cardiac rhythm. In: Zipes DP, Jalife J (Eds) Cardiac Electrophysiology. From Cell to Bedside. Saunders, Philadelphia 1991, 300–304.
37. Parati G, Casadei R, Groppelli A, Di Rienzo M, Mancia G. Comparison of finger and intra-arterial blood pressure monitoring at rest and during laboratory testing. *Hypertension* 1989;**13**:647–655.
38. Ernsting J, Parry DJ. Some observations on the effects of stimulating the stretch receptor in the carotid artery of man. *J Physiol* 1957; **137**:45P–46P.
39. Eckberg DL, Cavanaugh MS, Mark AL, Abboud FM. A simplified neck suction device for activation of carotid baroreceptors. *J Lab Clin Med* 1975; **85**:167–173.
40. Ludbrook J, Mancia G, Ferrari A, Zanchetti A. The variable-pressure neck-chamber method for studying the carotid baroreflex in man. *Clin Sci* 1977; **53**:165–171.
41. Sleight P, La Rovere MT, Mortara A, Pinna G, Maestri R, Leuzzi S, Bianchini B, Tavazzi L, Bernardi L. Physiology and physiopathology of heart rate and blood pressure variability in humans: is power spectral analysis largely an index of baroreflex gain? *Clin Sci* 1995; **88**:103–109.
42. Bernardi L, Bianchini B, Spadacini G, Leuzzi S, Valle F, Marchesi E, Passino C, Calciati A, Vigano M, Rinaldi M, Martinelli L, Finardi G, Sleight P. Demonstrable cardiac reinnervation after human heart transplantation by carotid baroreflex modulation of RR interval. *Circulation* 1995; **92**:2895–2903.
43. Eckberg DL. Adaptation of the human carotid baroreceptor-cardiac reflex. *J Physiol (London)* 1977; **269**:579–589.
44. Mancia G, Ferrari A, Gregorini L, Valentini R, Ludbrook J, Zanchetti A. Circulatory reflexes from carotid and extracarotid baroreceptor areas in man. *Circ Res* 1977; **41**:309–315.
45. Parati G, Di Rienzo M, Bertinieri G, Pomidossi G, Casadei R, Groppelli A, Pedotti A, Zanchetti A, Mancia G. Evaluation of the baroreceptor-heart rate reflex by 24-hour intra-arterial blood pressure monitoring in humans. *Hypertension* 1988; **12**:214–222.

46. Bertinieri G, Di Rienzo M, Cavallazzi A, Ferrari AU, Pedotti A, Mancia G. A new approach to the analysis of the arterial baroreflex. *J Hypertension* 1985; **3** (Suppl 3): s79–s81.
47. Pagani M, Somers V, Furlan R, Dell'Orto S, Conway J, Baselli G, Cerutti S, Sleight P, Malliani A. Changes in autonomic regulation induced by physical training in mild hypertension. *Hypertension* 1988; **12**:600–610.
48. Robbe HWJ, Mulder LJM, Ruddel H, Langewitz WA, Veldman JB, Mulder G. Assessment of baroreceptor reflex sensitivity by means of spectral analysis. *Hypertension* 1987; **10**:538–543.
49. Taylor JA, Eckberg DL. Fundamental relations between short-term RR interval and arterial pressure oscillations in humans. *Circulation* 1996; **93**:1527–1532.
50. Baselli G, Porta A, Ferrari G. Analysis of cardiovascular variability signals. In: M Malik, AJ Camm (Eds) Heart Rate Variability. Futura, Armonk 1995, 135–146.
51. Watkins LL, Grossman P, Sherwood A. Noninvasive assessment of baroreflex control in borderline hypertension: comparison with the phenylephrine method. *Hypertension* 1996; **28**:238–243.
52. Maestri R, Pinna GD, Mortara A, La Rovere MT, Tavazzi L. Assessing baroreflex sensitivity in post-myocardial infarction patients: comparison of spectral techniques and phenylephrine. *J Am Coll Cardiol* 1998; **31**:344–351.
53. La Rovere MT, Pinna GD. Physiological versus clinical value of spectral and nonspectral techniques in the evaluation of the baroreflex. *Ann Noninvas Electrocard* 1997; **2**:203–205.

PART THREE

CLINICAL PRACTICE

CHAPTER ELEVEN

AGEING AND CARDIAC AUTONOMIC STATUS

Bořivoj Semrád, Bohumil Fišer and Nataša Honzíková

Recent studies have shown the significance of autonomic changes for the diagnosis and prognosis of many cardiovascular diseases. A consensus exists that, in the setting of acute myocardial ischaemia, sympathetic hyperactivity facilitates the onset of malignant arrhythmias, whereas vagal activation can exert an antifibrillatory effect. Thus, evaluation of the activity of the autonomic nervous system can contribute to more refined post-myocardial infarction risk stratification. Decreased heart rate variability (HRV) which is a marker of lower parasympathetic and/or higher sympathetic activity and depressed baroreflex sensitivity (BRS) indicate an increased risk of sudden cardiac death (SCD) in these patients.[1–3] BRS is also decreased in patients with essential hypertension and therapeutic normalisation of blood pressure is not accompanied by BRS increase.[4,5] On the other hand, both BRS and HRV physiologically decline with age[6] which complicates the evaluation of cardiac autonomic status in the elderly.

Appropriate evaluation of BRS and HRV in the elderly is more complicated than in younger age groups, because of several factors. It is, for instance, necessary to account for the increased incidence of hypertension and ischaemic heart disease with increasing age. Moreover, some elderly subjects are unable to participate in measurements requiring active collaboration.

Malik M (Ed) Clinical guide to cardiac autonomic tests, 285–300.

This chapter reviews age dependent changes of the cardiovascular system, including alterations in autonomic nervous system regulation and morphological variations of the arterial tree. The advantages and disadvantages of methods which can be used for the evaluation of HRV and BRS in children, healthy elderly, and ischaemic patients are discussed with examples of post-myocardial risk stratification and attempts to evaluate the physiological and clinical relevance of individual measurements.

GENERAL AGE RELATED CHANGES

The most prominent age-related changes in the cardiovascular system are changes of the pulse wave.[7–10] With increasing age, aortic wave velocity increases subsequent to stiffening of the aortic wall. A high velocity causes the reflected wave to return early in systole, modifying the systolic pulse in a characteristic way related to the age of the subject. According to Langewouters, aortic mechanical properties change almost exclusively with age.[11] Although severe atherosclerosis tends to result in an aorta with slightly lower Young's elasticity modulus, this is compensated for by an enlarged cross-section and results in equal compliance and no significant difference in wave velocity. This is caused by lipid deposits and calcified plaques having little mechanical strength, certainly negligible compared to collagen fibres. Hypertension does not accelerate aortic ageing. However, pulse wave velocity increases linearly from approximately 5 m/s at 30 years of age to 14 m/s at 90, resulting in an age-dependent increase of systolic and pulse pressure.

Stiffening of aortic wall is also likely to decrease the sensitivity of aortic baroreceptors. Very probably, similar changes of the wall of the carotid sinus decrease the sensitivity of carotid baroreceptors. The decline of BRS with age was first described by Gribbin et al[6] and was confirmed in post-myocardial infarction patients[3] (Figure 1). The plausible explanation is the increase of arterial stiffness and decrease of baroreceptor sensitivity, as mentioned above. On the other hand, other conclusion Gribbin et al[6] that BRS is decreased by high blood pressure, is not fully acceptable. Long lasting therapy with ACE-inhibitor enalapril decreases blood pressure, normalises forearm blood flow after 5 minutes of ischaemia, which indicates the regression of hypertrophy of the resistance vessels. BRS remains low in patients treated with enalapril, reaching approximately 50% of age matched normotensive controls.[4] Similarly, treatment with Ca-antagonists lasting several weeks normalises blood pressure but not BRS.[5] The low BRS as a risk factor for the development of hypertension is a more plausible explanation in accordance with Sleight et al.[12]

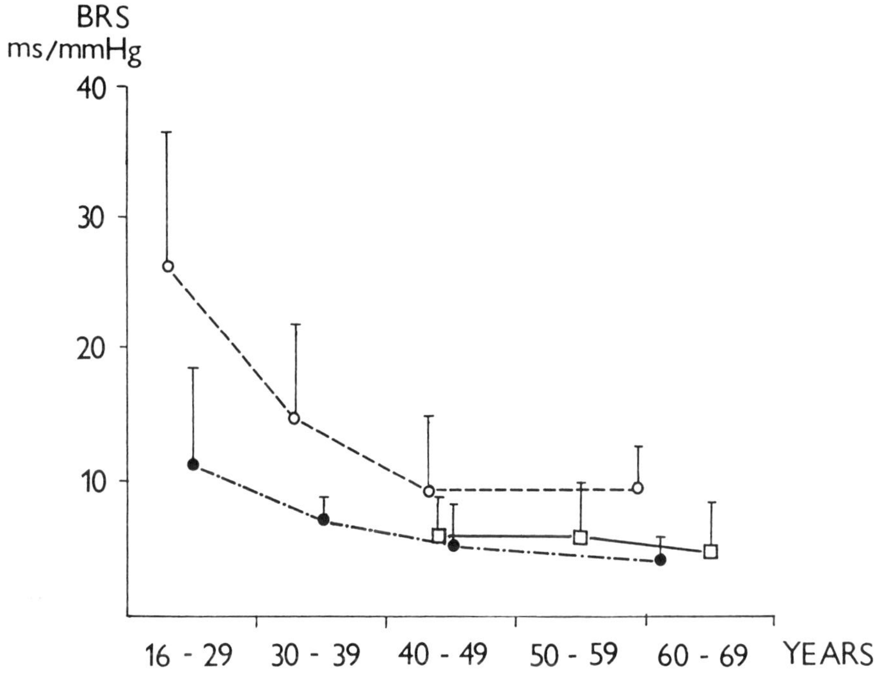

FIGURE 1. The relationship between BRS and age in normotensives with mean arterial pressure (MAP) < 100 mm Hg (circles) and hypertensives with MAP > 100 mm Hg (points) and in post-myocardial infarction patients (squares). Means ± SD are shown (Data for normotensives and hypertensives are from Gribbin et al[6], data of myocardial infarction patients from our laboratory).

BRS inversely correlates with an increase of blood pressure during psychological stress which can be a link between low BRS and increased blood pressure in hypertensive subjects.[13] The question whether age related decline of HRV is a consequence of decreased BRS or whether, vice versa, the decrease of BRS is partially a consequence of lower parasympathetic activity due to ageing, will be discussed at the end of this chapter.

With ageing, heart rate response to β-adrenergic agonists and to β-adrenergic antagonists declines. Muscarinic parasympathetic blockade with atropine results in a lesser increase of heart rate in healthy elderly vs. younger individuals. The decrease of low frequency (LF) and high frequency (HF) spectral components with age was observed in studies of HRV spectral analysis.

HEART RATE VARIABILITY

Pagani et al found that the LF/HF ratio does not change with age and suggested, that ageing is associated with a new equilibrium between sympathetic and parasympathetic activity.[14] Studying a broader age range, Shanon et al found that LF power declined linearly with age across the whole age range, while HF power declined significantly only between the ages of 9 and 30 years.[15] This was confirmed in a study with metronome controlled respiration showing that respiratory sinus arrhythmia at constant tidal volume declined rapidly between the ages of 22 and 35 years with no further decrease up to the age of 78 years.[16] A larger study by Schwartz et al showed a decline of both HF and LF spectral components with age during spontaneous and metronome controlled breathing in supine and standing positions.[17] The decline of HF, but not of LF, power with age was higher in a group of 27–48 vs. 48–70 years. Pfeffer et al[18] and Lipsitz et al[19] observed that the difference between young and elderly is mainly due to a decline in parasympathetic functions. Veernman et al[20] observed that changing the position from supine to standing, a higher increase of LF component was observed in younger subjects vs. older persons. A less marked influence of age on HRV was observed after myocardial infarction. Kleiger et al. found only a weak inverse correlation ($r = -0.2$) between HRV and age.[1]

The physiological interpretation of observed differences in the magnitude of LF and HF components is not simple. Increased parasympathetic tonic influence is probably accompanied by the increased variability in the high frequency range of spectra. On the other hand, the increase of sympathetic nervous activity can increase the variability in the LF component, e.g. in a reaction to standing in young people. The relationship between sympathetic activity and the LF component is also dependent on the average sympathetic activity. If it is too high, it can decrease the variability in the low frequency component, e.g. in patients with congestive heart failure.[21]

Different methods for HRV determination are based on 24-hours ECG Holter monitoring allowing to perform the measurement together with the diagnosis of arrhythmic events and ST segment analysis. While some of these measurement have been implemented within commercial equipment,[22] the determination of BRS is not commercially available.

BAROREFLEX SENSITIVITY

Data about BRS in various age groups are sparse and depend on the technology for BRS determination. BRS has traditionally been determined by a pharmacological method.[23] A pressor agent (usually phenylephrine), which does not affect heart rate directly, is administered as an intravenous bolus injection and the magnitude of the bradycardiac response to the pressure rise represents the BRS index. The development of reliable non-invasive blood pressure monitoring, coupled with on-line methods of analysis, has largely solved the technical problems.[24,25] However the need for an intravenous injection limits the use of this method in large populations and follow-up studies. Minor side effects are also common after administration of phenylephrine.[26] Moreover, the method enables to evaluate the cardiac but not the vasomotor response.

Several groups used phase 4 of the Valsalva manoeuvre to asses BRS showing a high correlation between baroreflex slopes derived from the phenylephrine method and those obtained by analysis the overshoot bradycardia of the Valsalva manoeuvre.[27–29] The range of blood pressure changes is comparable to changes elicited by phenylephrine administration but the phase of blood pressure increase is too short, 4 to 5 s compared to 30 s with the phenylephrine method. For a more reliable determination of BRS, several measurements are necessary and the psychological stress connected with these repeated measurements may influence the results[30].

The neck chamber technique is also often used.[31] The stimulus is pure, that is, confined to the baroreceptor areas. The suction is psychologically better tolerated compared to an increase of neck pressure which causes diminished baroreceptor stimulation. The main problem is that the transformation of neck chamber pressure to carotid sinus pressure is not possible to establish in individual subjects The psychological impact is large and the method cannot be practically used in children as parents are rarely willing to consent.

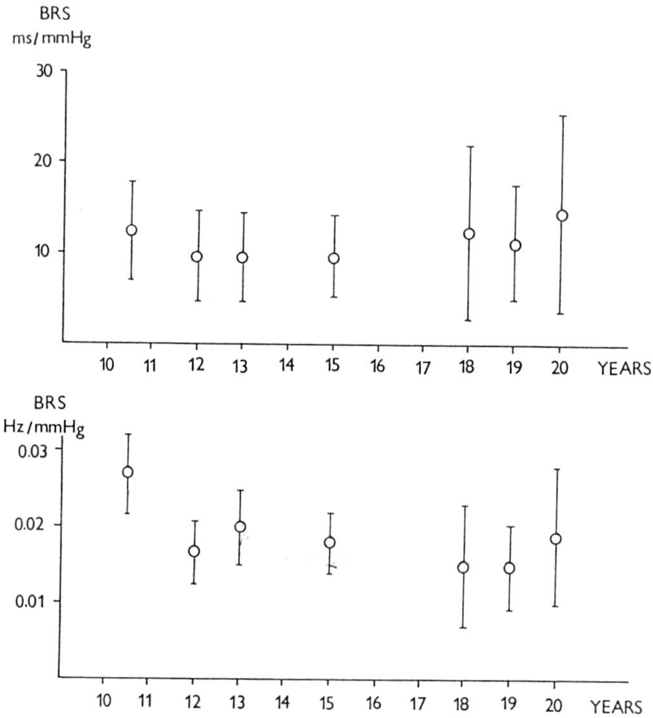

FIGURE 2. BRS [ms/mm Hg] and BRS_f [Hz/mm Hg] in children and young adults (mean±SD).

Analysis of spontaneous beat-to-beat fluctuations of arterial pressure and inter-beat interval has several advantages, mainly the comfort during measurement. Two basic approaches have been proposed and validated, based on 'time-domain' and on 'frequency-domain' measurements. With the time-domain approach, the sequences of three or more consecutive heart beats are identified characterised by a progressive increase in systolic blood pressure and progressive lengthening of RR or pulse interval, or characterised by a progressive reduction in systolic blood pressure and progressive shortening of RR or pulse interval. The sensitivity of baroreflex control of heart rate is obtained by computing the slope of the regression line between changes in RR or pulse intervals.[32] The frequency-domain approach uses the spectral analysis of simultaneously obtained sequences of RR intervals and blood pressure.[33–35]

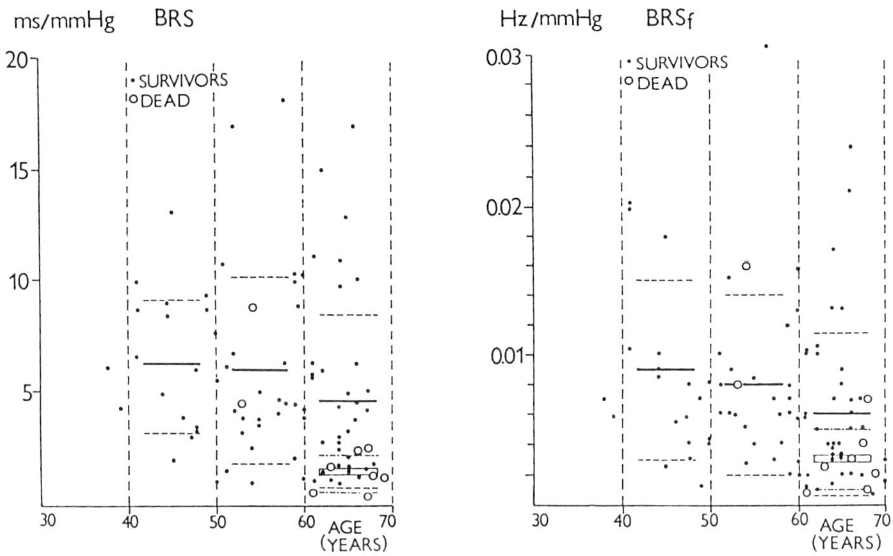

FIGURE 3. BRS and BRS_f in patients 7–14 days after first signs of myocardial infarction. Deceased patients died during one year. Means ±SD are indicated in survivors and deceased patients.

BRS can be expressed in ms/mm Hg or Hz/mm Hg.[12,37] During physical exercise, the absolute changes in heart rate elicited by a given change in pressure are, according to Bevegard and Shepherd, remarkably similar to those at rest. However, when these changes in heart rate are evaluated from the inter-beat interval changes, the data indicated a reduced gain for baroreflex control of heart rate.[38] We confirmed this finding in our laboratory studying BRS [ms/mm Hg] and BRS_f [Hz/mm Hg] during psychological stress (oral examinations at the medical school) and during mild physical exercise in healthy young adults. BRS decreased and BRS_f remained constant during psychological stress and exercise.[13]

One hundred and seventy seven subjects between 10 and 20 years of age were analysed. The results are seen in Figure 2. In this range of age, we observed a significant age related increase of systolic pressure (correlation coefficient with age r = 0.47, p<0.01), diastolic pressure (r = 0.20, p< 0.05) and pulse interval (r = 0.54, p<0.01) On the other hand, BRS remained constant (r = 0.12, ns). There was a non-significant trend of BRS_f to decrease. Large variation of BRS and BRS_f was noted. In 16.4% of our population, we observed BRS < 5 ms/mm Hg.

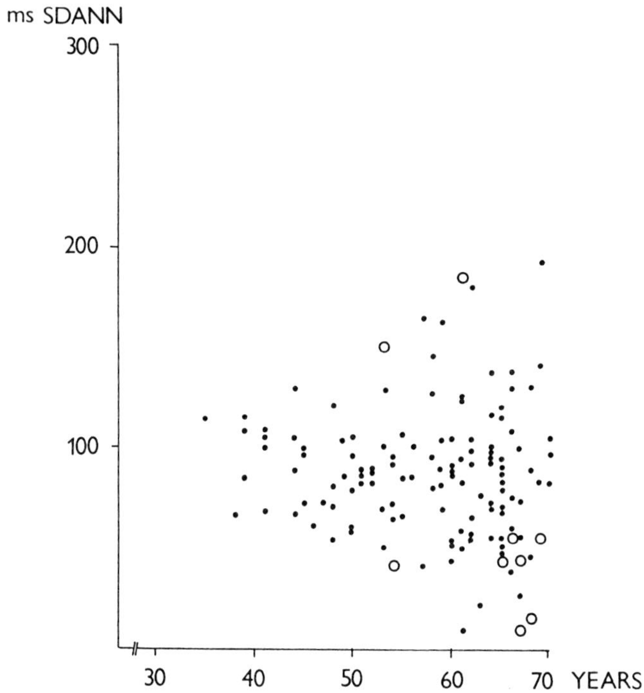

FIGURE 4. The relationship between SDANN (standard deviation of the average normal to normal intervals for all 5-minute segments of a 24-hour recording) and age in surviving patients monitored 7–14 days after first signs of myocardial infarction (points) and patients deceased during 1 year (circles).

We also examined 139 patients 7–14 days after first signs of myocardial infarction. In addition to BRS and BRS_f, other indexes of the autonomic status were assessed in these patients. HRV was expressed as SDANN and SDNN index from 24 hours Holter ECG. The peak of respiratory arrhythmia, that is the power of HF component in the HRV spectrum, was assessed at 0.33 Hz, which was the frequency of metronome controlled breathing. Five patients were younger than 40 years, 22 patients were between 40 and 50 years, 39 patients between 50 and 60 and 73 patients between 60 and 70. Nine patients died, seven of them were between 60 and 70 years old. Significant correlation with age was found in BRS, BRS_f and in SDANN and SDNN index only. (BRS: r = −0.312, BRS_f: r = −0.287, Figure 3, SDANN: r = −0.216, Figure 4, SDNN index: r = −0.375, Figure 5.) Correlation between age and HF power and between age and mean 24-hours RR interval was not observed.

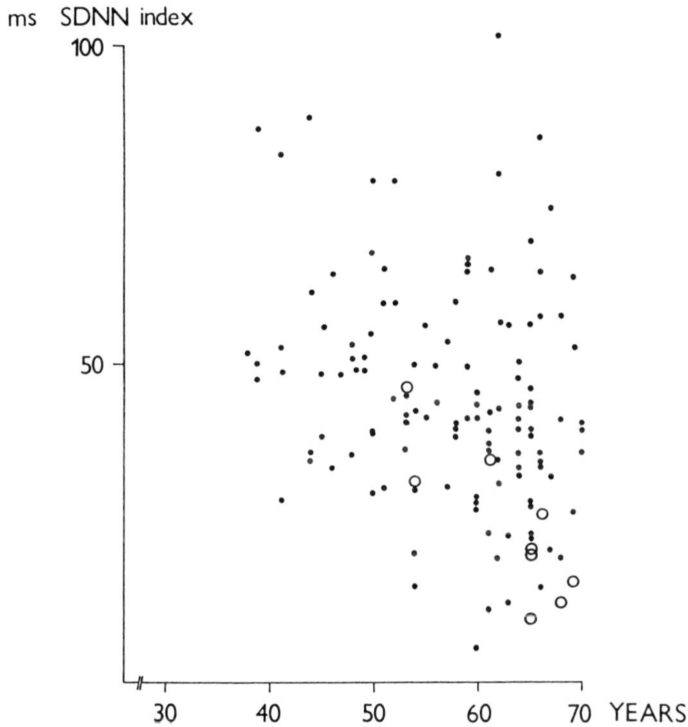

FIGURE 5. The relationship between SDNN index (mean of the standard deviation of all normal to normal intervals for all 5-minute segments of a 24-hour recording) and age. Layout as in Figure 4.

We found a mild correlation between BRS and SDANN r = 0.100, and between BRS and SDNN index r = 0.402. The information derived from ambulatory ECG recording concerns primarily vagal tone, whereas the BRS determination discloses the capability of the parasympathetic nervous system to react to a gross stimulus and thus concerns primarily vagal reflexes. The main conclusion of Schwartz at al.[2] that the two types of measures are not redundant and cannot be used to predict each other is fully justified.

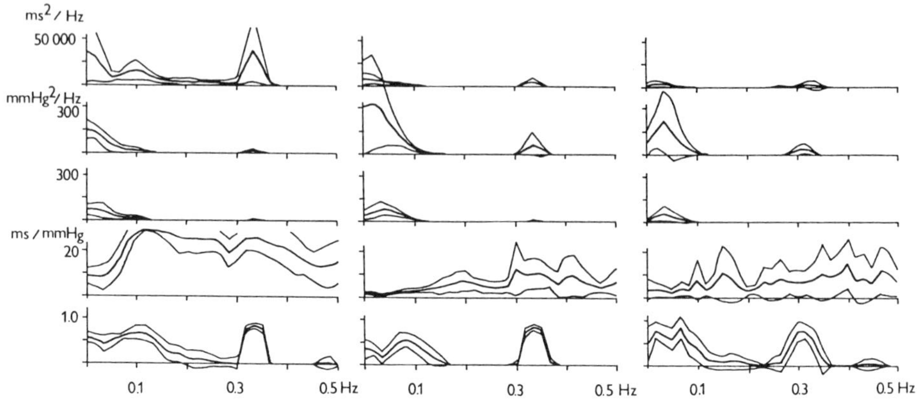

FIGURE 6. Spectral analysis of spontaneous fluctuation of circulatory variables in healthy young subjects with high BRS (left), in healthy young subjects with low BRS (middle) and in 65–70 years old patients 7–14 days after myocardial infarction. From above: pulse interval, systolic and diastolic blood pressure, modulus (at 0.1 Hz corresponds to BRS) and coherence. Notice that spectra from young adults with low BRS and aged patients are identical from 0.05 to 0.5 Hz.

It is self-evident that heart rate varies according to the influence of both limbs of the autonomic nervous system on the intrinsic heart rate (HR_0). HR_0 can be determined only when the autonomic nervous system influence is completely blocked by drugs. According to Jose and Colins, HR_0 decreases with age according to formula:[39]

$$HR_0 = 118 - (\text{age[years]} \times 0.56)$$

The decline of heart rate variability with increasing age is generally attributed of age-related reduction in parasympathetic activity. Sympathetic activity also declines with age, but at a slower rate. On this basis we can explain why heart rate at rest is not age-dependent but we cannot explain why the vagally mediated respiratory arrhythmia is not age dependent. Hence we propose an alternative hypothesis that the age-related decrease of baroreceptor sensitivity is based on the age-related change in the arterial wall. As a result, elderly have decreased response to baroreflex mediated changes of heart rate on blood pressure fluctuations. Experimental evidence of Mancia et al. that BRS is inversely related to blood pressure variation and positively related to heart rate variability supports this hypothesis.[40] Additional evidence is seen in Figure 6. Both mechanisms seem to play a role in the decrease of HRV with ageing.

Ageing of neuronal nets responsible for the baroreflex is not fully understood. We cannot estimate the relative contribution of the decline in baroreceptors sensitivity and the decline in the gain of the neuronal baroreflex net. Very little is known about the heredity of baroreflex gain, which can be important for hypertension development as well as risk of sudden cardiac death.

CONCLUSION

Our main concern is the cardiac death: Is the decreased vagal and increased sympathetic activity together with diminished baroreflex sensitivity the cause of cardiac death in the elderly or is it only a contributing factor whilst the cause is the damaged myocardium? The mortality in post-infarction patients is age-dependent, concentrated in the age group above 60 years. At the same time, the correlation between age and left ventricular ejection fraction is weak in our measurements ($r = -0.144$). The occurrence of late potentials in signal averaged ECGs depends on age only slightly (30 – 50 years age group: incidence 18.5 %, 50 – 60 years: 20.5 %, 60 – 70 years: 26.0 %). Hence, it is generally accepted that the activity of autonomic nervous system plays an important role.

The mechanism by which decreased vagal activity and increased sympathetic activity decrease the fibrillatory threshold is approximately known. The explanation of the relationship between reduced BRS and cardiac mortality is more complicated. The studies of the changes of BRS during exercise or psychological stress in individual healthy subjects indicate that BRS [ms/mm Hg] is heart rate dependent. On the other hand BRS_f [Hz/mm Hg] is heart rate independent in such studies. Our results from myocardial infarction patients revealed that BRS indicates the risk of death better than BRS_f. This can be interpreted as a result of two factors, the decrease of baroreceptor sensitivity and the high heart rate, both of which contribute to the increase in the risk of death in post-infarction patients (Figure 7).

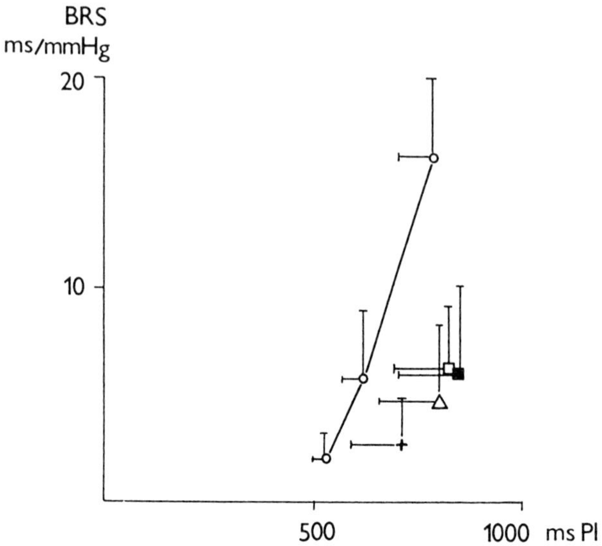

FIGURE 7. The relationship between BRS and pulse interval in young adults at rest and during bicycling[43] (0.5 and 1.0 W/kg of body weight, circles) and in various age groups of myocardial infarction patients (40–50 years: open squares; 50–60 years: closed squares; 60–70 years: triangles; deceased: crosses). Means ± SD are shown.

The mechanisms by which powerful vagal reflexes may exert a protective effect against life-threatening ventricular arrhythmias are not yet fully understood. It has been suggested that BRS measurement discloses the capability of the parasympathetic nervous system to react to a gross stimulus and thus concerns primarily vagal reflexes.[2] It has been shown that even a single premature ventricular contraction causes an increase in sympathetic outflow and that vagal activity exerts a restraint on the increase in sympathetic outflow evoked by the premature ventricular contraction.[41,42]

Patients with a marked depression of vagal reflexes may leave the increase in sympathetic activity completely unopposed and might be more susceptible to sustained ventricular arrhythmias.

Increased heart rate is a marker of increased sympathetic activity. This can explain why heart rate dependent BRS is a better marker of risk than the heart rate independent BRS_f. Another possibility is that the increased sympathetic activity increases blood pressure and the baroreflex can oppose this by decreasing the sympathetic activity. Thus both limbs of the efferent part of the baroreflex may contribute to the protective baroreflex function.

REFERENCES

1. Kleiger RE, Miller JP, Bigger JT Jr, Moss AJ, the Multicenter Post-Infarction Research Group. Decreased heart rate variability and its association with increased mortality after acute myocardial infarction. *Am J Cardiol* 1987; **59**:256–262.
2. Schwartz PJ, Vanoli E, Stramba-Badiale M, de Ferrari GM, Billman GE, Foreman RD. Autonomic mechanisms and sudden death: New insight from the analysis of baroreceptor reflexes in conscious dogs with and without a myocardial infarction. *Circulation* 1988; **78**:969–973.
3. La Rovere MT, Specchia G, Mortara A, Schwartz PJ. Baroreflex sensitivity, clinical correlates and cardiovascular mortality among patients with a first myocardial infarction: A prospective study. *Circulation* 1988; **78**:816–824.
4. Siegelová J, Fišer B, Dušek J, Al-Kubati M. Die Baroreflexsensitivitatsmessung bei Patienten mit essentieller Hypertonie: Einfluss von Enalapril. *Nieren und Hochdruckkrankheiten* 1995, **24**:20–22.
5. Siegelová J, Fišer B, Dušek J. Baroreflex heart rate sensitivity in essential hypertension: the effect of Isoptin SR 240 therapy. In: Varro V, de Chatel R (Eds) 22nd Congress of the International Society of Internal Medicine. Monduzzi Editore, Bologna 1994, 99–102.
6. Gribbin B, Pickering EG, Sleight P, Peto R. Effect of age and high blood pressure on baroreflex sensitivity in man. *Circ Res* 1971; **29**:424–431.
7. Kelly R, Hayward C, Avolio A, O Rourke MF. Noninvasive determination of age-related changes in the human arterial pulses. *Circulation* 1989; **80**:1652–1659.
8. De Jong-De Vos van Steenwijk CCE, Wieling W, Johannes JM, Harms MP, Kuis W, Wesseling KH. Incidence and hemodynamic characteristics of near-fainting in healthy 6- to 16-year old subjects. *J Am Coll Cardiol* 1995; **25**:1615–1621.
9. Idema RN, Van den Meiracker AH, Imholz BPM, Man in t Veld AJ, Settels JJ, Ritsema van Eck HJ, Schalekamp MADH. Comparison of Finapres non-invasive beat-to-beat finger blood pressure with intrabrachial artery pressure during and after bicycle ergometry. *J Hypertens* 1989; **7**:S58–S59.
10. Rongen GA, Bos WJW, Lenders JWM, Van Montfrans GA, Van Lier HJJ, Van Goudoever J, Wesseling KH, Thien T. Comparison of intrabrachial and finger blood pressure in healthy elderly volunteers. *Am J Hypertens* 1995; **8**:237–248.
11. Langewouters GJ, Wesseling KH, Goedhard WJA. The static elastin properties of 45 human thoracic and 20 abdominal aortas in vitro and the parameters of a new model. *J Biomechanics* 1984; **17**:425–435.

12. Sleight P. Baroreceptors and Hypertension. In: Persson PB, Kirchheim HR (Eds) Baroreceptor Reflexes. Springer, Berlin 1991, 271–292.
13. Al-Kubati MAA, Fišer B, Siegelová J. Baroreflex sensitivity during psychological stress. *Physiol Res* 1997; **46:**27–33.
14. Pagani M, Lombardi F, Guzzetti S, Rimoldi R, Furlan R, Pizzinelli P, Sandrone G, Malfatto G, Dell'Orto S, Piccaluga E, Turiel M, Baselli G, Cerutti S, Mlliani A. Power spectral analysis of heart rate and arterial pressure variabilities as a marker of sympatho-vagal interaction in man and conscious dog. *Circ Res* 1986; **59:**178–193.
15. Shannon DC, Carley DW, Benson H. Aging of modulation of heart rate. *Am J Physiol* 1987; **253:**H874–H877.
16. Hirsch JA, Bishop B. Respiratory sinus arrhythmia in humans: how breathing pattern modulates heart rate. *Am J Physiol* 1981; **241:**H620–H629.
17. Schwartz JB, Gibb WJ, Tran T. Aging effects on heart rate variability. *J Gerontol* 1991, **46:**M99–M106.
18. Pfeffer MA, Weinberg CR, Cook D, Best JD, Reenan A, Halter JB. Differential changes of autonomic nervous system function with age in man. *Am J Med* 1983, **75:**249–258.
19. Lipsitz LA, Mietus J, Moody GB, Goldberger AL. Spectral characteristics of heart rate variability before and during postural tilt: relations to aging and risk of syncope. *Circulation* 1990; **81:**1803–1810.
20. Veerman DP, Imholz BPM, Wieling W, Karemaker JM, van Montfrans GA. Effects of aging on blood pressure variability in resting conditions. *Hypertension* 1994; **24:**120–130.
21. Saul J, Arai Y, Berger RD, Lilly LS, Colucci WS, Cohen RJ. Assessment of autonomic regulation in chronic congestive heart failure by heart rate spectral analysis. *Am J Cardiol* 1988; **61:**1292–1299.
22. Task force of the European Society of Cardiology and the North American Society for Pacing and Electrophysiology. Heart rate variability. Standards of mesurement, physiological interpretation, and clinical use. *Circulation* 1996; **93:**1043–1065.
23. Smyth HS, Sleight P, Pickering GW. Reflex regulation of arterial pressure during sleep in man. A quantitative method of assessing baroreflex sensitivity. *Circ Res* 1969; **24:**109–121.
24. Penaz J. Photoelectric measurement of blood pressure, volume and flow in the finger. In: Digest 10th Int Conf Med Biol Eng, Dresden 1973, 104.
25. Wesseling KH, de Wit B, Settels JJ, Klawer WH. On the indirect registration of finger blood pressure after Penáz. *Funkt Biol Med* 1982; **1:**245–250.
26. Farrell TG, Odemuyiwa O, Bashir Y, Cripps TR, Malik M, Ward DE, Camm AJ. Prognostic value of baroreflex sensitivity testing after acute myocardial infarction. *Br Heart J* 1992; **67:**129–137.
27. Palmero HA, Caeiro TF, Iosa DJ, Bas J. Baroreceptor reflex sensitivity index derived from phase 4 of the Valsalva maneuver. *Hypertension* 1981; **3** (suppl II):134–137.

28. Smith SA, Stallard TJ, Salih MM, Littler WA. Can sinoaortic baroreceptor heart rate reflex be determined from phase IV of the Valsalva manoeuvre? *Cardiovasc Res* 1987, **21**:422–427.
29. Trimarco B, Volpe M, Ricciardelli B. Valsalva moneuver in the assessment of baroreflex responsiveness in borderline hypertensives. *Cardiology* 1983; **70**:6–14.
30. Al-Kubati MAA. Baroreflex sensitivity during the stimulation of low-pressure receptors. *Scripta Medica* 1996; **69**:191–196.
31. Eckberg DL, Cavanaugh MS, Mark AL, Abboud FM. A simplified neck suction device for activation of carotid baroreceptors. *J Lab Clin Med* 1975; **85**:167–173.
32. Parati G, Casadei R, Groppelli A, Rienzo M, Mancia G. Comparison of finger and intra-arterial blood pressure monitoring at rest and during laboratory testing. *Hypertension* 1989; **13**:647–655.
33. Robbe HWJ, Mulder LJM, Ruddel H, Langewitz WA, Veldman JBP, Mulder G. Assessment of baroreceptor reflex sensitivity by means of spectral analysis. *Hypertension* 1987; **10**:538–543.
34. Fišer B, Honzíková N, Penaz J. Power spectra of spontaneous variations of indirectly recorded blood pressure, heart rate and acral blood flow. *Automedica* 1978; **2**:143–147.
35. Pagani M, Somers V, Furlan R, Dell'Orto S, Conway J, Baselli G, Cerutti S, Sleight P, Malliani A. Changes in autonomic regulation induced by physical training in mild hypertension. *Hypertension* 1988; **12**:600–610.
36. Honzíková N, Fišer B, Honzík J. Noninvasive determination of baroreflex sensitivity in man by means of spectral analysis. *Physiol Res* 1992; **41**:31–37.
37. Elghozi JL, Girard A, Laude D, Valensi P, Weise F. Spectral analysis of short-term variability of blood pressure and pulse interval using finger blood pressure monitoring. *Homeostasis* 1995; **36**:153–159.
38. Bevegard BS, Shepherd JT. Regulation of the circulation during exercise in man. *Physiol Rev* 1967; **47**:178–213.
39. Jose AD, Collison D. The normal range and determinants of the intrinsic heart rate in man. *Cardiovasc Res* 1970; **4**:160–167.
40. Mancia G, Parati G, Pomidossi G, Casadei R, DiRienzo M, Zanchetti A. Arterial baroreflexes and blood pressure and heart rate variabilities in human. *Hypertension* 1986; **8**:147–153.
41. De Ferrari GM, Landolina M, Mantica M, Manfredini R, Schwartz PJ. Baroreflex sensitivity, but not heart rate variability, is reduced in patients with life-threatening ventricular arrhythmias long after myocardial infarction. *Am Heart J* 1995; **130**:473–480.
42. Lombardi F, Sandrone G, Pernpruner S, Sala R, Garimoldi M, Cerutti S, Baselli G, Pagani M, Malliani A. Heart rate variability as an index of sympathovagal interaction after acute myocardial infarction. *Am J Cardiol* 1987; **60**:1239–1245.

43. Honzíková N, Semrád B, Fišer B. Baroreflex sensitivity and heart rate variability in patients after myocardial infarction: effect of thrombolytic therapy. In: Man in t Veld AJ, van Montfrans GA, Langenwouters GJ, Lie KI (Eds) Measurement of heart rate and blood pressure variability in man. Van Zuiden, Amsterdam 1995, 113–117.

CHAPTER TWELVE

CONCOMITANT THERAPY AND AUTONOMIC TESTS

Xavier Copie, Louis Guize and Jean-Yves Le Heuzey

There is an increasing awareness of the role played by the autonomic nervous system in a diversity of cardiovascular diseases.[1,2] Autonomic tests were initially developed to improve our understanding of physiological and pathophysiological mechanisms in humans and animals. They subsequently evolved to become clinical tests that can be used in everyday practice.[3,4]

Simple autonomic tests are used for the diagnosis of autonomic neuropathy, especially diabetic neuropathy.[5] Baroreflex sensitivity was devised in the 1960s by the group of Oxford.[6] It is now an accepted, if not widely used, method for risk stratification after myocardial infarction.[7,8] Heart rate variability was studied as early as 1973 by Sayers,[9] and as a potential risk stratifier after myocardial infarction by Wolf in 1978.[10] However the physiological implications of spectral analysis of heart rate variability were widely recognised after Akselrod's paper in 1981.[11] The importance of heart rate variability for risk stratification after myocardial infarction was established after Kleiger's paper in 1987,[12] and has since gained widespread acceptance.[13–16]

Although the pharmacological environment can easily be controlled in the autonomic laboratory, concomitant therapies may greatly influence the results and significance of autonomic tests in clinical practice. Therefore it is important to understand which

Malik M (Ed) Clinical guide to cardiac autonomic tests, 301–330.

treatments influence the autonomic nervous system, what are the effects of these treatments on the blunt results of autonomic tests, but also if the clinical interpretation of autonomic tests is modified in the presence of various medications. Finally, one should ask if the "improvement" of autonomic tests with specific treatments can be considered as a surrogate proof of the clinical benefits of these treatments.

It is also important to have in mind that a treatment may have a different effect on the autonomic nervous system depending on the baseline conditions (Figure 1). In other words, the effect of a treatment in a patient may be influenced by the activity of the autonomic nervous system itself, and will therefore be different in a normal healthy volunteer, in a patient with uncomplicated hypertension, after myocardial infarction,[17,18] or in a patient with congestive heart failure.[19]

We will review the effects and implications of various medical therapy on "simple" cardiovascular autonomic function tests such as Valsalva manoeuvre or lower body negative pressure, tilt-testing, baroreflex sensitivity, and heart rate variability.

β-BLOCKERS

Among the treatments likely to modify the activity of the autonomic nervous system, β-blockers are probably the most frequently used. Notably, they are important as antihypertensive therapy,[20] in post-myocardial infarction patients,[21] and may be beneficiary in patients with heart failure.[22]

The effects of β-blockers on the autonomic nervous system have been studied extensively, using either autonomic tests such as the Valsalva manoeuvre, lower body negative pressure, cold pressure test, tilt-testing, baroreflex sensitivity, or heart rate variability. As stated previously, the population studied is important to consider when addressing the effects of β-blockers on the autonomic nervous system. It is likely that β-blockers will not have the same effects in patients with high sympathetic tone such as in heart failure, or in young healthy volunteers with a high parasympathetic tone. It is also important to distinguish the effects of β-blockers on heart rate and on the autonomic nervous system itself, as their effect on heart rate is important. Indeed, a change in autonomic testing with β-blockers should be distinguished from a simple decrease in heart rate. If a result of such test can simply be mathematically explained by the decrease in heart rate, it is likely that the result of the test is not clinically relevant.

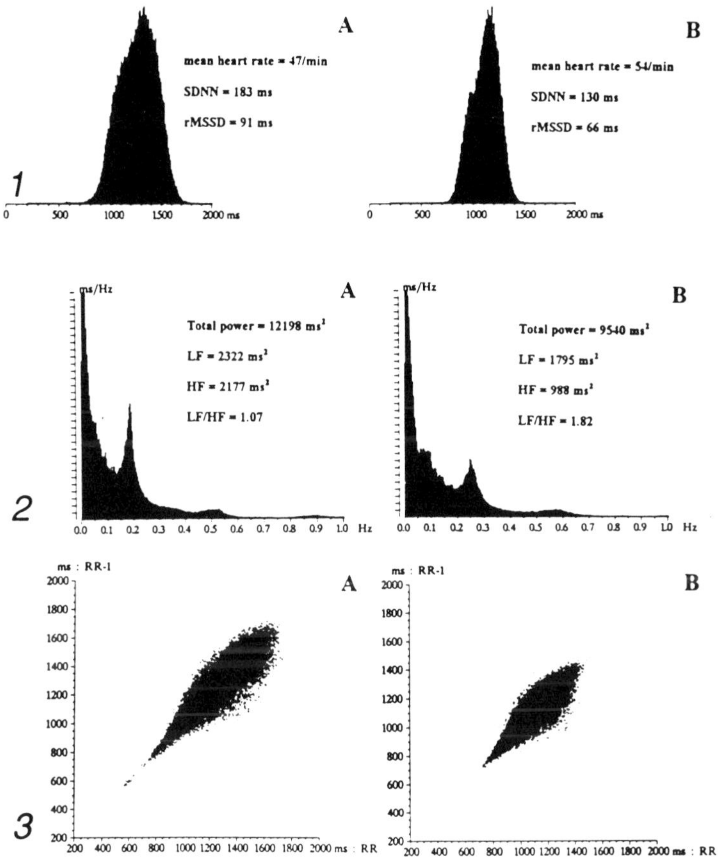

FIGURE 1. Effect of myocardial infarction on heart rate variability. A 48-year old male had a 24-hour Holter ECG recorded during hospitalisation for unstable angina (A). The patient returned ten weeks later with anterior myocardial infarction (B). This case report illustrates the influence of cardiovascular disease on autonomic tests. Although the patient received the same treatments during both Holter recordings (atenolol, aspirin, heparin), heart rate variability assessed in time domain (1), frequency domain (2), or using scatterplots of RR intervals (3) decreased significantly after myocardial infarction.

A few studies have addressed the effects of β-blockers on the results of Valsalva manoeuvre.[23–26] In a group of four subjects, Leon and colleagues failed to elicit significantly different rate responses after Valsalva strain performed during β-blocking therapy as compared to baseline.[23] This result was challenged by Spodick et al in a double-blind study enrolling ten normal healthy volunteers using a single dose of 100 mg alprenolol or placebo[24] As expected, they found a significant reduction in heart rate. The effects of Valsalva manoeuvre on heart rate, left ventricular ejection time, and ejection time index were essentially the same qualitatively with placebo or alprenolol but at a lower parallel set. However, measurements of blood pressure were not reported. Hoshino et al found a significant effect of propranolol on the result of Valsalva manoeuvre in a group of ten healthy volunteers.[25] The pressure overshoot characteristic of Valsalva manoeuvre phase IV was no longer present during β-blocking therapy. In asymptomatic post-infarction patients, Bennett et al found a significant effect of β-blocking therapy on maximum heart rates and maximum systolic pressures seen after the manoeuvre, but this effect was entirely attributable to the differences in resting state.[26] Thus, it is likely that β-blockers do not modify qualitatively heart rate in response to Valsalva manoeuvre, and that most differences seen during β-blocking therapy can be explained by differences in baseline heart rate. The pressure response, and specifically the pressure overshoot seen in phase IV of Valsalva manoeuvre, is likely to be blunted by β-blocking therapy. However, the decrease in pressure overshoot seen in some patients as a result of β-blocking therapy, may be explained solely by a lower baseline blood pressure.

Lower body negative pressure is often used to simulate orthostatic stress.[27] Low levels of lower body negative pressure are responsible for a selective inhibition of cardiopulmonary baroreceptors, and specifically left ventricular receptors. In a study performed in ten healthy volunteers, Ferguson et al demonstrated that acute administration of propranolol significantly attenuated the vasoconstrictor response to lower body negative pressure.[28] This effect of propranolol was not present when neck pressure, cold pressor test, and hand grip were performed. These results support the view that acute administration of β-adrenoreceptor antagonist selectively reduces the activity of left ventricular baroreceptors.

The patterns of cardiovascular response to lower body negative pressure was compared in asymptomatic post-myocardial infarction patients treated with propranolol, atenolol, or placebo by Bennett et al.[26] The slopes for the regressions of RR interval on systolic blood pressure during lower body negative pressure were not significantly different between placebo-treated and β-adrenoreceptor antagonist-treated patients. The vasoconstrictor response to lower body negative pressure as measured by the forearm vascular resistance tended to be higher in patients treated with β-blockers, but was not significantly different. Discrepancies between these two studies may be explained by the type of administration of the treatments (acute vs. chronic administration), and by the populations studied (healthy volunteers vs. post-myocardial infarction patients). It is possible that the negative inotropic effect of β-blockers seen with acute administration is responsible for the decrease in ventricular baroreceptor activity, and is no longer present during chronic administration.

Head-up tilt testing is now largely accepted for the diagnosis and management of patients with neurocardiogenic syncope. The effects of β-blockers on the outcome of tilt-testing have been investigated in several studies.[29–32] In most studies, β-blockers have been shown to improve the outcome of tilt testing and symptoms in a majority of patients. For Sra and co-workers, the effect of intravenous esmolol during tilt testing can predict the efficacy of long term oral β-adrenergic blocker therapy.[29] In one study, the efficacy of metoprolol was compared with that of verapamil.[32] Metoprolol was significantly more efficacious than verapamil in managing neurocardiogenic syncope.

A bradycardia in response to an increase in arterial pressure is one of the most consistent reflex mechanisms of cardiovascular regulation. Concurrent with the bradycardia, a decrease in peripheral vascular resistance occurs that complements the cardiac response. This important baroreflex is often studied using the "Oxford technique" first described in 1969.[6] Baroreflex sensitivity is known to be impaired in patients with hypertension,[33] after myocardial infarction,[17] and in congestive heart failure.[34] Improvement of baroreflex sensitivity using medical therapy is usually considered beneficial for the patients.[35]

Although there are some controversies regarding the effects of β-adrenoreceptor antagonists on baroreflex sensitivity, most studies have demonstrated an improvement of baroreflex with the use of β-blockers. Pickering et al demonstrated an increase in baroreflex sensitivity in healthy volunteers after an acute administration of i.v. propranolol.[36] This improvement in baroreflex sensitivity was evidenced at rest, but was no longer present during exercise. A similar result was found in a study by Eckberg et al using the graded neck suction technique in supine and standing healthy volunteers.[37] Propranolol significantly improved baroreflex sensitivity both in the supine and standing positions.

Several studies have addressed the effects of β-receptor antagonists on baroreflex sensitivity in hypertensive patients.[38–43] All studies used the Oxford technique to measure baroreflex sensitivity. Experiments were performed after acute treatment with β-blockers,[39,42] or chronic treatment,[40] or both.[38,41,43] These papers yield apparently conflicting results. Although most authors agree to say that the antihypertensive effect of β-receptor antagonists is not related to changes in baroreflex sensitivity, the effect of this treatment on baroreflex sensitivity is variously appreciated. Some authors found no change in baroreflex sensivity,[38,42] while others demonstrated an increase in baroreflex sensitivity,[39,43] or a different effect according to the type of patient.[40,41] From these studies, it appears that β-blockers may have an effect on baroreflex sensitivity in the less severe patients, namely young patients with borderline hypertension, but may not increase baroreflex sensitivity in older patients with a more severe hypertension.

Heart rate variability is now the test most often used to assess the effect of the autonomic nervous system on cardiovascular regulation. Heart rate variability is also a potent predictor of subsequent mortality after myocardial infarction,[44] and probably in patients with heart failure.[45] Because heart rate and heart rate variability are related, the effects of β-blockers are difficult to study.[46,47] Nevertheless, the effects of β-blockers on heart rate variability have been studied extensively in different experimental and clinical conditions.[11,48–62]

In the first experiments using heart rate variability measured in the frequency domain during sort-term recordings in control conditions in healthy volunteers or conscious dogs, it was shown that acute β-adrenergic receptor blockade increases the high frequency component of heart rate variability.[11,48–50] The effects of acute β-blockade on the low frequency component of heart rate variability at rest is less clear. Some authors found a non significant decrease in the low frequency peak,[49] while others did not.[50] However, concordant results were obtained after activation of the sympathetic system by tilt or standing. In these conditions, acute β-blockade reduces the low frequency peak of heart rate variability.[49,50] Consequently, in most experiments, acute β-blockade is associated with a decrease in the low to high frequency ratio. However, this result may depend on the population studied. In healthy volunteers, with a high level of parasympathetic tone at rest and a low sympathetic tone, the low to high frequency ratio may not be significantly increased by β-blockade. This was the result of the study by Ahmed and co-workers[57] who found no change in heart rate, nor in heart rate variability after intravenous administration of propranolol in healthy volunteers.

In the study by Pagani and co-workers, chronic β-blockade was responsible for more important changes in heart rate variability than acute β-blockade.[50] It is then important to take into account the duration of treatment with β-adrenoreceptor antagonists.

In a placebo-controlled study comparing the effects of atenolol and diltiazem on heart rate variability in healthy volunteers, Cook et al found a significant decrease in heart rate with atenolol, together with an increase in beat-to-beat heart rate variability (pNN50, RMSSD).[53] There was no change in global heart rate variability as assessed by SDNN. The decrease in the night/day difference in average NN interval was responsible for this lack of increase in SDNN despite an increase in beat-to-beat heart rate variability. There was also an increase in the low frequency power. From this experiment, one can conclude first that the low frequency power measured in healthy volunteers during a 24-hour Holter recording is not predominantly a measure of sympathetic tone, and second that the most prominent effect of β-blockade in healthy volunteers is to increase the measures of heart rate variability dependent on parasympathetic tone. Using a specific approach to measure heart rate variability, Coumel et al found a decrease in all components of heart rate variability (short, medium, and long oscillations) in healthy volunteers treated with acebutolol, a β-blocker with β-adrenergic intrinsic activity.[52] This change in heart rate variability was associated with a decrease in the long to short oscillations ratio (which can be assimilated to the low to high frequency ratio of the power spectra).

In patients with coronary artery disease and no recent myocardial infarction, Niemela et al compared the effects of metoprolol, atenolol, and placebo on heart rate variability measured during 24-hour Holter recordings.[56] Both treatments significantly increased all components of heart rate variability, excepted the low frequency power. The differences between this study and the study by Cook et al[53] may be related to the baseline activity of the autonomic nervous system. Baseline sympathetic tone was probably higher in patients with coronary artery disease than in healthy volunteers. Therefore modifications in heart rate variability with β-adrenergic blockade may have been easier to evidence in coronary artery disease. Also, it is possible that the low frequency power of heart rate variability measured during 24-hour Holter recordings has not the same signification in healthy volunteers and in patients with coronary artery disease. This low frequency peak may be in part if not largely mediated by parasympathetic tone in healthy volunteers, especially at night during low breathing rates.

In patients with a recent myocardial infarction, care should be taken to include a control group in all studies as heart rate variability changes rapidly in the weeks,[18] and months following myocardial infarction.[63] However, in an uncontrolled study, Bekheit et al found a significant increase in mean NN, with no change in SDNN, and a decrease in the low frequency power of heart rate variability one week after treatment with metoprolol.[51] In a placebo-controlled study, Molgaard et al found a significant decrease in heart rate, with no change in SDNN, but an increase in beat-to-beat heart rate variability as assessed by pNN50, with metoprolol in patients with a recent myocardial infarction.[54] Kontopoulos et al found an increase in global as well as beat-to-beat heart rate variability in patients with a recent myocardial infarction.[59] These two studies are difficult to compare, especially as the study by Kontopoulos et al do not provide the magnitude of modifications in heart rate in the patients receiving metoprolol.[59] Using a different approach to study heart rate variability, Keeley et al found no change in scatterplot measurements in patients receiving metoprolol after myocardial infarction.[62] However, they found an increase in beat-to-beat heart rate variability with no change in global heart rate variability as assessed by SDNN and SDANN. Thus, in patient with a recent myocardial infarction, the most consistent effect of beta-adrenergic blockade is a decrease in heart rate, and an increase in beat-to-beat heart rate variability. Changes in global heart rate variability (SDNN) are in most cases minimal.

These effects of β-blockers after myocardial infarction are important to consider because heart rate variability is now accepted as an important risk stratifier for these patients. Although, Kleiger et al did not find any effect of a treatment with β-blockers on the predictive value of SDNN,[12] this question has not been studied in detail. The other large database from which numerous studies on heart rate variability in post-myocardial infarction patients have been performed is from St George's hospital in London.[13–15] In this database all patients are studied after withdrawal of β-blocking therapy. A recent study from the Gissi trialists in Italy provided interesting results on heart rate variability following myocardial infarction,[64] but unfortunately the authors did not analyse the effect of treatment with β-blocking therapy on the prognostic value of heart rate variability.

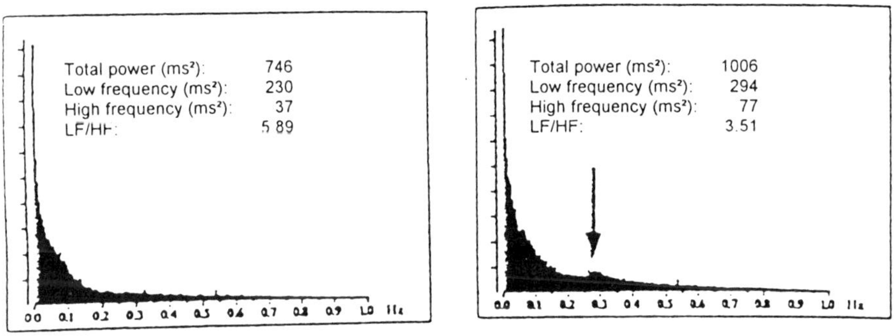

FIGURE 2. Effect of bisoprolol on heart rate variability in a patient with heart failure. Left panel, day power spectrum at baseline. Right panel, day power spectrum after two months of treatment with bisoprolol. Bisoprolol increased the high frequency power and decreased the LF to HF ratio (From Pousset et al[61]).

Given the lack of specific study on the subject, one has to speculate the effect of concomitant treatment on the prognostic value of heart rate variability in post-myocardial infarction patients. Two facts can be taken into account. The first one is an experimental study by Adamson and co-workers.[65] In this model, a myocardial infarction is created in dogs that are subsequently exercised with induction of myocardial ischemia at peak exercise. This results in ventricular fibrillation in more than 50% of the dogs. Heart rate variability measured before myocardial infarction was not different between animals later shown to be susceptible or resistant to ventricular fibrillation. However, treatment with β-blockers unmasked an important difference in heart rate variability between resistant and susceptible dogs. Animals who will not experience ischaemic ventricular fibrillation do improve their heart rate variability with β-blockers significantly more than dogs susceptible to ventricular fibrillation. Therefore, in this model, treatment with β-blocking agents does improve the predictive value of heart rate variability. The second one is the result of the studies in human on the effects of β-blockade on heart rate variability after myocardial infarction. As previously stated, β-blockers either improve heart rate variability, and especially beat-to-beat heart rate variability, or do not have a significant effect on heart rate variability. As patients treated with β-blockers after myocardial infarction have a better outcome than others,[66] it can be speculated that β-blockers either improve or have no significant effect on the predictive value of heart rate variability after myocardial infarction.

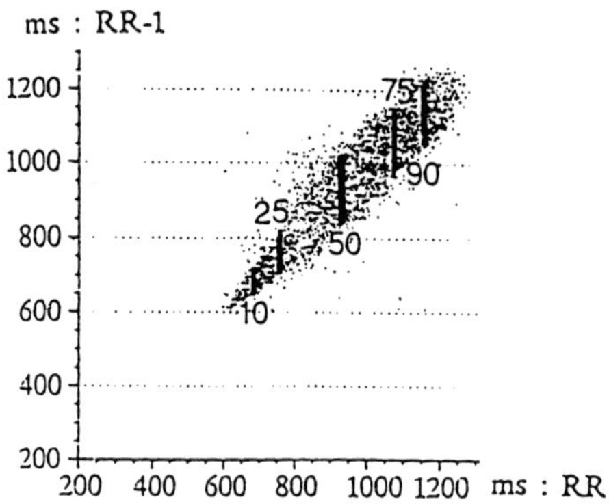

FIGURE 3. Method for measuring beat-to-beat heart rate variability at selected heart rates using scatterplots of RR intervals. Scatterplot dispersion is measured vertically at 10th, 25th, 50th, 75th, and 90th percentiles of total RR-interval distribution. (From Copie et al,[60] with permission).

The effects of β-blockers on heart rate variability have also been studied in patients with heart failure.[52,60,61] In the study by Coumel et al, β-blockade had a significantly more important effect on heart rate in patients with heart failure, than in healthy volunteers or patients with left ventricular hypertrophy and no heart failure. However, the effect of acebutolol on the long to short oscillations ratio was not significantly different in the three groups of patients.[52] In an ancillary study of the CIBIS trial, Pousset et al, and Copie et al, examined the effects of bisoprolol on heart rate variability in patients with severe heart failure (NYHA class III or IV).[60,61] The two studies used different approaches to measure heart rate variability. In the study by Pousset et al, conventional time-domain and frequency-domain measures of heart rate variability were used. Bisoprolol had a significant effect on heart rate, and increased beat-to-beat heart rate variability (RMSSD and pNN50) (Figure 2). Bisoprolol had no significant effect on global heart rate variability as assessed by SDNN. Using scatterplots of RR intervals,[67] Copie et al found an increase in beat-to-beat variability which was significant at the lowest heart rates (Figures 3 and 4).

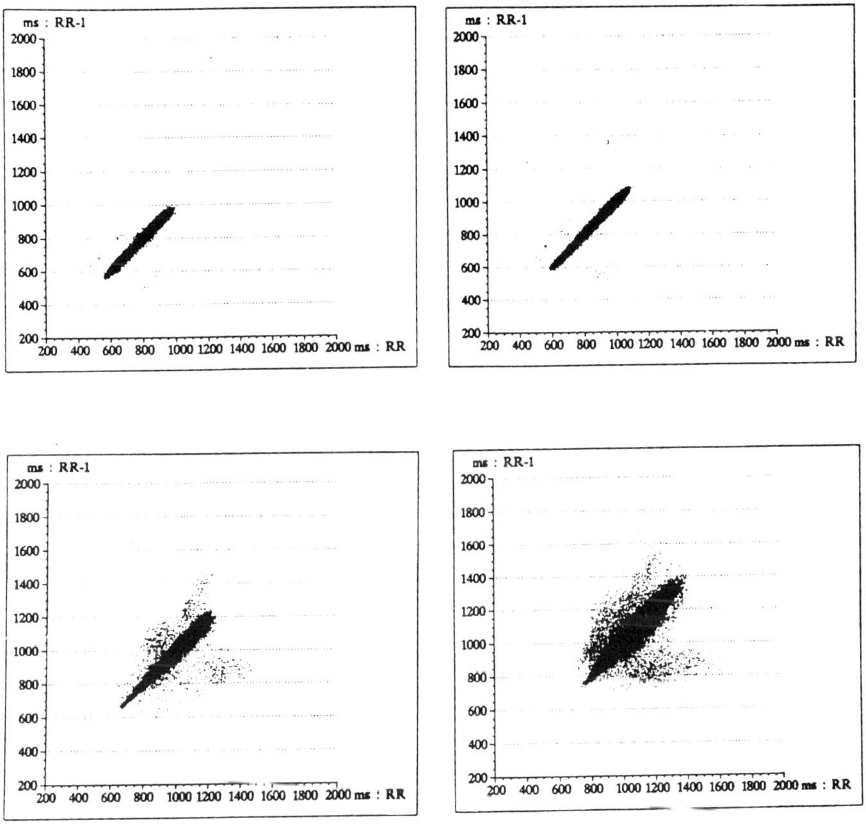

FIGURE 4. Effect of bisoprolol on heart rate variability in patients with heart failure. Scatterplot evolution in a patient who received placebo *(top)*, and a patient who received bisoprolol *(bottom)*. (From Copie et al,[60] with permission.)

In patients with frequent ventricular arrhythmias treated with sotalol, Hohnloser et al found an increase in beat-to-beat heart rate variability (RMSSD, pNN50, and high frequency power spectrum), together with an increase in SDNN and decrease in heart rate.[55] Contrary to these findings, Pitzalis et al found a decrease in heart rate, but no change in SDNN in patients with frequent ventricular premature beats treated with nadolol.[58] However, this study did not evaluate the effect of nadolol on beat-to-beat heart rate variability.

From all these studies on the effects of β-blockade on heart rate variability, it can be concluded that their most constant effect is a decrease in heart rate, and an increase in beat-to-beat heart rate variability, probably in relation with an increased parasympathetic tone. The effects of β-blockers on global heart rate variability as measured by SDNN is inconstant and may be dependent on the experimental conditions.

An increase in heart rate variability during treatment with a β-blocker may be considered beneficial for a patient, and is certainly encouraging. However, an increase in heart rate variability cannot be considered as a surrogate end-point, and ultimately the effect of the treatment on survival remains the only unquestionable end-point.

ANGIOTENSIN CONVERTING ENZYME INHIBITORS

The ACE inhibitors are widely used for the treatment of hypertension, congestive heart failure,[68] and in post-myocardial infarction patients.[69] Although the benefits of these treatments are largely related to their effects on cardiovascular haemodynamics,[70] their ability to modify the autonomic balance has been verified in several studies.[59,71–86]

Improvement of baroreflex sensitivity with ACE inhibition has been demonstrated in different studies.[72,74,77,78] However, a study found no improvement in baroreflex sensitivity with ACE inhibition.[76] The improvement of baroreflex sensitivity with ACE inhibitors is present in patients likely to have a depressed baroreflex sensitivity such as patients with heart failure,[74,77,83] or after myocardial infarction,[78] but may not be present in healthy volunteers.[76] For Osterziehl et al, improvement of baroreflex sensitivity in patients with heart failure was only present in the group of patients who showed an improvement in haemodynamics after treatment with captopril or enalapril.[77] The mechanisms responsible for an improvement in baroreflex sensitivity with ACE inhibitors are still unclear. A possible explanation is a removal of the sympatho-excitatory effect of angiotensin II, which has been shown to decrease the baroreceptor heart rate reflex.[87]

Results from direct muscle sympathetic nerve activity recordings also support the view that ACE inhibitors decrease sympathetic neural drive. This was demonstrated both in healthy volunteers[85] and in patients with heart failure.[83,86] The sympatholytic action of ACE inhibition has been recently demonstrated to be at a postsynaptic level.[84]

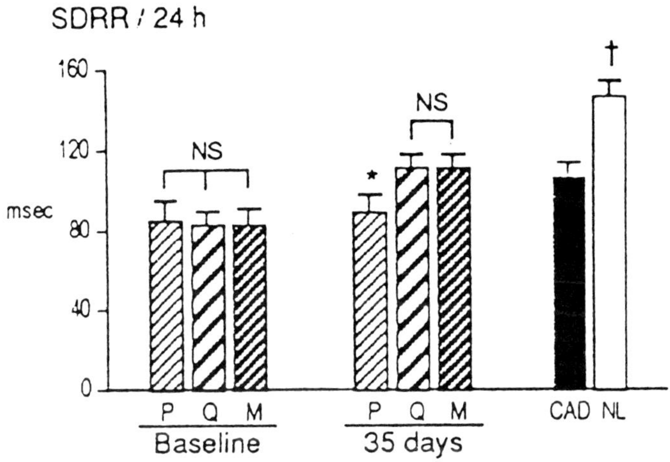

FIGURE 5. Comparison of the effect of metoprolol (M) and quinapril (Q) on heart rate variability in post-myocardial infarction patients. After 35 days of treatment, metoprolol and quinapril significantly increased heart rate variability as compared to placebo (P). There was no significant difference in heart rate variability between patients receiving metoprolol and quinapril. These patients had a heart rate variability similar to patients with stable coronary artery disease (CAD). NL: normal volunteers, $*p<0.05$ compared with quinapril and metoprolol, $+p<0.05$ compared with stable coronary artery disease (From Kontopoulos et al[59]).

The study of the effects of ACE inhibitors on heart rate variability is an other approach to understand their effect on the autonomic nervous system. In their principal paper, Akselrod and co-workers demonstrated a marked effect of converting enzyme inhibition on the power spectrum of heart rate fluctuations in conscious dogs.[11] Converting enzyme inhibition had little or no effect on heart rate or blood pressure, but was associated with a marked increase in the area under the very low frequency peak of heart rate fluctuations. This effect of ACE inhibitors was subsequently demonstrated in patients with a recent acute myocardial infarction[59,82] (Figure 5), and in patients with heart failure.[79,80] However, one study did not find a significant effect of enalapril on heart rate variability in healthy volunteers.[81] There is a controversy regarding the effect of converting enzyme inhibition on beat-to-beat heart rate variability estimated either by the high frequency peak of the power spectral density or the beat-to-beat indices in time-domain analysis (RMSSD, or pNN50). In other words, there are conflicting results on the effects of ACE inhibitors on vagal tone

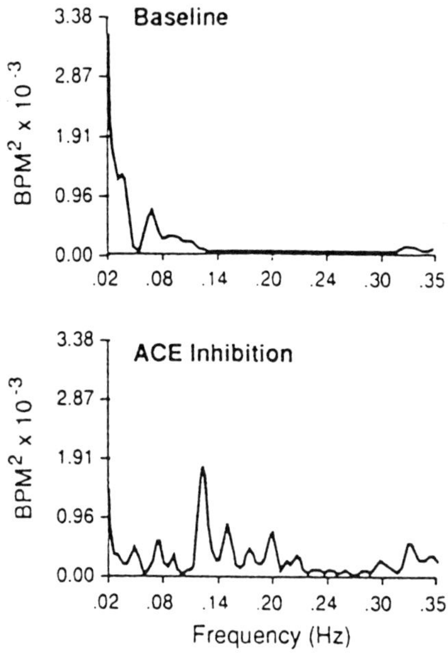

Figure 6. Power spectral density of heart rate variability recorded in a patient with congestive heart failure at baseline study *(top)*, and after 12 weeks of therapy with zofenopril *(bottom)*. There is marked augmentation of heart rate variability after therapy with angiotensin-converting enzyme inhibitor (From Binkley et al[80]).

as assessed by beat-to-beat heart rate variability. Explanations for these conflicting results are not straightforward. As previously explained, angiotensin II inhibits the parasympathetic outflow through a central mechanism.[87] Therefore blockade of the renin-angiotensin system is expected to increase vagal tone. However no change in markers of vagal activity was found in different studies.[11,81,82] An increase in the high frequency heart rate variability was found by Binkley and co-workers in heart failure patients receiving zofenopril[80] (Figure 6). A careful monitoring substantiated the fact that changes in respiratory rates were not responsible for this increase in high frequency heart rate variability.

There is no study on the effect of treatment with ACE inhibitors on the prognostic value of heart rate variability, specifically after myocardial infarction. As these treatments have clearly been shown to improve heart rate variability and are also known to improve survival, our understanding is that heart rate variability should be interpreted in the same way in patients after myocardial infarction whether they are treated or not with ACE inhibitors. However, there is no definite evidence that the same cut-off value for heart rate variability can be used in patients with and without ACE inhibitors after myocardial infarction.

DIGITALIS

Although a recent study showed that digoxin does not improve survival in patients with heart failure, digitalis are widely used in this clinical setting.[88] They remain one of the only orally available treatment with positive inotropic properties which does not decrease survival. The clinical benefits observed with digitalis in patients with heart failure can be attributed to the hemodynamic improvements produced by these agents. However, the effects of digitalis on the autonomic nervous system may account for an important part of their beneficial clinical effects.

The modifications of the control of the heart by the autonomic nervous system induced by digitalis treatment have been evaluated in numerous studies.[89–100] The effects of digitalis on the autonomic nervous system have been studied either by direct recording of cardiac vagal fibres during digitalis administration,[101] or through the effects of digitalis on sympathetic neural recordings,[99] baroreflex sensitivity,[94,96,97] or heart rate variability.[81,100]

Ferguson and co-workers demonstrated a different effect of digitalis on muscle sympathetic nerve activity in patients with heart failure and healthy volunteers.[99] In this study, intravenous lanatoside C had no effect on muscle sympathetic nerve activity in healthy volunteers but evoked an early, profound, and sustained decrease in muscle sympathetic nerve activity in patients with heart failure. This effect of digitalis in patients with heart failure was not found with dobutamine despite comparable inotropic effect. Therefore digitalis glycosides are capable of a marked sympatho-inhibitory action in patients with heart failure. Part of this sympatho-inhibitory action may be due to mechanical activation of afferent arterial baroreflex mechanisms, but an increase in baroreflex sensitivity is probably the main mechanism.

A potentiation of baroreceptor reflexes by digitalis was indeed demonstrated by Ferrari and co-workers.[96,97] Therapeutic doses of lanatoside C in healthy volunteers or patients with uncomplicated essential hypertension increased the reflex effects of baroreceptor stimulation but had no significant effect on baroreceptor deactivation.[97] This differ from animal studies in which large doses of digitalis are responsible for an enhanced heamodynamic response after common carotid occlusion.[91] This selective effect of digitalis studied at therapeutic doses in humans may be beneficial because excessive tachycardia is avoided during reductions in blood pressure.

Using heart rate variability to study the effect of digoxin on the autonomic nervous system, Krum et al,[100] and Kaufman et al,[81] demonstrated an increase in indexes of vagal modulation of heart rate in healthy volunteers[81] and patients with heart failure.[100] In the study by Kaufman et al performed in healthy volunteers, there was no significant change in heart rate, but a significant increase in all spectral components of heart rate variability. This change was markedly important for the high frequency component of heart rate variability, which was corroborated by similar increases in beat-to-beat time domain indexes of heart rate variability (i.e. RMSSD and pNN50). This effect of digoxin on heart rate variability was compared with that of enalapril or placebo. Contrary to digoxin, enalapril had no effects on heart rate variability in these healthy volunteers (Figure 7). In patients with heart failure treated with digoxin, Krum and co-workers found a significant decrease in mean RR interval and an increase in all spectral components of heart rate variability. In the same study, digoxin was also responsible for a decrease in plasma norepinephrine levels. Interestingly the increase in plasma norepinephrine did not correlate with a decrease in the low frequency spectral component of heart rate variability, a measure often considered to reflect sympathetic tone. This study illustrates the fact that the low frequency peak derived from 24-hour Holter recordings is not a direct measure of sympathetic tone, but is in part vagally mediated and is modified by baroreflex influences on heart rate variability.

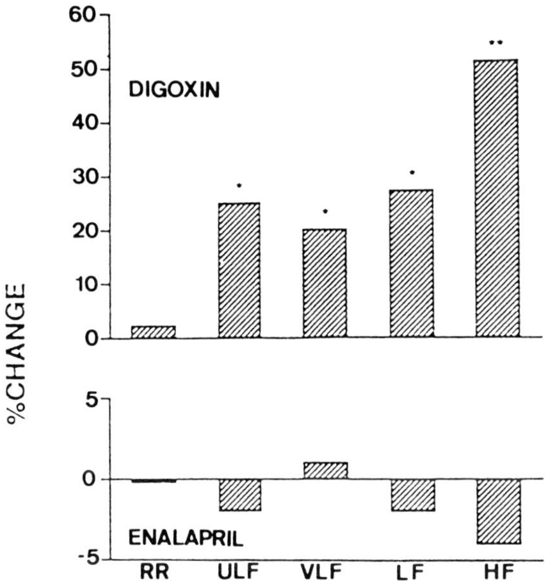

FIGURE 7. Comparative effect of digoxin and enalapril on heart rate and heart rate variability in healthy volunteers. Digoxin was associated with a significant increase in heart rate variability, while enalapril had remarkedly no effect on heart rate variability. RR – mean RR interval, ULF – ultra low frequency, VLF – very low frequency, LF – low frequency, HF – high frequency, *$p<0.01$, **$p<0.001$ (From Kaufman et al[81]).

Some questions concerning autonomic nervous system and digitalis remain poorly understood. The first one is the effect of a treatment with digitalis on the predictive value of baroreflex sensitivity or heart rate variability, specifically after myocardial infarction or in heart failure.[45] As stated before there is the same interrogation concerning β-blockers and ACE inhibitors. Is heart rate variability still a good predictor of subsequent mortality and arrhythmic events, and should we use the same dichotomy limits in patients receiving digitalis or not ? Although the question was partially answered by Kleiger and co-workers in their principal study, we believe that it has not been fully answered.[12] The second question is the apparent paradox between an increase in heart rate variability and the lack of effect on total mortality in patients with heart failure. However, this paradox was not demonstrated in the same patients and it is possible to speculate that patients with an increase in heart rate variability during digitalis treatment will benefit from it, whereas others would not. This remains purely speculative and should be demonstrated in a prospective study.

ANTIARRHYTHMIC THERAPY

Since the results of the CAST study,[102] antiarrhythmic therapy is much less popular for the treatment of patients with heart disease. Recently completed trials with amiodarone also cast doubt on the efficiency of antiarrhythmic therapy in various clinical settings.[103,104] Studies on the effects of antiarrhythmic therapy on the autonomic nervous system may help understand their clinical effects.[55,105–109]

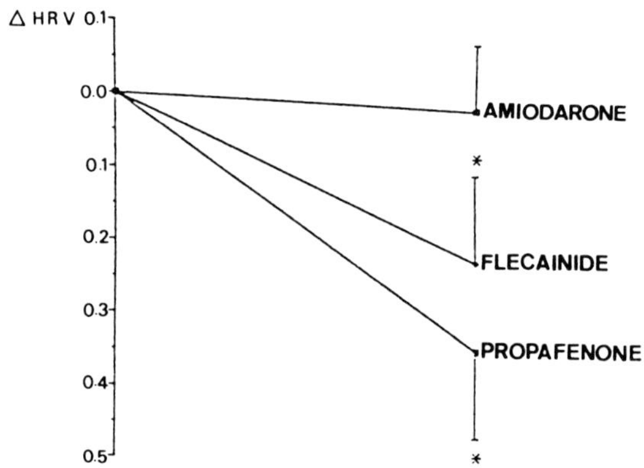

FIGURE 8: Effect of antiarrhythmic therapy on heart rate variability (From Zuanetti et al[107]).

As stated previously, Hohnloser and co-workers demonstrated an increase in heart rate variability in patients with ventricular arrhythmias treated with sotalol,[55] a β-blocker with class III antiarrhythmic activity. Sotalol increased global heart rate variability as assessed by SDNN, and beat-to-beat heart rate variability as assessed by RMSSD and pNN50.

In an invasive electrophysiological study in the basal state and after autonomic blockade with propranolol and atropine, Alboni and co-workers demonstrated a possible vagolytic effect of oral flecainide (200 to 250 mg daily).[105]

In a study comparing the effects of three antiarrhythmic drugs (amiodarone, flecainide, and propafenone) on heart rate variability, Zuanetti and co-workers found a decrease in beat-to-beat heart rate variability, as assessed by NN50, with flecainide and propafenone, whereas there was no change in heart rate variability with amiodarone[107] (Figure 8). A decrease in heart rate variability with propafenone was also evidenced by Lombardi and co-workers.[108] However, these authors also demonstrated an increase in the high-frequency component of heart rate variability together with a decrease in the low-frequency component. These changes in the spectral components of heart rate variability with propafenone were consistent with a β-blocking effect of propafenone.

These results may explain some of the proarrhythmic effects seen with Class I agents, and the relative absence of such effects with amiodarone.

CALCIUM CHANNEL BLOCKERS

Although the safety and efficacy of calcium channel blockers are still controversial, they are used in a diversity of cardiac diseases.[110] Their effects on autonomic nervous system have been evaluated in a limited number of studies.[111–113]

Dihydropyridines such as nifedipine act prominently as arterial vasodilators. Therefore the acute antihypertensive response to such agents is accompanied by reflexly mediated tachycardia, which is probably the effect of increased sympathetic tone.[112] In a study comparing the acute and chronic effects of nifedipine therapy in patients with mild to moderate essential hypertension, Kiowski and co-workers found no change in baroreflex sensitivity both in the acute and chronic settings.[113]

In healthy volunteers, Cook and co-workers did not find any significant effect of diltiazem (240 mg per day) on heart rate and heart rate variability measured from 24-hour Holter recordings.[53] This result contrasted with the effect of atenolol in the same subjects (see above). In patients with a recent myocardial infarction, Bekheit and co-workers found no significant effect of nifedipine on low-frequency energy during tilt, a parameter related to sympathetic tone. On the contrary, both diltiazem and metoprolol reduced the low frequency energy during tilt.[51]

LOW DOSE ANTIMUSCARINIC AGENTS

Defective parasympathetic control of the heart may in itself trigger ventricular arrhythmias and sudden arrhythmic death.[114] Therefore there is a potential interest in a treatment that would selectively increase vagal activity.[115,116] Although antimuscarinic agents are normally used for their anticholinergic properties, at very low doses they can induce a paradoxical increase in cardiac vagal activity. This paradoxical increase in vagal activity is thought to result from a central action on inhibitory fibres projecting from the ventrolateral respiratory reticular formation and making synaptic connections with vagal motoneurons.[117] Atropine, by blocking these inhibitory mechanisms, may increase the activity of vagal motoneurons. In addition to central actions, peripheral mechanisms, such as the presynaptic muscarinic modulation of acetylcholine release may contribute to the effect of low doses of antimuscarinic agents.[118] This is suggested by the evidence that pirenzepine, which does not appear to have central actions, increases vagal activity.

Several studies have addressed the effects of treatment with low doses antimuscarinic agents on indices of vagal activity in experimental animal models,[119] healthy volunteers,[120–123] in post-myocardial infarction patients,[124–128] and in patients with heart failure.[129] In all studies, indices of vagal activity as assessed by heart rate variability or reflex vagal tone as assessed by baroreflex sensitivity, increased with low doses of antimuscarinic agents. This effect was evidenced in most populations studied. Therefore, these treatments are likely to improve the autonomic environment in patients at risk of sudden death.

However, these treatments have not been tested at a large scale in patients with a high arrhythmic risk. To our knowledge, they represent a unique type of treatment for which we know their effect on heart rate variability or baroreflex sensitivity without any knowledge of their effect on mortality or arrhythmic risk in man. However, caution must be applied before extrapolating the results of these studies to the possible prevention of arrhythmic death. Such caution is even more warranted after the results of the experimental study by Hull and co-workers.[119] In this study, treatment with scopolamine significantly increased heart rate variability at rest but prevented ventricular fibrillation in only one out of seven high risk dogs. Clinicians should be aware of these results to avoid extrapolating the effect of a treatment on autonomic indices as a proof of its efficacy. However, it is also possible that the model of ischaemic ventricular fibrillation in dogs does not reflect perfectly the mechanism of sudden death in humans, and that low doses of antimuscarinic agents will prove useful in patients at risk of sudden death.

OTHER AGENTS

Other treatments or therapeutic interventions have been shown to influence the control of the heart by the autonomic nervous system. Among these, clonidine has been shown to increase baroreflex sensitivity in rabbits, but also in man.[130] Nitrates activate the sympathetic nervous system, a result in marked contrast with the effects of ACE inhibitors.[85] Pedretti and co-workers found a higher heart rate variability in patients who underwent thrombolysis within 6 hours of the onset of myocardial infarction, than in patients who did not.[131] Successful coronary angioplasty also increased heart rate variability in patients with a reversible abnormal regional wall motion.[132]

The effects of non cardiovascular drugs on autonomic tests have also been studied extensively, but would be beyond the scope of this review. Notably, treatments used in psychatric disorders such as imipraminc are known to decrease heart rate variability.[133]

CONCLUSION

Autonomic tests are useful to evaluate the effects of a treatment on the control of the cardiovascular system by the autonomic nervous system. The activation of the sympathetic nervous system is likely to be deleterious and probably indicates that a treatment will not improve survival and may even increase the risk of sudden death. However, a decrease in sympathetic tone and/or increase in parasympathetic tone cannot be automatically translated in a beneficial effect of the treatment.

REFERENCES

1. Barron HV, Lesh MD. Autonomic nervous system and sudden cardiac death. *J Am Coll Cardiol* 1996; **27**:1053–1060.
2. Packer M. The neurohormonal hypothesis: a theory to explain the mechanism of disease progression in heart failure. *J Am Coll Cardiol* 1992; **20**:248–254.
3. van Ravenswaaij-Arts CMA, Kollée LAA, Hopman JCW, Stœlinga GBA, van Geijn HP. Heart rate variability. *Ann Int Med* 1993; **118**:436–447.
4. Le Heuzey JY. La variabilité sinusale: intérêt en rythmologie. *Arch Mal Cœur* 1992; **85**:37–43.
5. Ewing DJ, Martyn CN, Young RJ, Clarke BF. The value of cardiovascular autonomic function tests: 10 years experience in diabetes. *Diabetes Care* 1985; **8**:491–498.
6. Smyth HS, Sleight P, Pickering GW. Reflex regulation of arterial pressure during sleep in man: a quantitative method of assessing baroreflex sensitivity. *Circ Res* 1969; **24**:109–121.
7. Billman GE, Schwartz PJ, Stone HL. Baroreceptor reflex control of heart rate: a predictor of sudden cardiac death. *Circulation* 1982; **66**:874–880.
8. Farrell TG, Paul V, Cripps TR, Malik M, Bennett ED, Ward D, Camm AJ. Baroreflex sensitivity and electrophysiological correlates in patients after acute myocardial infarction. *Circulation* 1991; **83**:945–952.
9. Sayers B Mc A. Analysis of heart rate variability. *Ergonomics* 1973; **16**:85–97.
10. Wolf MM, Varigos GA, Hunt D, Sloman JG. Sinus arrhythmia in acute myocardial infarction. *Med J Aust* 1978; **2**:52–53.
11. Akselrod S, Gordon D, Ubel FA, Shannon DC, Barger AC, Cohen RJ. Power spectrum analysis of heart rate fluctuations: a quantitative probe of beat-to-beat cardiovascular control. *Science* 1981; **213**:220–222.
12. Kleiger RE, Miller JP, Bigger JT Jr, Moss AJ, and the Multicenter Post-Infarction Research Group. Decreased heart rate variability and its association with increased mortality after acute myocardial infarction. *Am J Cardiol* 1987; **59**:256–262.
13. Farrel TG, Bashir Y, Cripps T, Malik M, Poloniecki J, Bennett ED, Ward DE, Camm AJ. Risk stratification for arrhythmic events in postinfarction patients based on heart rate variability, ambulatory electrocardiographic variables and the signal-averaged electrocardiogram. *J Am Coll Cardiol* 1991; **18**:687–697.
14. Copie X, Hnatkova K, Staunton A, Lü Fei, Camm AJ, Malik M. Predictive power of increased heart rate versus depressed left ventricular ejection fraction and heart rate variability for risk stratification after myocardial infarction. A two-year follow-up study. *J Am Coll Cardiol* 1996; **27**:270–276.
15. Copie X, Hnatkova K, Lü Fei, Staunton A, Camm AJ, Malik M. Gender specificities in risk stratification after myocardial infarction. *Ann Noninvas Electrocardiol* 1997; **2**:59–68.

16. Casolo GC, Stroder P, Signorini C, Calzolari F, Zucchini M, Balli E, Sulla A, Lazzerini S. Heart rate variability during the acute phase of myocardial infarction. *Circulation* 1992; **85**:2073–2079.
17. Schwartz PJ, Zaza A, Pala M, Locati E, Beria G, Zanchetti A. Baroreflex sensitivity and its evolution during the first year after myocardial infarction. *J Am Coll Cardiol* 1988; **12**:629–636.
18. Copie X, Le Heuzey JY, Iliou MC, Pousset F, Lavergne T, Guize L. Evolution de la variabilité de la fréquence cardiaque après infarctus du myocarde. Apport du diagramme de Poincaré. *Arch Mal Cœur* 1995,88:1621–1626.
19. Sharpey-Schafer EP. Effects of Valsalva's manœuvre on the normal and failing circulation. *Br Med J* 1955; **1**:693–695.
20. Prichard BNC, Gillam MS. Propranolol in hypertension. *Am J Cardiol* 1966; 387–391.
21. ISIS-1 (First International Study of Infarct Survival) Collaborative Group. Randomised trial of intravenous atenolol among 16027 cases of suspected acute myocardial infarction: ISIS-1. *Lancet* 1988; **i**:921–923.
22. CIBIS Investigators and Committees. A randomized trial of β-blockade in heart failure. The Cardiac Insufficiency Bisoprolol Study (CIBIS). *Circulation* 1994; **90**:1765–1573.
23. Leon DF, Shaver JA, Leonard JJ. Reflex heart rate control in man. *Am Heart J* 1970; **80**:729–739.
24. Spodick DH, Meyer MB, Quarry-Pigott VM. Effect of β-adrenergic blockade on beat-to-beat response to Valsalva maneuver. *Br Heart J* 1974; **36**:1082–1086.
25. Hoshino PK, Blaustein AS, Gaasch WH, Johnson A. Effect of propranolol on left ventricular response to the Valsalva maneuver in normal subjects. *Am J Cardiol* 1988; **61**:400–404.
26. Bennett T, Wilcox RG, Hampton JR. Cardiovascular reflexes in patients after myocardial infarction. Effects of long-term treatment with β-adrenoreceptor antagonists. *Br Heart J* 1980; **44**:265–270.
27. Zoller RP, Mark AL, Abboud FM, Schmid PG, Heistad DD. The role of low pressure baroreceptors in reflex vasoconstrictor responses in man. *J Clin Invest* 1972; **51**:2967–2972.
28. Ferguson DW, Thames MD, Mark AL. Effects of propranolol on reflex vascular responses to orthostatic stress in humans. Role of ventricular baroreceptors. *Circulation* 1983; **67**:802–807.
29. Sra JS, Murthy VS, Jazayeri MR, Shen YE, Troup PJ, Avitall B, Akhtar M. Use of intravenous esmolol to predict efficacy of oral beta-adrenergic blocker therapy in patients with neurocardiogenic syncope. *J Am Coll Cardiol* 1992; **19**:402–408.
30. Müller G, Deal BJ, Stasburger JF, Benson DW. Usefulness of metoprolol for unexplained syncope and positive response to tilt testing in young persons. *Am J Cardiol* 1993; **71**:592–595.
31. Cohen MB, Snow JS, Grasso V, Lehnert L, Goldner BG, Jadonath RL, Cohen TJ. Efficacy of pindolol for treatment of vasovagal syncope. *Am Heart J* 1995; **130**:786–790.

32. Jhamb DK, Singh B, Sharda B, Kaul U, Goel P, Talwar KK, Wasir HS. Comparative study of the efficacy of metoprolol and verapamil in patients with syncope and positive head-up tilt test response. *Am Heart J* 1996; **132**:608–611.
33. Bristow JD, Honour AJ, Pickering GW, Sleight P, Smyth HS. Diminished baroreflex sensitivity in high blood pressure. *Circulation* 1969; **39**:48–54.
34. Eckberg DL, Drabinsky M, Braunwald E. Defective cardiac parasympathetic control in patients with heart disease. *New Engl J Med* 1971; **285**:877–883.
35. Eckberg DL, Sleight P. Baroreflex malfunction in cardiovascular patients: general principles. In: Eckberg DL, Sleight P (Eds) Human baroreflexes in health and disease. Oxford University Press, New York 1992, 303–326.
36. Pickering TG, Gribbin B, Strange Petersen E, Cunnigham DJC, Sleight P. Effects of autonomic blockade on the baroreflex in man at rest and during exercise. *Circ Res* 1971; **30**:177–185.
37. Eckberg DL, Abboud FM, Mark AL. Modulation of carotid baroreflex responsiveness in man: effects of posture and propranolol. *J Appl Physiol* 1976; **41**:383–387.
38. Simon G, Kiowski W, Julius S. Effect of beta adrenoreceptor antagonists on baroreceptor reflex sensitivity in hypertension. *Clin Pharm Therap* 1977; **22**:293–298.
39. Takeshita A, Tanaka S, Nakamura M. Effects of propranolol on baroreflex sensitivity in borderline hypertension. *Cardiovasc Res* 1978; **12**:148–151.
40. Krediet RT, Dunning AJ. Baroreflex sensitivity in hypertension during beta-adrenergic blockade. *Br Heart J* 1979; **41**:106–110.
41. Watson RDS, Stallard TJ, Littler WA. Effects of β-adrenoreceptor antagonists on sino-aortic baroreflex sensitivity and blood pressure in hypertensive men. *Clin Sci*1979; **57**:241–247.
42. Takata S, Ogawa J, Okuwa H, Yagi S, Iwase N, Hattori N. Effect of propranolol on baroreflex control of heart rate in young patients with essential hypertension. *Angiology* 1986; **37**:106–111.
43. Floras JS, Vann Jones J, Hassan MO, Sleight P. Effects of acute and chronic β-adrenoreceptor blockade on baroreflex sensitivity in humans. *J Auton Nerv Syst* 1988; **25**:87–94.
44. Malik M, Camm AJ. Heart rate variability and clinical cardiology. *Br Heart J* 1994; **71**:3–6.
45. Binder T, Frey B, Porenta G, Heinz G, Wutte M, Kreiner G, Gössinger H, Schmidinger H, Pacher R, Weber H. Prognostic value of heart rate variability in patients awaiting cardiac transplantation. *Pacing Clin Electrophys* 1992; **15**:2215–2220.
46. van Hoogenhuyze D, Weistein N, Martin GJ, Weiss JS, Schaad JW, Sahyouni N, Fintel D, Remme WJ, Singer DH. Reproducibility and relation to mean heart rate of heart rate variability in normal subjects and in patients with congestive heart failure secondary to coronary artery disease. *Am J Cardiol* 1991; **68**:1668–1676.
47. Coumel P, Maison-Blanche P, Catuli D. Heart rate and heart rate variability in normal young adults. *J Cardiovasc Electrophysiol* 1994; **5**:899–911.

48. Coker R, Koziell A, Oliver C, Smith SE. Does the sympathetic nervous system influence sinus arrhythmia in man? Evidence from combined autonomic blockade. *J Physiol* 1984; **356**:459–464.
49. Pomeranz B, Macaulay RJB, Caudill MA, Kutz I, Adam D, Gordon D, Kilborn KM, Barger C, Shannon DC, Cohen RJ, Benson H. Assessment of autonomic function in humans by heart rate spectral analysis. *Am J Physiol* 1985; **248**:H151–H153.
50. Pagani M, Lombardi F, Guzzetti S, Rimoldi O, Furlan R, Pizzinelli P, Sandrone G, Malfatto G, Dell'Orto S, Piccaluga E, Turiel M, Baselli G, Cerutti S, Malliani A. Power spectral analysis of heart rate and arterial pressure variabilities as a marker of sympatho-vagal interaction in man and conscious dog. *Circ Res* 1986; **59**:178–193.
51. Bekheit S, Tangella M, el-Sakr A, Rasheed Q, Craelius W, el-Sherif N. Use of heart rate spectral analysis to study the effects of calcium channel blockers on sympathetic activity after myocardial infarction. *Am Heart J* 1990; **119**:79–85.
52. Coumel Ph, Hermida JS, Wennerblöm B, Leenhardt A, Maison-Blanche P, Cauchemez B. Heart rate variability in myocardial hypertrophy and heart failure, and effects of beta-blocking therapy: A non-spectral analysis of heart rate oscillations. *Eur Heart J* 1991; **65**:1345–1350.
53. Cook JR, Bigger JT, Kleiger RE, Fleiss JL, Steinman AB, Rolnitzky LM. Effect of atenolol and diltiazem on heart period variability in normal persons. *J Am Coll Cardiol* 1991; **17**:480–484.
54. Molgaard H, Mickley H, Pless P, Bjerregaard P, Moller M. Effects of metoprolol on heart rate variability in survivors of acute myocardial infarction. *Am J Cardiol* 1993; **71**:1357–1359.
55. Hohnloser SH, Klingenheben T, Zabel M, Just H. Effect of sotalol on heart rate variability assessed by Holter monitoring in patients with ventricular arrhythmias. *Am J Cardiol* 1993; **72**:67A–71A.
56. Niemela MJ, Airaksinen KEJ, Huikuri HV. Effect of β-blockade on heart rate variability in patients with coronary artery disease. *J Am Coll Cardiol* 1994; **23**:1370–1377.
57. Ahmed MW, Kadish AH, Parker MA, Goldberger JJ. Effect of physiologic and pharmacologic adrenergic stimulation on heart rate variability. *J Am Coll Cardiol* 1994; **24**:1082–1090.
58. Pitzalis MV, Mastropasqua F, Massari F, Totaro P, Passantino A, Rizzon P. The effect of nadolol on heart rate and the standard deviation of the RR intervals. *Eur Heart J* 1995; **16**:269–275.
59. Kontopoulos AG, Athyros VG, Papageorgiou AA, Papadopoulos GV, Avramidis MJ, Boudoulas H. Effect of quinapril or metoprolol on heart rate variability in post-myocardial infarction patients. *Am J Cardiol* 1996; **77**:242–246.
60. Copie X, Pousset F, Lechat Ph, Jaillon P, Guize L, Le Heuzey JY, and CIBIS investigators. Effects of β-blockade with bisoprolol on heart rate variability in patients with advanced heart failure: analysis of scatterplots of RR intervals at selected heart rates. *Am Heart J* 1996; **132**:369–375.

61. Pousset F, Copie X, Lechat Ph, Jaillon P, Boissel JP, Hetzel M, Remme W, Guize L, Le Heuzey JY. Effects of bisoprolol on heart rate variability in heart failure. *Am J Cardiol* 1996; **77**:612–617.
62. Keeley EC, Lange RA, Hillis D, Joglar JA, Page RL. Correlation between time-domain measures of heart rate variability and scatterplots in patients with healed myocardial infarcts and the influence of metoprolol. *Am J Cardiol* 1997; **79**:412–414.
63. Bigger TJ, Kleiger RE, Fleiss JL, Rolnitzky LM, Steinman RC, Miller JP. Components of heart rate variability measured during healing of acute myocardial infarction. *Am J Cardiol* 1989; **61**:208–215.
64. Zuanetti G, Neilson JMM, Latini R, Santoro E, Maggioni AP, Ewing DJ, on behalf of GISSI-2 Investigators. Prognostic significance of heart rate variability in post-myocardial infarction patients in the fibrinolytic era. The Gissi-2 results. *Circulation* 1996; **94**:432–436.
65. Adamson PB, Huang MH, Vanoli E, Foreman RD, Schwartz PJ, Hull SS. Unexpected interaction between β-adrenergic blockade and heart rate variability before and after myocardial infarction. A longitudinal study in dogs at high and low risk for sudden death. *Circulation* 1994; **90**:976–982.
66. Hartikainen JEK, Malik M, Staunton A, Poloniecki J, Camm AJ. Predictive power of heart rate variability used as a stratifier of cardiac mortality after myocardial infarction in patients discharged with and without beta-blocker therapy. *Annals Noninvas Electrocard* 1996; **1**:12–18.
67. Copie X, Le Heuzey JY, Iliou MC, Khouri R, Lavergne T, Pousset F, Guize L. Correlation between time-domain measures of heart rate variability and scatterplots in post-infarction patients. *PACE* 1996; **19**:342–347.
68. The CONSENSUS Trial Study Group. Effects of enalapril on mortality in severe congestive heart failure: results of the Cooperative North Scandinavian Enalapril Survival Study (CONSENSUS). *N Engl J Med* 1987; **316**:1429–1435.
69. The ISIS-4 Collaborative group. International Study of Infarct Survival-4. A randomized trial assessing early oral captopril, oral mononitrate, and intravenous magnesium sulfate in 58,050 patients with suspected acute myocardial infarction. *Lancet* 1995; **345**:669–685.
70. Levine TB, Olivari MT, Garberg V, Sharkey SW, Cohn JN. Hemodynamic and clinical response to enalapril, a long acting converting-enzyme inhibitor, in patients with congestive heart failure. *Circulation* 1984; **69**:548–553.
71. Cody RJ, Bravo EL, Fouad FM, Tarazi RC. Cardiovascular reflexes during chronic converting enzyme inhibition and sodium depletion: the response to tilt in hypertensive patients. *Am J Med* 1981; **71**:422–426.
72. Ebert TJ. Captopril potentiates chronotropic baroreflex responses to carotid stimuli in humans. *Hypertension* 1985; **7**:602–606.
73. Campbell BC, Sturani A, Reid JL. Evidence of parasympathetic activity of an angiotensin-converting enzyme inhibitor captopril in normotensive men. *Clin Sci* 1985; **68**:49–56.
74. Osterziehl KJ, Rohring N, Dietz R, Manthey J, Hecht J, Kuber W. Influence of captopril on the arterial baroreceptor reflex in patients with heart failure. *Eur Heart J* 1988; **9**:1137–1145.

75. Vogt A, Unterberg C, Kreuzer H. Acute effects of the new angiotensin converting enzyme inhibitor ramipril on hemodynamics and carotid sinus baroreflex activity in congestive heart failure. *Am J Cardiol* 1987; **59:**149D–154D.
76. Volpe M, Lembo G, Condorelli G, De Luca N, Lamenza F, Indolfi C, Trimarci B. Converting enzyme inhibition prevents the effects of atrial natriuretic factor on baroreflex responses in humans. *Circulation* 1990; **82:**1214–1221.
77. Osterziehl KJ, Dietz R, Schmid W, Mikulaschek K, Kuber W. ACE inhibition improves vagal activity in patients with heart failure. *Am Heart J* 1990; **120:**1120–1129.
78. Bonaduce D, Petretta M, Morgano G, Attisano T, Bianchi V, Arrichiello P, Rotondi F, Condorelli M. Effects of converting enzyme inhibition on baroreflex sensitivity in patients with myocardial infarction. *J Am Coll Cardiol* 1992; **20:**587–593.
79. Flapan AD, Nolan J, Neilson JMM, Ewing DJ. Effect of captopril on parasympathetic activity in chronic cardiac failure secondary to coronary artery disease. *Am J Cardiol* 1992; **69:**532–535.
80. Binkley PF, Haas JG, Starling RC, Nunziata E, Hatton PA, Leier CV, Cody RJ. Sustained augmentation of parasympathetic tone with angiotensin-converting enzyme inhibition in patients with congestive heart failure. *J Am Coll Cardiol* 1993; **21:**655–661.
81. Kaufman ES, Bosner MS, Bigger JT, Stein PK, Kleiger RE, Rolnitzky LM, Steinman RC, Fleiss JL. Effects of digoxin and enalapril on heart period variability and response to head-up tilt test in normal subjects. *Am J Cardiol* 1993; **72:**95–99.
82. Bonaduce D, Marciano F, Petretta M, Migaux ML, Morgano G, Bianchi V, Salemme L, Valva G, Condorelli M. Effects of converting enzyme inhibition on heart period variability in patients with acute myocardial infarction. *Circulation* 1994; **90:**108–113.
83. Grassi G, Cattaneo BM, Seravalle G, Lanfranchi A, Pozzi M, Morganti A, Carugo S, Mancia G. Effects of chronic ACE inhibition on sympathetic nerve trafic and baroreflex control of circulation in heart failure. *Circulation* 1997; **96:**1173–1179.
84. Lyons D, Roy S, O'Byrne S, Swift CG. ACE Inhibition. Postsynaptic adrenergic sympatholytic action in men. *Circulation* 1997; **96:**911–915.
85. Noll G, Wenzel RR, de Marchi S, Shaw S, Lüscher TF. Differential effects of captopril and nitrates on muscle sympathetic nerve activity in volunteers. *Circulation* 1997; **95:**2286–2292.
86. Dibner-Dunlap ME, Smith ML, Kinugawa T, Thames MD. Enalapril augments arterial and cardiopulmonary baroreflex control of sympathetic nerve activity in patients with heart failure. *J Am Coll Cardiol* 1996; **27:**358–364.
87. Scroop GC, Lowe RD. Central pressor effect of angiotensin mediated by parasympathetic nervous system. *Nature* 1968; **220:**1331–1335.
88. The Digitalis Investigation Group. The effect of digoxin on mortality and morbidity in patients with heart failure. *N Engl J Med* 1997; **336:**525–533.

89. Chai CY, Wang HH, Hoffman BF, Wang SC. Mechanisms of bradycardia induced by digitalis substances. *Am J Physiol* 1967; **212**:26–34.
90. Ten Eick RE, Hoffman BF. The effect of digitalis on the excitability of autonomic nerves. *J Pharmacol Exp Ther* 1969; **169**:95–108.
91. Quest JA, Gillis RA. Carotid sinus reflex changes produced by digitalis. *J Pharmacol Exp Ther* 1971:177:650–661.
92. Quest JA, Gillis RA. Effect of digitalis on carotid sinus baroreceptor activity. *Circ Res* 1974; **35**:247–255.
93. Gillis RA, Pearle DL, Levitt O. Digitalis: a neuroexcitatory drug. *Circulation* 1975; **52**:739–741.
94. Cokkinos DV, Haralambakis AG, Thalassinos NC, Heinomas ET, Demopoulos JN, Voridis EM, Gardikas CD. Effects of digitalis on diminished baroreceptor sensitivity in diabetics. *Angiology* 1979; **30**:549–557.
95. Gillis RA, Quest JA. The role of the nervous system in the cardiovascular effects of digitalis. *Pharmacol Rev* 1980; **31**:19–97.
96. Ferrari A, Gregorini L, Ferrari MC, Preti L, Mancia G. Digitalis and baroreceptor reflexes in man. *Circulation* 1981; **63**:279–285.
97. Ferrari A, Bonazzi O, Gregorini L, Gardumi M, Perondi R, Mancia G. Modification of the baroreceptor control of atrio-ventricular conduction induced by digitalis in man. *Cardiovasc Res* 1983; **17**:633–641.
98. Thames MD, Miller BD, Abboud FM. Sensitization of vagal cardiopulmonary baroreflex by chronic digoxin. *Am J Physiol* 1982; **243**:H815–H818.
99. Ferguson DW, Berg WJ, Sanders JS, Roach PJ, Kempf JS, Kienzle MG. Sympathoinhibitory responses to digitalis glycosides in heart failure patients. Direct evidence from sympathetic neural recordings. *Circulation* 1989; **80**:65–77.
100. Krum H, Bigger JT, Goldsmith RL, Packer M. Effect of long-term digoxin therapy on autonomic function in patients with chronic heart failure. *J Am Coll Cardiol* 1995; **25**:289–294.
101. McLain PL. Effects of cardiac glycosides on spontaneous efferent activity in vagus and sympathetic nerves of cats. *Int J Neuropharmacol* 1969; **8**:379–387.
102. The Cardiac Arrhythmia Suppression Trial (CAST) Investigators. Preliminary report: effect of encainide and flecainide on mortality in a randomized trial of arrhythmia suppression after myocardial infarction. *N Engl J Med* 1989; **321**:406–412.
103. Cairns JA, Connolly SJ, Roberts R, Gent M, for the Canadian Amiodarone Myocardial Infarction Arrhytmia Trial Investigators. Randomised trial of outcome after myocardial infarction in patients with frequent or repetitive ventricular premature depolarisations: CAMIAT. *Lancet* 1997; **349**:675–682.
104. Sopher M, Camm AJ. Adverse effects of amiodarone at low dose: Plus ça change. *J Am Coll Cardiol* 1997; **30**:799–801.
105. Alboni P, Paparella N, Cappato R, Candini GC. Direct and automatically mediated effects of oral flecaïnide. *Am J Cardiol* 1988; **61**:759–763.
106. Alboni P, Paparella N. Vagolytic effect of propafenone (letter). *Am Heart J* 1988; **116**:1648.

107. Zuanetti G, Latini R, Neilson JMN, Schwartz PJ, Ewing DJ, and the antiarrhythmic drug evaluation group (ADEG). Heart rate variability in patients with ventricular arrhythmias: effect of antiarrhythmic drugs. *J Am Coll Cardiol* 1991; **17**:604–612.
108. Lombardi F, Torzillo D, Sandrone G, Dalla Vecchia L, Cappiello E. Beta-blocking effect of propafenone based on spectral analysis of heart rate variability. *Am J Cardiol* 1992; **70**:1028–1034.
109. Lombardi F, Torzillo D, Sandrone G, Dalla Vecchia L, Cappiello E. Autonomic effects of antiarrhythmic drugs and their importance. *Eur Heart J* 1992; **13**:F38–F43.
110. Packer M. Pathophysiological machanisms underlying the adverse effects of calcium channel-blocking drugs in patients with chronic heart failure. *Circulation* 1989; **80**:59–67.
111. Adinoff B, Mefford I, Waxman R, Linnoila M. Vagal tone decreases following intravenous diazepam. *Psychiatry Res* 1992; **41**:89–97.
112. Lund-Johansen P, Omvik P. Haemodynamic effects of nifedipine in essential hypertension at rest and during exercise. *J Hypertension* 1983; **1**:159–163.
113. Kiowski W, Erne O, Bolli P, Bühler F. Acute and chronic sympathetic reflex activation and antihypertensive response to nifedipine. *J Am Coll Cardiol* 1986; **7**:344–348.
114. Lown B, Verrier RL. Neural activity and ventricular fibrillation. *N Engl J Med* 1976; **294**:1165–1170.
115. Vanoli E, De Ferrari GM, Stramba-Badiale M, Hull SS, Foreman RD, Schwartz PJ. Vagal stimulation and prevention of sudden death in conscious dogs with a healed myocardial infarction. *Circ Res* 1991; **68**:1471–1481.
116. De Ferrari GM, Vanoli E, Curcuruto P, Tommasini G, Schwartz PJ. Prevention of life-threatening arrhythmias by pharmacologic stimulation of the muscarinic receptors with oxotremorine. *Am Heart J* 1992; **124**:883–890.
117. Mc Closkey DI. Cardiorespiratory integration. In: Levy MN, Schwartz PJ (Eds) Vagal Control of the Heart: Experimental Basis and Clinical Implications. Futura, Armonk 1994, 481–508.
118. Wetzel GT, Brown JH. Presynaptic modulation of acetylcholine release from cardiac parasympathetic neurons. *Am J Physiol* 1985; **248**:H146–H151.
119. Hull SS, Vanoli E, Adamson PB, De Ferrari GM, Foreman RD, Schwartz PJ. Do increases in markers of vagal activity imply protection from sudden death? The case of scopolamine. *Circulation* 1995; **91**:2516–2519.
120. Raczkowska M, Eckberg DL, Ebert TJ. Muscarinic cholinergic receptors modulate vagal cardiac responses in man. *J Auton Nerv Syst* 1983; **7**:271–278.
121. Dibner-Dunlap ME, Eckberg DL, Magid NM, Cintrón-Treviño NM. The long term increase of baseline and reflexly augmented levels of human vagal-cardiac nervous activity induced by scopolamine. *Circulation* 1985; **71**:797–804.
122. Vybiral T, Bryg RJ, Maddens ME, Bhasin SS, Cronin S, Boden WE, Lehmann MH. Effects of transdermal scopolamine on heart rate variability in normal subjects. *Am J Cardiol* 1990; **65**:604–608.

123. Raeder EA, Stys A, Cohen RJ. Effect of low-dose scopolamine on autonomic control of the heart. *Ann Noninvas Electrocard* 1997; **2**:236–241.
124. Casadei B, Pipilis A, Sessa F, Conway J, Sleight P. Low doses scopolamine increase vagal tone in the acute phase of myocardial infarction. *Circulation* 1993; **88**:353–357.
125. Pedretti R, Colombo E, Braga SS, Carù B. Influence of transdermal scopolamine on cardiac sympathovagal interaction after acute myocardial infarction. *Am J Cardiol* 1993; **72**:384–392.
126. De Ferrari GM, Mantica M, Vanoli E, Hull SS, Schwartz PJ. Scopolamine increases vagal tone and vagal reflexes in patients after myocardial infarction. *J Am Coll Cardiol* 1993; **22**:1327–1334.
127. Vybiral T, Glaeser DH, Morris G, Hess KR, Yang K, Francis M, Pratt CM. Effects of low dose transdermal scopolamine on heart rate variability in acute myocardial infarction. *J Am Coll Cardiol* 1993; **22**:1320–1326.
128. Pedretti RFE, Colombo E, Braga SS, Ballardini L, Carù B. Effects of oral pirenzepine on heart rate variability and baroreceptor reflex sensitivity after acute myocardial infarction. *J Am Coll Cardiol* 1995; **25**:915–921.
129. La Rovere MT, Mortara A, Pantaleo P, Mestri R, Cobelli F, Tavazzi L. Scopolamine improves autonomic balance in advanced congestive heart failure. *Circulation* 1994; **90**:838–843.
130. Sleight P, West MJ, Korner PI, Oliver JR, Chalmers JP, Robinson JL. The action of clonidine on the baroreflex control of heart rate in conscious animals and man, and on single aortic baroreceptor discharge in the rabbit. *Arch Int Pharmacodyn* 1975; **214**:4–11.
131. Pedretti RFE, Colombo E, Braga SS, Carù B. Effect of thrombolysis on heart rate variability and life-threatening ventricular arrhythmias in survivors of acute myocardial infarction. *J Am Coll Cardiol* 1994; **23**:19–26.
132. Bonaduce D, Petretta M, Piscione F, Indolfi C, Migaux ML, Bianchi V, Esposito N, Marciano F, Chiarello M. Influence of reversible segmental left ventricular dysfunction on heart period variability in patients with one-vessel coronary artery disease. *J Am Coll Cardiol* 1994; **24**:399–405.
133. Yeragani VK, Pohl R, Balon R. Effect of imipramine treatment on heart rate variability measures. *Neuropsychobiology* 1992; **26**:27–32.

CHAPTER THIRTEEN

INFLUENCE OF SMOKING, ALCOHOL, CAFFEINE AND RECREATIONAL DRUGS ON CARDIAC AUTONOMIC TESTS

Azad V Ghuran and Marek Malik

The assessment of the autonomic nervous system is an important area of clinical cardio-neurophysiology. The individual tests are being used mainly to quantify the status of the sympathetic and parasympathetic balance in a variety of clinical conditions including a history of syncope, diabetes, cardiac failure, risk stratification after acute myocardial infarction, etc.

The majority of the presently used tests are non invasive and clinically fully practical. Consequently, it is not surprising that these tests are being utilised more and more frequently. As new clinical settings appear in which the autonomic tests are considered appropriate, the clinical populations to which the tests are applied broaden. Together with this trend, difficulties are emerging in interpreting the results of the tests in unusual conditions that may not be directly associated with the primary diagnosis. Since the autonomic nervous system is, by its very essence, rather sensitive practically to every disturbance of the organism, particular care is needed to distinguish true autonomic pathologies from abnormalities due to coincident factors. As described elsewhere in this book, special standards are

Malik M (Ed) Clinical guide to cardiac autonomic tests, 331–356.

needed to evaluate autonomic tests in patients on autonomically active therapy. Similar to pharmaceutical agents, autonomic tests may be seriously distorted by a number environmental and recreational agents, such as alcohol, tobacco, caffeine, and recreational drugs.

The purpose of this chapter is to review how some of these recreational substances may influence the autonomic nervous system tests. However, before reviewing the effects of the recreational substances on autonomic tests, the general changes in autonomic and cardiovascular physiology induced by these substances have to be summarised.

RECREATIONAL SUBSTANCES AND CARDIOVASCULAR SYSTEM

SMOKING

As repeatedly discussed at various levels of the society, tobacco, and in particular cigarette, smoking is the largest single preventable cause of illness and premature death in developed countries.

Nicotine is the main pharmacological active agent in tobacco smoke, apart from the polycyclic aromatic hydrocarbons, tobacco glycoproteins and carbon monoxide. The amount of nicotine absorbed from an average cigarette is about 1.5 grams which raises the plasma nicotine concentration to 130–200 mmol/l. After smoking a single cigarette, the plasma concentration of nicotine exhibits the standard double-compartmental exponential decay and falls to 50% within 10 minutes and diminishes more slowly over the next 1–2 hours. The rapid decline results from redistribution between blood and other tissues; the slower decline is due to hepatic metabolism.

At a cellular level, nicotine acts on nicotinic acetylcholine receptors and causes neuronal excitation. The overall central nervous system changes are a mixture of inhibitory and excitatory effects. The peripheral effects of small doses of nicotine result from stimulation of autonomic ganglia and peripheral sensory receptors in the heart and lungs. Stimulation of these receptors elicits several autonomic reflex responses causing tachycardias, increased cardiac output and arterial pressure, reduction of gastrointestinal motility, and sweating. Tolerance develops with repeated dosage. Secretion of adrenaline and noradrenaline from the adrenal medulla and a reduction in its clearance contribute to the effects on the cardiovascular system.[1] Smoking also stimulates antidiuretic hormone from the posterior pituitary gland.[2]

Smokers experience a 70 % greater coronary heart disease mortality than non-smokers; among heavy smokers, this figure increases to 200 %. This excess mortality is due to both myocardial infarction and sudden cardiac death.[3] These pathologies are caused by direct and indirect influences on atherosclerotic lesions per se, and by promoting coronary occlusive phenomena through a complex interaction of many factors that include:

- Vasoconstriction of epicardial and small coronary arteries which lead to a reduction in coronary flow and increase vessel resistance[4] (endothelium vasodilatation is impaired probably from an imbalance of prostacyclin / thrombaxane A 2 ratio and catecholamine secretion[5]),
- Endothelial injury and dysfunction which promote atherosclerosis,[6]
- Increase platelet aggregation and adhesiveness, shortened platelet survival, increase fibrinogen levels, decrease clotting time and increased blood viscosity (chronic carbon monoxide exposure causes a relative hypoxia which increases haematocrit and, consequently, blood viscosity),
- Pulmonary dysfunction producing arterial hypoxia that results in further tissue hypoxia,
- Increased in heart rate and electrical instability of the myocardium (probably due to adrenergic stimulation and catecholamine release),
- Adverse effects on blood cholesterol,[7] and
- Possible carbon monoxide induced reduction of ventricular fibrillation threshold, which is known to occur in animals and which may also contribute to atherogenesis.[8,9]

Although some older epidemiological surveys of blood pressure measurement in out-patients reported that smokers have a lower blood pressure than non smokers,[10] more recent studies of ambulatory blood pressure monitoring demonstrate that long term cigarette smoking increases average heart rate and blood pressure throughout the whole day.[11]

ALCOHOL

The consumption of alcohol far exceeds that of any other recreational substance. Moreover, in the western world, the consumption has been steadily increasing over the past decades.

At a cellular level, the action of alcohol involves several mechanisms, including inhibition of calcium entry through voltage gated calcium channels, and enhancement of gamma-amino-butyric-acid (GABA) mediated synaptic inhibition and antagonism of excitatory amino acids.[2] Above all, alcohol is a depressant of the central nervous system and its acute effects of slurred speech, motor incoordination, euphoria, self confidence, etc., are well known. Alcohol also inhibits ADH secretion and causes a self- limiting diuresis.

Chronic effects include cerebral and cerebellar degeneration, together with peripheral and autonomic neuropathy. It is believed that alcohol exerts a direct toxic effect causing degeneration of the myelinated fibres of the vagus nerve and thoracic paravertebral sympathetic trunk.[12,13] The depression in sympathetic function may also involve higher centres.[13]

The acute effects of alcohol on the cardiovascular system include mainly arrhythmias (sinus tachycardia, atrial fibrillation, atrial flutter, supraventricular and ventricular ectopics), and vasodilatation. The cutaneous vasodilatation increases heat loss, but paradoxically causes a feeling of increase warmth. Lowering blood pressure by acute alcohol consumption has also been described.[14,15]

The chronic cardiovascular effects of alcohol are associated with several pathologies; including:

- Hypertension. A general association of chronic alcoholism with hypertension has been demonstrated in large epidemiological studies.[16,17] At the same time, however, some studies show a lower prevalence of hypertension in moderate drinkers compared to totally alcohol abstaining individuals. This may suggest that chronic alcohol consumption and increased blood pressure are non-linearly related with a substantial hypertension only in heavy drinkers ('J shape' relationship).[16,18]
- A low incidence of ischaemic heart disease in moderate drinkers. A moderate chronic alcohol intake seem to result in increased high-density lipoprotein concentration and decreased platelet aggregation.
- Dilated Cardiomyopathy. Chronic consumption of alcohol may cause myocardial damage by three mechanisms, namely by:
    - direct toxic effect of alcohol or its metabolites,
    - negative nutritional effects seen mainly in association with thiamine deficiency (although it is now clear that alcoholic

cardiomyopathy can occur in the absence of nutritional deficiencies), and
- rarely, toxic effects due to additives in the alcoholic beverage such as cobalt.[19]

Alcohol and especially its first metabolite are sympathomimetics and are believed to facilitate the release of catecholamines and glucocorticoid from the adrenal gland. Consequently, heavy alcohol ingestion can produce an imbalance in peripheral sympathetic and parasympathetic neural discharge, a mechanism which intensifies the cardiotoxic effect of catecholamines by a deleterious steroid / catecholamine interaction. This is believed to have proarrhythmic properties, facilitate myocytolysis, and, therefore, increase the risk of sudden cardiac death.[20,21] At the same time, however, alcohol triggered catecholamine release is still being debated. A recent study by Heikkonen et al[22] found no effect of acute alcohol intake on the adrenergic system when measuring plasma catecholamines and lymphocytic beta adrenergic receptor density in healthy subjects.

Liver cirrhosis per se (independent of aetiology) can cause autonomic dysfunction.[23,24] The extent of autonomic dysfunction correlates with increasing severity of liver disease.[21] Dillon et al have shown that the administration of captopril causes a partial recovery of autonomic dysfunction in liver cirrhosis.[25]

CAFFEINE

Caffeine, a naturally occurring xanthine, is contained in various beverages such as coffee, tea and chocolate drinks. The average daily consumption is approximately 200mg.

Several cellular actions of caffeine have been observed, but it is not entirely clear how the individual mechanisms contribute to its pharmacological profile. Among the most important cellular effects are inhibition cyclic nucleotide phosphatase, antagonism of adenosine (A1,A2) receptors and modulation of intracellular calcium handling.[2]

During the last two decades, the scientific interest in the cardiovascular effects of caffeine has increased dramatically. Particularly, the effects on hypertension, possible association with coronary artery disease and serum lipids, and arrhythmogenic effects of caffeine have been investigated extensively.[26,27]

A review by Myers[28] of 17 separate studies revealed that a single dosage of 200–300 mg of caffeine increases blood pressure in non-users and abstainers. However, no clear consensus exists in respect of the effects of caffeine on blood pressure as presently available studies provide discordant results. Robertson et al[29] demonstrated that in normotensive subjects who abstained from drinking coffee for 3 weeks, caffeine at a dose level of 250 mg increases blood pressure (by 10–20 mm Hg) and lowers heart rate (by 3–5 beats per minute). The duration of elevated blood pressure after caffeine intake was as much as 4 hours. These changes were not seen in subjects who were treated 1–3 days before with caffeine, which suggests tolerance to the circulatory effects.[30,31] In contrast, other studies showed no significant correlation between caffeine consumption and blood pressure.[32,33] Dusseldorp et al[34] investigated the effects of switching from regular coffee to decaffeinated coffee in normotensive, non-smoking adults and observed a decline by 1.5 mm Hg and 1 mm Hg in systolic and diastolic blood pressure, respectively. Hence, this study suggests that caffeine abstinence might marginally lower blood pressure. In contrast, a similar 2 months study by Rosmarin et al[35] did not detect a significant fall in blood pressure after a period of prolonged abstinence in 21 male moderate coffee drinkers.

The cardiovascular effects seen in these studies raise the question as to what degree the adreno-cortical system and catecholamines might be involved. Caffeine ingestion induces a 2–3 times fold increase in adrenaline[27,36,37] and to a lesser extent in nor adrenaline. Smits et al[38] compared the effects of caffeine in healthy volunteers and adrenalectomised subjects and found an increase in blood pressure in both groups following caffeine ingestion. An increase in adrenaline and noradrenaline concentration was seen in the healthy volunteers but not in the adrenalectomised subjects. These findings suggest that the pressor response to drinking coffee is not purely related to catecholamine changes.

The effects of caffeine in precipitating arrhythmias, particularly supraventricular arrhythmias, are well acknowledged. Adenosine was reported to prolong the AV nodal refractory time in man.[39] There is evidence suggesting that caffeine competitively antagonises endogenous adenosine[40] and caffeine has actually been shown to shorten the refractory period of the AV node, right atrium, and right ventricle, but, at the same time, an increase of left atrial refractory period was noted.[41] Prineas et al showed a positive association between tea and coffee consumption and ventricular ectopic activity.[42] It has also been suggested that caffeine can have a ventricular proarrhythmic effect but only in subjects with frequent ventricular ectopics beats.[43] On the contrary, data reassuring the safety of coffee drinking in cardiac patients also exist. Harris et al found no increase in ventricular arrhythmias after 300 mg of caffeine in a group of 70 post myocardial infarction patients.[44]

The Tromso heart study group observed an increase in serum cholesterol concentration in healthy volunteers after the ingestion of boiled coffee.[45] They also demonstrated that abstinence from boiled coffee but not from filtered coffee lowered plasma cholesterol in patients with hypercholesterolemia.[46] In general, these studies suggest that there is a tendency for coffee to cause an increase in cholesterol. The increase in cholesterol is seen mainly with boiled coffee and is probably related to other compounds other than caffeine as tea (which also contains caffeine) has no effect on serum cholesterol.[42]

The effects of caffeine on blood pressure and plasma lipids raise the question whether these and other effects of coffee may contribute to the risk of developing ischaemic heart disease. The current studies are conflicting.[47–49] To a large extent, the discrepancies between published reports may be partly explained by the lack of controlled variables between the studies (smoking behaviour, alcohol consumption, body mass index, physical activity, socio-economic class).

COCAINE

Cocaine use and abuse have reached epidemic proportions and continue to overwhelm urban economic, social and health-care systems. Cocaine and its free base form 'crack' are extracted from the coca plant. Cocaine is absorbed from all mucous membranes of the body. It may also be smoked, inhaled or injected intravenously, intramuscularly or subcutaneously. Onset of its action ranges from 3 seconds to 5 minutes depending on the route of administration. Peak effect varies from 1–20 minutes while duration of action ranges from 5–90 minutes.[50]

Cocaine is an indirectly acting sympathomimetic drug, which inhibits the re-uptake of noradrenaline and dopamine at the preganglionic sympathetic nerve endings.[2] This increases the synaptic concentration of these monoamines and their increased concentration can explain some of the cardiovascular effects, such as vasoconstriction, hypertension, tachycardia and predisposition to ventricular arrhythmias.

Cocaine has been shown to directly release dopamine in the brain and there is evidence that cocaine acting via a central mechanism causes the release of noradrenaline from the adrenal medulla.[51,52] Cocaine in high doses can acutely depress myocardial contractility probably through direct local anaesthetic effect[53,54] (it decreases sodium influx across cell membranes which lowers the amount of sodium available for sodium/calcium exchange). Consequently, the cardiac effects of cocaine are complex and depend on a balance between competing mechanisms, namely the local anaesthetic effect versus the inotropic effect of the sympathomimetic stimulation.

Cocaine leads to myocardial ischaemia and infarction by multiple factors. It may promote diffuse or local coronary artery spasm in normal or atherosclerotic coronary arteries[55–57] which results in stasis of blood with thrombus formation. In vitro studies have shown that cocaine activates platelets, increases platelet aggregation and potentiates thromboxane production.[58,59] The potential of cocaine to induce a pro-coagulant effect by decreasing the concentration of protein C and anti-thrombin 3 has also been shown.[60] On discontinuation of cocaine, coagulation activity returns to normal. Cocaine also increases oxygen demand by increasing heart rate, blood pressure and myocardial contraction. Long-term use may cause repetitive episodes of coronary spasms and paroxysms of hypertension causing endothelial damage and subsequent acceleration of atherosclerosis.

Dilated cardiomyopathy is associated with prolonged cocaine toxicity.[61] The cause of this association is unknown but may result from a direct toxic effect on cardiac muscle, myocarditis or from profound sympathetic stimulation which can cause subendocardial ischaemia and fibrosis characterised by contraction band necrosis.[62,63] Myocarditis has also frequently been seen in cocaine abusers.[64] The cause is again unknown and may be related to infectious agents and contaminants, e.g. heavy metals (manganese has been found in street preparations of cocaine).

Circulating levels of catecholamines in cocaine users can be elevated as much as 5 fold. It is well known that an increase in catecholamine levels reduces cardiac electrical stability and predisposes to lethal arrhythmias, particularly under conditions of ischaemia. The presence of contraction bands can act as a substrate for lethal re-entrant arrhythmias. Chronic cocaine use is associated with left ventricular hypertrophy which is a known risk factor for ventricular arrhythmias and may also provide a substrate for myocardial ischaemia. Ventricular arrhythmia, polymorphic ventricular tachycardia, supraventricular tachycardia, accelerated idioventricular rhythm, sino-atrial node and AV block, prolonged PR, QRS and QT intervals have all been reported.[65–67]

Cocaine abuse has also been associated with acute aortic rupture. This is probably related to substantial increases in blood pressure. In comparison with other recreational drugs, the use of cocaine has been reported to be an independent risk factor for the development of endocarditis.[68] Interestingly, in contrast to other illicit drug use, left sided cardiac lesions may be more commonly infected in cocaine abusers who develop endocarditis.[50]

MARIJUANA

Biologically active products derived from the hemp plant cannabis sativa are collectively referred to as cannabis. The main active constituent is delta 1-tetrahydrocannabinol (THC) though pharmacologically active metabolites may be important.[2] (Note, that delta 1-THC is often termed delta 9-THC, according to a different ring numbering system.)

The typical effect of marijuana in man is tachycardia with no change or a slight increase in blood pressure,[70–72] whilst in animals, the effect is bradycardia and hypotension.[73] The tachycardia is dose-related[74] and is associated with an increase in limb blood flow.[70] Maximum tachycardia can be reached within 30 minutes and can persist for between 90 minutes to 12 hours. ECG abnormalities in the form of P and T wave changes, ST segment depolarisation changes, premature ventricular and atrial ectopics have all been reported.[71,75,76] These findings are reversible and do not persist. Whether marijuana specifically caused these changes independent of its effect on heart rate is not clear.

There is an increase frequency of angina pectoral symptoms at a lower level of exercise after smoking marijuana.[80] This may be secondary to an increase in myocardial oxygen demand due to an increase in heart rate.

The tachycardia induced by marijuana may be caused by an increase in sympathetic tone as this effect is prevented by beta adrenoreceptor blockade. This mechanism is reflected by the ability of cannabis to shorten the pre-ejection period, lengthen the left ventricular ejection time and increase stroke volume, that is by the ability to enhance left ventricular performance.[81] In contrast, Tashkind et al[69] showed that propranolol did not prevent the marijuana induced tachycardia. They also showed that after smoking marijuana, cardiac output and pulse rate increased by 24% and 53%, respectively, while stroke volume and blood pressure did not change significantly. Therefore the increase in cardiac output after marijuana smoking seems to be mainly related to an increase in heart rate, which appears to be caused by a mechanism other than beta stimulation.

The influence of marijuana on exercise tolerance has also been studied. Shapiro et al[82] showed that after smoking marijuana, the duration of exercise was shorter at any given workload level. The heart rate was higher at rest and at all workloads. This premature achievement of maximum heart rate during exercise may explain why marijuana diminishes maximum exercise capability.

When marijuana is taken for long chronic periods or in large single doses, the effects are different.[77–79] Both bradycardia and hypotension develop, probably because of a decrease in central nervous system sympathetic discharge[83] which results in parasympathetic dominance.[84–86] The bradycardia is reversed by vagotomy, ganglia blocking drugs and anticholinergic drugs.[84] Marijuana also reduces venous tone[84] which explains the orthostatic hypotension. Fluid retention and weight gain is also observed with chronic use, probably due to a compensatory response to chronic blood pressure lowering.[86] Thus, the effects of marijuana on the autonomic nervous system are complex and appear biphasic.

HEROIN

Heroin (diacetylmorphine) is a semisynthetic analogue of morphine, which is commonly abused. Heroin and methadone use has increased by 900% between 1974–1992.[87]

When given intravenously, half-life of heroin is 3 minutes. It is slowly metabolised to morphine, but heroin itself is more active because of its greater lipid solubility and hence, its ability to cross the blood brain barrier. Heroin has sedative effects on the central nervous system causing inability to concentrate, drowsiness and respiratory depression in addition to inducing a euphoric state. It can cause acute noncardiogenic pulmonary oedema.[88] The oedema fluid contains protein concentrations similar to plasma and pulmonary capillary wedge pressures are normal when measured. This therefore suggests that the cause for pulmonary oedema may be due to an alveolar-capillary membrane leak.

Cardiac arrhythmias have also been documented.[88,89] These include premature atrial and ventricular ectopics, atrial fibrillation, idioventricular rhythm, bradyarrhythmias and QT prolongation. Sudden death presumed secondary to arrhythmias can also occur. Hypotension can result from a combination of depression of the vasomotor centre in the medulla[90,91] and histamine release from mast cell degranulation. Bacterial endocarditis mainly affecting right sided cardiac structures is a well know complication of intravenous drug abusers. Heroin can also cause stimulation of anti diuretic hormone.

AMPHETAMINE

Amphetamine is another drug that is commonly abused. It is readily absorbed from the gastrointestinal tract and freely penetrates the blood brain barrier. It is also readily absorbed from the nasal mucosa and often taken by snorting. It is mainly excreted unchanged in the urine. The plasma half life varies from about 5 hours to 20–30 hours depending on urine flow and pH (acidic urine increases the rate of excretion).

Amphetamine is a sympathomimetic drug which releases noradrenaline, dopamine and serotonin from the nerve terminals. Amphetamines have greater stimulant activity than catecholamines. Peripherally, they cause stimulation of alpha and beta-receptors. Centrally, they stimulate the cerebral cortex, medullary respiratory centre, and reticular activating centre.

Amphetamine can cause myocardial infarction in patients with normal coronary arteries.[92,93] The mechanism is uncertain; it may be related to the excessive catecholamine excess resulting in an increase myocardial oxygen demand, coronary artery spasm, platelet aggregation and thrombus formation. Arrhythmias including sinus, supraventricular, and ventricular tachycardias, and ventricular fibrillation have been reported, as were acute and chronic cases of cardiomyopathy.[93–95] Amphetamine abuse has also been implicated as a cause of necrotising angiitis.[96] Both hypertension and hypotension can occur and a case report of severe orthostatic hypotension exists proposing a suppressed central sympathetic vasomotor drive being the main cause.[97]

BENZODIAZEPINE

Benzodiazepine is a synthetic compound discovered by an unplanned reaction in 1961.[2] It acts by binding to specific regulatory site on the GABA receptors. It is well absorbed orally and reaches a peak concentration in about an hour. Even in overdoses, the patients remain drowsy but arousable without serious depression in respiration or cardiovascular function. The drug may have electrostabilising antifibrillatory effects in the early phase of local myocardial ischaemia.[98]

RECREATIONAL SUBSTANCES AND AUTONOMIC TESTS

HEART RATE VARIABILITY

As discussed elsewhere in this book, substantial advances[99] in the measurement and interpretation of heart rate variability (HRV) were made since the first report[100] on cyclical changes in heart rate that was made by Hales in 1733.

Alcohol

In healthy subjects without cardiovascular disorders, acute alcohol ingestion caused a significant decrease in HRV and an increase in heart rate.[15,101] A significant change in HRV, but not in heart rate has also been reported.[102] Because of alcoholic neuropathy, HRV is also significantly decreased in patients with chronic alcoholic dependency.[102,103]

Weise et al[104] investigated alcohol withdrawal and found a significant increase in heart rate and a decrease in HRV, mainly in the first week. Consequently, they concluded that the testing for autonomic cardiac neuropathy in alcoholics is of limited diagnostic value within the first week of alcohol withdrawal. Weise et al[104] also showed an increase in HRV, interpreted as an improvement in vagal neuropathy in patients with short to moderate drinking history or a functional rather than an organic vagal damage. Hirsch et al[105] showed that HRV alcoholics remained abnormal after an abstinence of 6 months. Further studies are required to prove whether decreased HRV associated with alcoholism is reversible or not.

Alcohol is hepatotoxic and causes cirrhosis. Dillon et al showed a reduction in HRV in liver cirrhosis (independent of aetiology).[23]

Smoking

Both Hayano et al[106] and Niedermaier et al[107] showed that acute smoking causes a transient but significant decrease in the HRV high frequency component (a decrease in vagal modulation) and an increase in the low frequency component (an increase in sympathetic modulations). Hayano also compared the short and long term effects of smoking and concluded that heavy smoking causes long term reduction in vagal cardiac control and blunted postural responses in autonomic cardiac regulation.[106]

Caffeine

Published data on the effects of caffeine and recreational drugs on HRV are limited. In a study by Kolodiichuk et al involving women alone, caffeine was shown to increase HRV.[108] As discussed above, caffeine is a stimulant and can cause an increase in catecholamines and possibly blood pressure. Therefore, it is plausible to speculate that the increase in HRV may be a compensatory measure with an increase in vagal tone, and consequently vagal modulation, to counteract the increase in blood pressure or circulating catecholamines.

Cocaine

Cocaine in lethal doses was given intramuscularly to conscious ferrets and their heart rate dynamics were assessed using time series and spectral analysis techniques.[109] In these animals, HRV was significantly decreased prior to sudden death. Transient low frequency oscillations in heart rate were also noted but, interestingly, there was no significant change in heart rate.

Oriol et al[110] investigated the heart rate dynamics in neonates exposed to cocaine via maternal use during pregnancy. Spectral analysis showed a significant decrease in HRV during ten minutes periods of quiescent sleep. However, other studies are not conclusive: HRV was both observed increased[111] and with no change[112] in cocaine exposed infants.

Marijuana, Heroin and Amphetamine

The involvement of the sympathetic and parasympathetic systems in modulating the acute effects of marijuana remains debatable as the present studies are conflicting, as already discussed above. Currently, there are no data on the effects of marijuana on HRV.

Heroin has inhibitory effects on the vasomotor centre, which can cause a decrease in sympathetic activity. Higher pulse rates are also seen in addicts[90] reflecting an inhibition of the cardioinhibitory centre as well. Specific effects on HRV are not known.

The effects of amphetamine on heart rate are variable with both increases and decreases being reported.[113–115]

Benzodiazepine

The effects of benzodiazepine on HRV were studied by Tulen et al[116] who investigated the effects of intravenous lorazepam (in variable accumulated doses) on cardiac vagal tone during rest and stress. The results showed a dose dependent decrease in heart rate but an increase in all frequency bands of the HRV spectral analysis. These responses occurred during the rest period but not during the stress test. Benzodiazepine can therefore exert a dose dependent increase in cardiac vagal tone, resulting in decreased heart rate and increased HRV but only during periods of rest.

In contrast Adinoff et al[117] reported an opposite effect in a similar uncontrolled study in healthy subjects after intravenous administration of diazepam. They reported a dose dependent increase in heart rate and a dose dependent attenuation of cardiac vagal tone.

VALSALVA MANOEUVRE

The various phases and the mechanisms involved in Valsalva manoeuvre are well known.[118] Investigations of blood pressure changes during the Valsalva manoeuvre tests the integrity of the sympathetic adrenergic cardiovascular function whereas heart rate changes test the parasympathetic cholinergic function (cardiovagal response).

Alcohol

Chida et al[119] described cardiovascular regulation in alcoholic neuropathy and found that during Valsalva manoeuvre, blood pressure similarly decreased in phase two but did not increase in phase four. The reflex bradycardia in phase four was not apparent. The data therefore suggested sympathetic dysfunction. The Valsalva ratio (cardiovagal test) has also been shown to be significantly decreased.[120]

Smoking

The effects of smoking on the Valsalva manoeuvre have been repeatedly investigated.[121–123] Kotamaki et al[121] showed that smokers tended to have a lower Valsalva ratio and higher tachycardia ratio. Piha and Seppanen[122] reported similar and significant differences in these ratios. Both studies concluded that in patients with a smoking history, there may be a dysfunction in some part of the autonomic reflex arch mediating the Valsalva manoeuvre.

Niedermaier et al showed a significant increase in systolic pressure in stage four compared to non-smokers.[107]

Other Substances

Studies researching the influence of caffeine, cocaine and benzodiazepines on the Valsalva manoeuvre are not available. It is know (see above) that caffeine is associated with an increase in circulating catecholamines. Cocaine is also an indirectly acting sympathomimetic. In a study of the effects of intranasal cocaine on humans, there was a normal response to the Valsalva manoeuvre.[124] Therefore we may speculate that coffee will also have none or minimum effects. However, until further studies are available clarifying this point, it may be recommended not to perform the manoeuvre following caffeine intake..

Benowitz et al[86] reported the effects of prolonged use of tetrahydrocannabinol. They found a significant impairment of heart rate slowing during phase four of the Valsalva manoeuvre. In addition the time taken from the end of Valsalva to the onset of bradycardia was delayed.

As mentioned earlier heroin can inhibit the vasomotor centre. Therefore impairment of the sympathetic component of the Valsalva manoeuvre is likely to occur. However, no direct investigations are available.

A case report was published on Valsalva manoeuvre after taking amphetamine.[97] During the Valsalva straining period, the blood pressure decreased progressively. However the overshoot in blood pressure that normally occurs in phase four was not present. This probably implies a failure of vasomotor sympathetic control.

RESPONSES TO STANDING, TILTING AND BARORECEPTOR REFLEX

Standing is associated with haemodynamic changes which induces an exercise reflex increase in heart rate. This is due to an increase in sympathetic and withdrawal of parasympathetic tones. At the same, time blood pressure remains maintained or is elevated, due to an increase in sympathetic activity and mechanical squeeze (by postural muscles) on both venous capacitance and arterial resistance vessels.[125] These changes are mediated by stimulation of baroreceptors in the aortic arch, carotid sinus and various thoracic arteries.

In tilt testing, the patient is passively moved from a supine position to 60–80° vertically. This results in a decrease in venous return and cardiac output which stimulates the baroreceptors (the mechanical squeeze is absent in passive tilting). Overall cardiac output decreases by 20% and blood pressure is maintained.[125]

Alcohol

In a study of acute intake of alcohol in normal subjects, the index of sensitivity of the baroreceptor reflex decreased significantly.[15] This was attributed to an inhibition of parasympathetic control. In chronic alcoholics, several studies showed a significant impairment in heart rate[120,121,126] but not to blood pressure[120,121] responses on standing.

Smoking

Mancia et al showed that smoking significantly decreased baroreceptor reflex sensitivity.[127] In contrast, Niedermaier et al reported that smoking resets the baroreceptor function to operate at a higher level to arterial pressure and did not affect its operational function.[107] In orthostatic testing, Kotamaki et al showed that the systolic blood pressure at 5 minutes after standing, and resting diastolic blood pressure was significantly higher in smokers.121

Other Substances

Published data on the effects of caffeine or marijuana influencing baroreceptor reflex sensitivity are not available. There are a number of animal studies, which show that cocaine inhibits baroreceptor reflex sensitivity.[128,129] This was originally thought secondary to the local anaesthetic effect of cocaine. However, Jacobsen et al recently investigated the effects of intranasal cocaine in humans and found no detectable effect on baroreflex control on sympathetic nerve discharge.[125] The reasons for these conflicting findings are not known.

In a group of heroin addicts, change in posture from supine to standing failed to show an increase in blood pressure.[90] Looking also at other measures of autonomic function, the same study concluded that heroin causes a decrease in sympathetic activity and inhibition of baroreceptor reflexes.

Amphetamines can cause marked orthostatic hypotension and inhibition of the baroreceptor reflex through its selective inhibition of central sympathetic outflow.[97] Analgosedation (a combination of benzodiazepine and an opioid analgesic) was found to raise baroreceptor reflex sensitivity in dogs.[98]

Deep Breathing

Heart rate variation with respiration, i.e. an increase with inspiration and a decrease with expiration, is also used as a cardiovagal test.

Alcohol

The acute intake of alcohol does not affect respiratory arrhythmia in deep breathing.[15] In long-standing chronic alcoholics with neuropathy, there is a significant decrease in the heart rate variation during both normal and deep breathing.[120,121]

Other Substances

The investigations of effects of smoking on respiratory arrhythmia due to deep breathing have shown no statistical differences.[120–122]

In a case report, amphetamine had no effect on deep breathing patterns.[97] Data on the effects of caffeine, cocaine, marijuana, heroin and benzodiazepines on deep breathing and respiratory arrhythmia are not available.

Cold Pressor Test

The cold pressor test evaluates the cardiovascular response to immersion of one or both hands in cold water. This manoeuvre induces an increase in heart rate, arterial pressure and vascular resistance. It tests the integrity of the sympathetic nervous system.

Alcohol

In chronic alcoholics the pressor response is significantly decreased.[119] There may also be a paradoxical decrease in blood pressure[119] attributed to alcoholic autonomic neuropathy.

Other Substances

Smoking and caffeine are associated with an increase in circulating catecholamines. Cocaine is an indirectly acting sympathomimetic drug. Therefore, we may expect an increase in the pressor response with the use of these substances. Published data to verify this hypothesis do not exist.

Benowitz et al[86] found a non significant impairment of the cold pressor response with prolonged tetrahydrocannabinol use. Beaconsfield et al[70] also showed an inhibition of normal limb vasoconstriction response after the immersion of the contralateral limb in cold water, when tetrahydrocannabinol was given in large doses.

An impaired test result is likely with heroin due to an impairment of the vasomotor centre. Amphetamine can inhibit the pressor response.[97]

ISOMETRIC HAND GRIP

Sustained muscle contraction causes blood pressure and heart rate to increase. The mechanism involves the exercise reflex, which withdraws parasympathetic tone and increases sympathetic activity.

Alcohol and Smoking

The extent of the rise in blood pressure in response to isometric exercise in patients with alcoholic neuropathy is significantly lower than in healthy controls.[119]

Smokers were reported to have a significantly higher increase in diastolic blood pressure than non smokers during the hand grip test .[121] This may reflect an increase in sympathetic activity or circulating catecholamines in smokers.

Other Substances

Caffeine and cocaine may also exert a similar effect to smoking, due to the association with a high catecholamine levels. The effects of marijuana, benzodiazepines and heroin on isometric hand grip have not been reported. However due to an impairment of the vasomotor centre by heroin and amphetamine, an impaired test is likely.

CONCLUSION

The autonomic nervous system has a key role in homeostasis, but its regulation involves numerous and complex reflex systems which are influenced by a number of environmental factors.

This chapter has highlighted that in the presence of recreational substances, some of the common methods of autonomic assessment may have their results influenced either directly or indirectly, that is through their effects on haemodynamic changes or by blocking or stimulating effects on the parasympathetic and/or sympathetic limbs. These confounding variables therefore need to be carefully documented prior to autonomic testing. To avoid possible bias, special care should be taken when interpreting the results of autonomic tests in alcoholics, heavy smokers and drub abusers.

Further attention should be directed to other environmental factors (e.g. pollutants) which are likely to affect the autonomic nervous system and its testing.

REFERENCES

1. Grassi G, Seravalle G, Calhoun D, Bolla G. Cigarette Smoking and the adrenergic nervous system. *Clin and Exper Hyperten* 1992, **A14**:251–260.
2. Rang H, Dale M. Pharmacology, Churchill Livivingstone, New York 1995.
3. AHA Task Force. Cigarette smoking and cardiovascular disease. *Circulation 1984*; **70**:1114A–1117A.
4. Zhu B Q, Parmley W. Haemodynamic and vascular effects of acute and passive smoking. *Am Heart J* 1995; **130**:1270–1275.
5. Klein L. Cigarette smoking, atherosclerosis and the coronary dynamic response: a unifying hypothesis. *J Am Coll Cardiol* 1984, **4**:972–974.
6. Zimmerman M, McGeachie J. The effects of nicotine on aortic endothelium: a quantitative ultrasound study. *Atherosclerosis*1984; **63**:33–41.
7. Wells AJ. Passive smoking as a cause of heart disease. *J Am Coll Cardiol* 1994; **24**:546–554.
8. Debias DA, Banerjee CM, Birkhead NC, Greene CH. Effects of carbon monoxide inhalation on ventricular fibrillation. *Arch Eviron Health* 1976; **31**:38–42.
9. Turino GM, Effect of carbon monoxide toxicity: physiology and biochemistry. *Circulation* 1981; **63**:253–259.
10. Green MS, Jucha E, Luz Y. Blood Pressure in smokers and non smokers: epidemiologic findings. *Am Heart J* 1986; **111**:932–940.
11. Pickering T, Schwartz JE, James JD. Ambulatory blood pressure monitoring for evaluating the relationship between lifestyle, hypertension and cardiovascular risk, *Clin Exp Pharmacol Physio* 1995; **22**:226–231.
12. Novak DJ, Victor M, The vagus and sympathetic nerve in alcoholic polyneuropathy. *Arch Neurol* 1974; **30**:273–284.
13. Guo Y, McLeod JG, Baverstock J. Pathological changes in the vagus nerve in diabetes and chronic alcoholism. *J Neurol Neurosurg Psychiat* 1987; **50**:1449–1453.
14. van de Borne P, Mark A, Montano N, Mion D, Somers V. Effects of alcohol on sympathetic activity, haemodynamics and chemoreflex sensitivity. *Hypertension* 1997; **29**:1278–1283.
15. Koskinen P, Virolainen J Kupari M. Acute alcohol intake decreases short term heart rate variability in healthy subjects. *Clin Sci* 1994; **87**:225–230.
16. Kanel WB, Sorlie P. Hypertension in Framingham, Epidemiology and control of hypertension. Stratton, New York 1975, 553–592.

17. Arkwright P, Beilin LJ, Rouse IL, Armstrong BK. Effects of alcohol use and other aspects of lifestyle on blood pressure levels and prevalence of hypertension in working populations. *Circulation* 1982; **66**:60–66.
18. Klatsky AL, Friedman GD, Siegeland AB, Geraard MJ. Alcohol Consumption and blood pressure. *New Eng J Med* 1977; **296**:1194–1200.
19. Braunnwald E. Heart disease, 5th Edn, Saunders, New York 1997, 1412–1414.
20. Adachi J, Mizoi Y. Acetaldehyde-mediated alcohol sensitivity and elevation of plasma catecholamine in man. *Japan J Pharmacol* 1983; **33**:531–539.
21. Guaza C, Borrell. Adrenomedullary response to acute and chronic ethanol administration to rats. *J Biochem Pharmaol* 1983; **32**:3091–3095.
22. Heikkonen E, Maki T, Kontula K, Ylikhari R, Harkonen M. Physical exercise after alcohol intake: effect on plasma catecholamine and lymphocytic beta adrenergic receptors. *Alcohol Clin Exp Res* 1991; **15**:291–294. .
23. Dillon J, Nolan J, Plevris J, Ewing D, Neilson J, Bouchier I. Autonomic function in cirrhosis assessed by cardiovascular reflex tests and 24 hour heart rate variability. *Am J Gastroenterol* 1994; **89**:1544–1547.
24. Gentile S, Marmo R, Peduto A, Montella F. Autonomic neuropathy in liver cirrhosis: relationship with alcoholic aetiology and severity of disease. *Ital J Gastroenterol* 1984; **26**:53–58.
25. Dillon J, Nolan J, Thomas H, Williams BC. The correction of autonomic dysfunction in cirrhosis by captopril. *J Hepatol* 1997; **26**:331–335.
26. Battig K. Coffee, cardiovascular and behavioural effects, current research trends, *Rev Environ Health* 1991; **9**:53–84.
27. Smits P, Thien T. Coffee and the human cardiovascular system. *Netherland J Med* 1987; **31**:36–45.
28. Myers MG. Effects of caffeine on blood pressure. *Arch Inter Med* 1988; **148**:1189–1193.
29. Robertson D, Frolich JC, Carr RK, Watson JT. Effects of caffeine on plasma renin activity, catecholamine and blood pressure. *New Engl J Med* 1978; **298**:181–186.
30. Robertson D, Wade D, Workman R, Woolsey R. Tolerance to the humoral and haemodynamic effects of caffeine. *J Clin Invest* 1981; **67**:1111–1117.
31. Ammon H, Biek PR, Mandelaz D, Verspohl EJ. Adaptation of blood pressure to continuous heavy coffee drinking in young volunteers. *Br J Clin Pharmacol* 1983; **15**:701–706.
32. Periti M, Salvaggio A, Quaglia G, Dimarzio L. Coffee consumption and blood pressure: an Italian Study. *Clin Sci* 1987; **72**:443–447.
33. Shirlow MJ, Berry G, Stokes G. Caffeine consumption and blood pressure: An epidemiological study. *Intl J Epidem* 1988; **17**:90–97.
34. Dusseldorp M, Smits P, Thien T, Katan MB. Effects of decaffeinated vs. regular coffee on blood pressure, a 12 wk. double blind trial. *Hypertension* 1989; **14**:563–569.
35. Rosmarin PC, Applegate WB, Somes GW. Coffee consumption and blood pressure; A randomised cross over clinical trial. *J Gen Intern Med* 1990; **5**:211–213.

36. Smits P, Thien T. The cardiovascular effects of regular and decaffinated coffee. *Br J Pharmacol* 1985; **53**:918–922.
37. Izzo J, Ghosal A, Kwong K, Freeman RB. Age and prior caffeine use alter the cardiovascular and adrenomedullary responses to oral caffeine. *Am J Cardiol* 1983; **52:** 769–773.
38. Smits P, Pieters G, Thien T. The role of epinephrine in the circulatory effects of coffee in man. *Clin Pharm Therap* 1986; **40**:431–37.
39. Di Marco JP, Sellers T, Berne RM, West GA. Adenosine: electrophysiological effects and therapeutic use for terminating paroxysmal SVT. *Circulation* 1983; **68**:1254–1263.
40. Fredholm B, The mechanism of action of theophylline and caffeine. *Acta Med Scand* 1985; **217**:149–153.
41. Dobmeyer DJ, Stine RA, Leier CV, Greenberg R. The arrhythmogenic effects of caffeine on human beings. *New Engl J Med* 1983; **309**:814–816.
42. Pirenas R, Jacobs DJ, Crow RS, Blackburn H. Coffee, tea and ventricular premature beats. *J Chronic Dis* 1980; **33**:67–72.
43. Sutherland DJ, McPherson DD, Reuton KW, Spencer CA. The effect of caffeine on cardiac rate, rhythm and ventricular repolarisation. *Chest* 1985; **87**:319–324.
44. Harris L, Myers MG, Leenen FHM, Grant DM. The Cardiovascular effects of caffeine in post myocardial infarction, *Circulation* 1985; **72**(Suppl III):116.
45. Arnesen E, Forde OH, Thelle DS. Coffee and serum cholesterol. *Br Med J* 1984; **288:** 1960.
46. Forde OH, Kuutsen SF, Arnesen E, Thelle DS. The Tromso Heart Study Group, Coffee consumption and serum lipid concentration in men with hypercholesterolemia; a randomised intervention study. *Br Med J* 1985; **290**:893–895.
47. Jick H, Miettinen OS, Neff RK, Shapiro S, Heinonen OP. Coffee and myocardial infarction. *New Engl J Med* 1973; **289**:63–67.
48. La Croix AZ, Mead LA, Liang KY, Thomas CB. Coffee consumption and the incidence of coronary heart disease. *New Engl J Med* 1986; **315**:977–982.
49. Dawber T, Kannel WB, Gordon T. Coffee and Cardiovascular disease; observation from The Framingham Study. *New Engl J Med* 1974; **291**:871–874.
50. Mouhaffet AH, Madu EC, Satmary WA, Fraker TD. Cardiovascular complication of cocaine. *Chest* 1995; **107**:1426–1434.
51. Moore K, Chiueh C, Zeldes G. Release of neurotransmitter from the brain in vivo by amphetamine, methylphenidate and cocaine. In: Ellinwood E, Kilbey M (Eds) Cocaine and other stimulants. Plenum, New York 1977, 143–160.
52. Chiueh C, Kopin J. Centrally mediated release of cocaine by endogenous epinephrine and norepinephrine from the sympathoadrenal medullary system of unanaesthetised rats. *J Pharmacol Exp Therap* 1978; **205**:148–154.
53. Hale SL, Alker KJ, Rezkalla SH, Eisenhauer AC, Kloner RA. Nifedipine protects the heart from the acute deleterious effects of cocaine if administered before but not after cocaine. *Circulation* 1991; **83**:1437–1443.

54. Perreault CL, Hague NL, Ransil BJ, Morgan JP. The effect of cocaine on intracellular calcium handling and myofilament calcium responsiveness in ferret myocardium. *Br J Pharmacol* 1990; **101**:679–685.
55. Smith H, Liberman HA, Broody SL, Battey LL, Donohue BC, Morris DC. Acute myocardial infarction temporally related to cocaine use. *Ann Intern Med* 1987; **107**:13–18.
56. Isner J, Estes NM, Thompson PD, Costanzo-Nordin MR, Subramanian R, Miller G, Katsas G, Swenney K, Sturner WQ. Acute cardiac events temporarily related to cocaine abuse. *New Engl J Med* 1986; **315**:1438–1443.
57. Cregler L, Mark H. Relation of acute myocardial infarction to cocaine abuse. *Am J Cardiol* 1985; **56**:794.
58. Zucker MB. Platelet function. In: Williams W, Butler E, Erslev AJ (Eds), Haematology. McGraw Hill, New York 1983, 1179.
59. Tonga G, Tempesta E. Platelet responsiveness and biosynthesis of thrombaxane and prostacyclin in response to in vitro cocaine treatment. *Homeostasis* 1985; **15**:100–107.
60. Chokshi S, Rongione A, Miller G. Cocaine and cardiovascular disease: the leading edge. *Cardiology* 1989; **111**:1–11.
61. Weiner R, Lockhart JT, Schwartz RG. Dilated Cardiomyopathy and cocaine abuse: report of 2 cases. *Am J Med* 1986; **81**:699–701.
62. Tazelaar HD, Karch SB, Stephens BG, Billingham ME. Cocaine and the heart. *Human Pathophysiol* 1987; **18**:195–199.
63. Karch S. The pathology and aetiology of cocaine induced heart disease. *Arch Pathol Lab Med* 1988; **112**:225–230.
64. Viramani R, Robinowitz M, Smialek JE, Smyth DF. Cardiovascular effects of cocaine: an autopsy of 40 patients. *Am Heart J* 1988; **115**:1068–1076.
65. Boag F, Havard CW. Cardiac arrhythmias and myocardial ischemia related to cocaine and alcohol consumption. *Post Grad Med J* 1985; **61**:957–989.
66. Ascher E, Stauffer J, Gaasch J WH. Coronary artery spasm, cardiac arrest, transient electrocardiographic Q waves and stopped myocardium in cocaine associated myocardial infarction. *Am J Cardiol* 1988; **61**:939–941.
67. Inou H, Zipes DP. Cocaine induced suprasensitivity and arrhythmogenesis. *J Am Coll Cardiol* 1988; **11**:867–874.
68. Chambers H, Morris L, Tanber MG. Cocaine use and the risk for endocarditis in intravenous drug abusers. *Ann Intern Med* 1987; **106**:833–836.
69. Tashkin D. Cannabis. *Ann of Intern Med* 1978; **89**:539–549.
70. Beaconsfield P, Ginsburg P, Rainsbury R. Marijuana smoking: Cardiovascular effects in man and possible mechanisms. *New Engl J Med* 1972; **287**:209–212.
71. Johnson S, Domino E. Some cardiovascular effects of marijuana smoking in normal volunteers. *Clin Pharmacol Therap* 1971; **12**:762–768.
72. Weiss J, Watanabe A, Lemberger L. Cardiovascular effects of delta 9 tetrahydrocannabinol in man. *Clin Pharmacol Therap* 1972; **13**:671–684.
73. Graham J, Li DM. Cardiovascular effects of cannabis in cats and rats. *Br J Pharmacol* 1973; **49**:1–10.

74. Domino EF, Rennick P, Pearl JH. Dose-effect relations of marijuana smoking on various physiological parameters in experienced male users. *Clin Pharmacol Therap* 1974; **15**:514–520.
75. Benowitz N, Jones RT. Prolonged delta 9 tetrahydrocannabinol ingestion. Effects of sympathomimetic amines and autonomic blockades. *Clin Pharmacol Therap* 1977; **21**:336–342.
76. Kochar M, Hosko MJ. Electrocardiographic effects of marijuana. *J Am Med Ass* 1973; **225**:25–27.
77. Hollister L, Actions of various marijuana derivatives in man. *Pharmacol Rev* 1971; **23**:349–357.
78. Perez-Reyes M, Lipton MA, Timmons MC, Wall ME, Brine DR, Davis KH. Pharmacology of orally administered delta 9 tetrahydrocannabinol. *Clin Pharmacol Therap* 1973; **14**:48–55.
79. Renault P, Schuster C, Freedman. Repeat administration of marijuana smoke to humans. *Arch Gen Psychiat* 1974; **31**:95–102.
80. Arnow W, Cassidy J. Effect of marijuana and placebo-marijuana smoking on angina pectoris. *New Engl J Med* 1974; **291**:65–67.
81. Kanakis C, Pouget C, Rosen KM. The effect of delta 9 tetrahydrocannabinol on cardiac performance with and withwout beta blockade. *Circulation* 1976; **53**:703–707.
82. Shapiro BJ, Reiss S, Sullivan SF, Tashkin DP. Cardiopulmonary effects of marijuana smoking during exercise. *Chest* 1976; **70**:441.
83. Hardman H, Domino E, Seevers. General pharmacological actions of some synthetic delta tetrahydrocannabinol derivatives. *Pharmacol Rev* 1971; **23**:295–315.
84. Cavero I, Buckley JP, Jandhyala BS. Haemodynamic and myocardial effects of delta 9 tetrahydrocannabinol in anaesthetised dogs. *Eur J Pharmacol* 1973; **24**:243–251.
85. Vollmer R, Cavero I, Ertel RJ, Solomon TA. Role of the central nervous system in the hypotension and bradycardia induced by delta 9 tetrahydrocannabinol. *J Pharmac Pharmacol* 1974; **26**:186–192.
86. Benowitz N, Jones RT. Cardiovascular effects of prolong delta 9 tetrahydrocannabinol ingestion. *Clin Pharmacol Therap* 1975; **18**:287–297.
87. Neelman J, Farrell M. Fatal methadone and heroin overdoses: time trends in England and Wales. *J Epidemiol Comm Health* 1997; **51**:435–437 .
88. Duberstein J, Kaufman DM. A clinical study of an epidemic of heroin intoxication and heroin induced pulmonary odema. *Am J Med* 1971; **51**:704–714.
89. Lipski J, Stimmel B, Donoso E. The effect of heroin and multiple drug abuse on the ECG. *Am Heart J* 1973; **86**:663–668.
90. Kapoor R, Singh S, Gandhi A. Autonomic functions and audovisual reaction time in heroin addicts. *Indian J Physiol Pharmacol* 1993; **37**:209–212.
91. Arif A, Westermeryer I. Manual of drug and alcohol abuse. Plenum, London 1988, 131–132.
92. Kalant H, Kalant OJ. Death in amphetamine users causes and rates. *Can Med Ass* 1975; **112**:299–304.

93. Carson P, Oldroyd K. Myocardial infarction due to amphetamine. *Br Med J* 1987; **294:**1525–1526.
94. Call T, Hartneck J, Dickinson WA. Acute Cardiomyopathy secondary to intravenous amphetamine abuse. *Ann Intern Med* 1982; **97:**559–560.
95. Smith H, Roche AH, Jaqusch MF. Cardiomyopathy associated with amphetamine administration. *Am Heart J* 1976; **91:**792–797.
96. Citron B, Halperm M, McCarron M. Necrotising angiitis associated with drug abuse, *New Engl J Med* 1970; **283:**1003–1011.
97. Smit A, Wieling W, Voogel A, Koster R. Orthostatic hypotension due to suppression of vasomotor flow after amphetamine intoxication. *Mayo Clin Proceed* 1996; **71:**1067–1070.
98. Vrana M, Fejfar Z, Hess L, Blazek Z. Analgosedation raises baroreflex sensitivity in acute myocardial ischemia. *Cor et Vasa* 1992; **34:**421–427.
99. Task Force of the European Society of Cardiology and the North American Society of Pacing and Electrophysiology: Heart rate variability – Standards of measurement, physiological interpretation, and clinical use. *Circulation* 1996; **93:**1043–1065.
100. Hales. Statistical Essays, Vol. 11 Haemostaticks, Innings Manby, London 1733.
101. Murata K, Araki S, Yokoyama K, Sata F, Yamasihta K. Autonomic neurotoxicity of alcohol assessed by heart rate variability. *J Auton Nerv Syst* 1995, **48:**105–111.
102. Weise F, Krell D, Brinkhoff N. Acute alcohol ingestion reduces heart rate variability. *Drug Alcohol Depend* 1986; **17:**89–91.
103. Yokoyama A, Takagi T, Muramatsu T, Hori S. Impaired autonomic nervous system in alcoholics assessed by heart rate variation. *Alcohol Clin Exp Res* 1991; **15:**761–765.
104. Weise F, Muller D, Krell D, Kielstein V. Heart rate variability in chronic alcoholics: a follow up study. *Drug Alcohol Depend* 1986; **17:**365–368.
105. Hirsch JA, Bishop B, Yook JL. Recovery of respiratory sinus arrhythmia in detoxified alcohol subjects. *J Appl Physiol* 1993; **74:**1816–1823.
106. Hayano J, Yamada M, Sakakibara Y, Fujinami T, Takata K. Short and long term effects of cigarette smoking on heart rate variability. *Am J Cardiol* 1990; **65:**84–88.
107. Niedermaier O, Smith M, Beightol M, Zukowska-Grojec Z, Goldstein G. Influence of cigarette smoking on human autonomic nervous function. *Circulation* 1993; **88:**562–571.
108. Kolodiichuk E, Arushanian EB. The effect of caffeine on the cardiac intervalogram indices depending on the ovarian cycle phase in women. *Farmakol Toksikol* 1991; **54:**28–30.
109. Stambler BS Morgan J, Mietus J, Moody G. Cocaine alters heart rate dynamics in conscious ferrets. *Yale J Bio Med* 1991; **64:**143–153.
110. Oriol N, Bennett F, Rigney D, Goldberger A. Cocaine effects on Neonatal Heart Rate Dynamics: preliminary findings and methodological problems. *Yale J Bio Med* 1993; **66:**75–84.

111. Regaldo M, Schechtman V, Bean X, Del Angel P. Cardiac and respiratory patterns during sleep in cocaine exposed neonates. *Early Human Develop* 1996; **44**:187–200.
112. Mehta S, Finkelhor R, Anderson R, Harcar-Sevcik, Wasser T. Transient myocardial ischemia in infants prenatally exposed to cocaine. *J Paed* 1993; **122**:945–949.
113. Liu W, Varner K, Characterisation of the sympathetic nerve responses to Amphetamine: role of central α-adrenergic receptors. *J Cardiov Pharmacol* 1996; **28**:712–722.
114. Shindler C, Zheng JW, Tella SR, Goldberg SR. Pharmacological mechanisms in the cardiovascular effects of methamphetamine in monkeys. *Pharmacol Biochem Behav* 1992; **42**:791–796.
115. Sharma U, Sheth UK. Hypotensive response to repeated doses of amphetamine and mephentermine in exp. Animals. *Indian J Med Res* 1977; **65**:101–104.
116. Tulen J, Mulder G, Moleman P, Pepplinkhuizen L, Man In 'T Veld, Van Steenis H. Effects of lorazepam on cardiac vagal tone during rest and mental stress: assessment by means of spectral analysis. *Psychopharmacol* 1994; **114**:81–89.
117. Adinoff B, Mefford I, Waxman R, Linnoila M. Vagal tone decreases following intravenous diazepam. *Psychiatry Res* 1992; **41**:89–87.
118. Sharpey-Shafer EP. Effects of Respiratory acts on the circulation. In: Hamilton WF, Dow P (Eds) Handbook of Physiology. American Physiology Society, Washington 1964, Sec. 2, Vol. 3, 1875–1886.
119. Chida K, Takasu T, Mori N, Tokunaga K, Komatasu K. Sympathetic dysfunction mediating cardiovascular regulation in alcoholic neuropathy. *Funct Neurol* 1994; **2**:65–73.
120. Monforte R, Estruch R, Valls Sole J, Nicolas J, Villalta J, Urbano-Marquez A. Autonomic and peripheral neuropathies in patients with chronic alcoholism. *Arch Neurol* 1995; **52**:45–51.
121. Kotamaki M. Smoking induced differences in autonomic responses in military pilot candidates. *Clin Autonom Res* 1995; **5**:31–36.
122. Piha SJ, Seppanen A. Attenuated Valsalva heart rate response in subjects with smoking history: possible role of autonomic dysfunction. *Clin Physiol* 1993; **13**:51–55.
123. Piha S. Cardiovascular autonomic reflexes in heavy smokers. *J Autonom Nerv Syst* 1994; **48**:73–77.
124. Jacobsen T, Grauburn P, Snyder R, Hansen J, Landau C, Hillis D. Effects of intranasal cocaine on sympathetic nerve discharge in humans. *J Clin Inv* 1997; **99**:628–634.
125. Ravits J. AAEM minimonograph: 48: autonomic nervous system testing. *Muscle and Nerve* 1997; **20**:919–937.
126. Rechlin T, Orbes I, Weiss M, Kaschka W. Autonomic cardiac abnormalities in alcohol-dependent patients admitted to a psychiatric department. *Clin Autonom Res* 1996; **6**:119–122.
127. Mancia G, Gropelli A, Rienzo D, Castiglioni P, Parati G. Smoking impairs baroreflex sensitivity in humans. *Am J Physiol* 1997; **273**:H1555–H1560.

128. Andresen MC, Yang M, Nelson SH, Steinsland OS, Cocaine inhibits baroreflex control of blood pressure by actions at arterial baroreceptors. *Am J Physiol* 1990; **258**:H1244–H1249.
129. Trouve RG, Nahas G, Latour C. Inhibition of cocaine on the baroreflex in the rat. *Proc Soc Exp Biol Med* 1992; **201**:215–218.

CHAPTER FOURTEEN

CARDIAC AUTONOMIC SYNDROMES

Roy Freeman

Neurological disorders of both the central and peripheral nervous system frequently involve the cardiovascular system. Cardiovascular autonomic dysfunction is often the most significant feature of these disorders and may result in substantial disability, morbidity and even death. The features of cardiovascular autonomic dysfunction include orthostatic hypotension, poor exercise tolerance, autonomic dysreflexia, baroreflex failure, silent myocardial ischemia and infarction, cardiac arrhythmias and sudden death.[1,2] This chapter will provide a neuro-anatomical approach to classify the neurological disorders that result in cardiovascular autonomic dysfunction. We will describe the features of these disorders and the associated autonomic test abnormalities.

ANATOMY OF THE AUTONOMIC NERVOUS SYSTEM

The autonomic nervous system has central, afferent and efferent pathways. The central areas of autonomic control are distributed throughout the neuraxis. The efferent pathways are organised in two major outflows, the sympathetic and parasympathetic nervous system. Visceral afferent neurones travel to the central nervous system in both sympathetic and parasympathetic nerves.

Malik M (Ed) Clinical guide to cardiac autonomic tests, 357–391.

CENTRAL AUTONOMIC PATHWAYS

Areas of the cerebral cortex, basal forebrain, hypothalamus, midbrain, pons and medulla are involved in autonomic control. These areas are interconnected and constitute a functional unit, referred to as the 'central autonomic network' (CAN). The CAN integrates visceral, humoral, and environmental information and co-ordinates autonomic with endocrine and behavioural responses to various stimuli.[3–5] Within this CAN there is well-established regional specialisation for the control of autonomic functions.

Cerebral Cortex

Studies in mammalian species suggest the insula function as a 'visceral sensory cortex.' Anatomic and physiologic studies have defined a topographically organised visceral map in this region, with localisation of neurones responsive to cardio-respiratory, baroreceptor, and chemoreceptor-responsive neurones in the caudal granular insular region.[3,4,6]

Stimulation of the medial prefrontal cortex (including the anterior cingulate gyrus and the inferior and prelimbic cortices) produces profound changes in blood pressure and heart rate.[3–5] Both its position on the medial bank of the cerebral hemisphere near the frontal motor cortex and its pattern of anatomic connectivity (namely extensive limbic sensory input with efferent connections to diencephalic and brainstem autonomic nuclei) suggest that this region might function as a 'visceral motor cortex.'[7,8]

The amygdala receives exteroceptive and interoceptive information that provides it with emotional significance. Information from these regions is then integrated and the appropriate autonomic, endocrine, and behavioural responses are generated through interconnections between the amygdala and the neocortex, basal forebrain, ventral striatum, hypothalamus, and autonomic and motor nuclei of the brainstem. The amygdala thus plays an important role in the coordination of the autonomic and behavioural responses to emotion.[9]

Hypothalamus

The autonomic output of the hypothalamus originates from the paraventricular nucleus, dorsomedial nucleus, and lateral hypothalamic area, and sends direct projections to autonomic centres in the brainstem and spinal cord. The hypothalamic autonomic pathways descend predominantly ipsilaterally in the dorsomedial and ventrolateral tegmentum of the brainstem.

Brainstem

The nucleus of the solitary tract is the first relay station of cardiovascular afferents. This nucleus integrates multiple autonomic reflexes and projects to other central autonomic regions. The reticular formation of the ventrolateral medulla contains distinct groups of neurones that maintain cardiomotor, vasomotor and respiratory functions. Neurones of the rostral ventrolateral medulla project directly to sympathetic preganglionic neurones and maintain basal arterial blood pressure. The nucleus ambiguous and neurones in its region provide the parasympathetic innervation to the heart.[5,10–12]

EFFERENT PATHWAYS

Reciprocally interconnected cortical, subcortical, and brainstem regions regulate the preganglionic sympathetic and parasympathetic efferent pathways. These pathways consist of a two-neurone pathway with one synapse in the autonomic ganglia. Sympathetic preganglionic neurones are located in the intermediolateral cell column of the spinal cord at the T1-L3 segments (thoracolumbar outflow). Parasympathetic preganglionic neurones are located both in the brainstem and in the S2-S4 segments of the spinal cord (craniosacral outflow). Sympathetic and parasympathetic preganglionic axons emerge from the brainstem or spinal cord as small myelinated fibres and release acetylcholine at their synapse on neurones of the autonomic ganglia.

Sympathetic and parasympathetic ganglion neurones send postganglionic unmyelinated varicose axons to the peripheral effector organs. The primary postganglionic sympathetic neurotransmitter is norepinephrine, which acts on alpha and beta adrenergic receptors. The primary postganglionic parasympathetic neurotransmitter is acetylcholine, which acts on muscarinic receptors. Postganglionic sympathetic terminals to the heart and smooth muscles also release ATP and neuropeptide Y, whereas the postganglionic parasympathetic terminals may release vasoactive intestinal polypeptide or nitric oxide.[13,14]

CARDIOVASCULAR SYSTEM

The heart receives parasympathetic and sympathetic innervation. The cell bodies of parasympathetic preganglionic neurones innervating the heart are located in the medulla. The vagus nerve provides brainstem cardiovascular parasympathetic output. Vagal preganglionic neurones are located in dorsal motor nucleus, which controls respiratory function, and in the region of the nucleus ambiguous, which provides cardioinhibitory innervation to the heart. Vagal neurones join the cardiac neural plexus and synapse with neurones in intracardiac ganglia. From these ganglia, short postganglionic parasympathetic neurones emerge to innervate the myocardial tissue. Vagal innervation is particularly abundant in the S-A node and the A-V conduction system. The right vagus nerve predominantly innervates the S-A node while the left vagus nerve predominantly innervates the A-V node. The response to vagal nerve activity has a short latency because the muscarinic receptor is coupled directly to potassium channels by a G protein and does not require activation of a second messenger system.[15,16]

The sympathetic innervation of the heart is derived from neurones in the intermediolateral columns of the spinal cord from the lower cervical and at the T1 to T6 thoracic levels. Axons of these neurones synapse in the superior, middle and inferior (stellate) cervical ganglia which are the origin of postganglionic sympathetic neurones that reach the heart in a neural plexus mixed with preganglionic parasympathetic neurones. Sympathetic axons innervate the S-A and A-V node, the conduction system and myocardial muscle fibres most prominently in the ventricles. The heart rate response to sympathetic stimulation is more gradual as activation of the secondary messenger cyclic AMP is required.[15,16]

DISEASES OF THE AUTONOMIC NERVOUS SYSTEM

CENTRAL AUTONOMIC DISORDERS

Generalised Central Neurodegenerative Diseases

Multiple system atrophy (MSA) with autonomic failure (the Shy-Drager syndrome) and Parkinson's disease are the most prevalent neurodegenerative diseases that result in autonomic failure. These disorders are often phenotypically similar and a definitive diagnosis may only be made at autopsy. MSA is a degenerative disorder of the central nervous system that affects extrapyramidal, paramidal cerebellar and autonomic neurones. Patients with MSA typically present with autonomic symptoms such as orthostatic hypotension, bowel and bladder dysfunction, anhidrosis and impotence, together with motor dysfunction that can involve the extrapyramidal, cerebellar and pyramidal systems.[17–19] The symptoms of orthostatic hypotension usually lead patients to seek medical attention. This disorder is frequently misdiagnosed as idiopathic Parkinson's disease. In contrast to idiopathic Parkinson's disease, MSA is more likely to have a symmetrical onset and is less responsive to levodopa and dopamine agonists. The resting tremor that is characteristic of Parkinson's disease is rarely present in MSA, while inspiratory stridor, vocal cord paralysis and antecollis commonly occur. The median survival of MSA is usually less than ten years. Pathological studies have demonstrated cell loss and gliosis that involves the striatonigral, olivopontocerebellar and autonomic systems. Autonomic dysfunction in MSA is due to loss of preganglionic neurones in the brainstem and spinal cord. An argyrophilic, intracytoplasmic oligodendriglial inclusion may be a specific pathological hallmark of multiple system atrophy.[20]

In contrast, the autonomic features of Parkinson's disease are usually not as severe as those seen in patients with MSA. They characteristically occur late in the course of the illness and are often associated with levodopa and dopamine agonist therapy. Nevertheless, autonomic dysfunction is frequently the source of significant morbidity for the Parkinsonian patient. In patients with Parkinson's disease, Lewy bodies are found not only in central but also in peripheral sympathetic ganglia and neurones suggesting that autonomic dysfunction in this disorder may be caused by both pre and postganglionic neuron dysfunction.[21]

Cardiovascular autonomic testing has been performed in a number of patients with MSA and Parkinson's disease.[18,22–31] A comparison of cardiovascular autonomic function test results between patients with Parkinson's disease and MSA with similar extrapyramidal features, reveals that the autonomic test abnormalities are quite marked in patients with MSA and generally milder in patients with Parkinson's disease.[28–30,32] For example, the reflex responses to standing[25,33] and to the Valsalva manoeuvre[22,25,33] were found to be preserved in patients with Parkinson's disease but to be deficient or absent in all patients with the MSA syndrome.[25] These autonomic test abnormalities in patients with MSA are present in tests measuring both the parasympathetic and sympathetic nervous system.[28] Using a scale that normalises for age and gender, test abnormalities in patients with MSA are of similar severity to patients with peripheral neuropathies that have manifestations of autonomic failure. In contrast, patients with Parkinson's disease had autonomic test abnormalities of magnitude that is similar to patients with a peripheral neuropathy without autonomic symptoms.[32] Cardiac parasympathetic nervous system abnormalities, as determined by measures of heart rate variability, are present in the majority of MSA patients (81% in one study)[18] In patients with MSA the heart rate response to deep respiration appears more severely impaired than the heart rate response to a Valsalva manoeuvre (determined by measuring the Valsalva ratio).[28] When carrying out such studies it is important to take into account the effects of the motor disability that is associated with these neurological disorders on the test performance.

There are several other generalised central neurodegenerative diseases such as progressive supranuclear palsy, amyotrophic lateral sclerosis (motor neuron disease), Alzheimer's disease, Huntington's disease and familial olivopontocerebellar atrophy (see Table 1). The autonomic features of these disorders are usually of secondary importance in comparison to the motor and cognitive manifestations. Mild or clinically insignificant autonomic dysfunction may be present in these disorders and cardiovascular autonomic function testing may demonstrate subclinical abnormalities.[33–37]

Telencephalic Disorders

Temporo-limbic seizures may induce changes in heart rate, rhythm and blood pressure. Seizure-induced ventricular tachycardia and fibrillation have been implicated in sudden death.[38–40] Sinus tachycardia, the most frequent cardiac concomitant of seizures, has been reported in several case studies.[41–44] Bradyarrhythmias occur less frequently. Nevertheless, sinus bradycardia, sinus arrest, A-V block, and prolonged asystole have been the subject of a number of case reports.[43–45,45–49] Other cardiac arrhythmias have been reported in individual case studies.[47,48,50,51] In the prospective study of Blumhardt, documented arrhythmias were present in 42% of patients during seizures and included repeated acceleration and deceleration of heart rate, AV nodal escape, paroxysmal supraventricular tachycardia, ventricular ectopy and bradycardia.[52] In contrast to this study, Keilson et al[53,54] noted no ventricular ectopy or cardiac conduction defects. Preliminary investigations of cardiovascular autonomic function during the interictal state have revealed minor abnormalities. These abnormalities may, in part, be secondary to antiepileptic medications.[55,56]

Ischaemic damage to the cerebral hemispheres, particularly the insula and prefrontal cortical regions, may result in autonomic dysfunction including a predisposition to cardiac arrhythmias. [57,58] Several investigators have carried out cardiac autonomic function testing of patients with hemisphere strokes. Results to date are inconclusive. Preliminary evidence suggests that cerebro-vascular disease of either hemisphere is associated with reduced heart rate modulation.[59,60] The changes in cardiac parasympathetic nervous system control may be more marked in patients with right hemisphere lesions[59] although not all evidence supports the lateralisation of parasympathetic control.[57,61]

Diencephalon

Hypothalamic disorders frequently produce disturbances of thermoregulation, osmotic balance, endocrine function, and state of alertness.[62–65] Cardiovascular manifestations of diencephalic disorders are less prominent. Hypertension may be a feature of diencephalic autonomic 'epilepsy' – a term introduced by Penfield to describe a constellation of autonomic signs and symptoms that occurred in a patient with an tumor at the foramen of Munro that impinged upon the thalami. The patient presented with a syndrome of paroxysmal flushing, diaphoresis, mydriasis, hypertension, lacrimation, shivering, hiccoughing, and respiratory changes that Penfield suggested was secondary to mechanical irritation by the tumor. Other authors have reported similar episodes associated with head trauma,[66] neoplasms,[67] neuronal degeneration,[68] and hydrocephalus,[66] and without obvious pathology.[69] Although the pathogenesis of diencephalic epilepsy is uncertain, it is unlikely that the manifestations are a consequence of a true epileptiform disorder. This disorder is more likely to represent a release phenomenon or a periodic functional hypothalamic alteration. Cardiovascular autonomic dysfunction has not been studied systematically in this disorder.

Sympathetic hyperactivity also occurs in familial fatal insomnia – a prion disease characterised by severe atrophy of the anteroventral

TABLE 1.
Central autonomic nervous system disorders

GENERALIZED CENTRAL NEURODEGENERATIVE DISORDERS
- Multiple system atrophy with autonomic failure (Shy-Drager Syndrome)
- Parkinson's disease
- Progressive supranuclear palsy
- Huntington's disease
- Familial olivopontocerebellar atrophy
- Amyotrophic lateral sclerosis (motor neuron disease)
- Alzheimer's disease

TELENCEPHALIC DISORDERS
- Cerebrovascular disease
- Temporolimbic seizures (autonomic epilepsy)

DIENCEPHALIC DISORDERS
- Wernicke's disease
- Episodic hyperhidrosis with hypothermia
- Paroxysmal autonomic hyperactivity (diencephalic epilepsy)
- Fatal familial insomnia
- Iatrogenic
  - Neuroleptic malignant syndrome
  - Serotonin syndrome

BRAINSTEM DISORDERS
- Tumors
- Vertebrobasilar ischemia
- Syringobulbia
- Arnold-Chiari type I malformation
- Demyelinating disease (e.g. multiple sclerosis)
- Infectious disease (e.g. poliolmyelitis)

SPINAL CORD DISORDERS
- Trauma
- Syringomyelia
- Demyelinating disease
- Tetanus
- Stiff man syndrome

and dorsomedial nuclei of the thalamus. This autosomal dominant disorder presents with disruption of endocrine circadian rhythms, motor dysfunction and progressive intractable insomnia.[70–74] Evaluation of cardiovascular autonomic reflexes in patients with familial fatal insomnia reveals intact baroreflex pathways but exaggerated blood pressure and biochemical responses to postural change, Valsalva manoeuvre and isometric handgrip. The autonomic test results suggest preserved parasympathetic function with increased basal and provoked sympathetic nervous system activity.[73]

Brainstem Disorders

Vertebrobasilar ischemia is the most common brainstem disorder that results in autonomic dysfunction. Transient ischaemic attacks in the basilar artery territory may present with paroxysmal hypertension even before any focal neurological deficit.[75] The autonomic manifestations of the lateral medullary infarction (Wallenberg's syndrome) include a Horner's syndrome and, occasionally, more severe autonomic abnormalities such as profound bradycardia, supine hypotension, or central hypoventilation.[76] Time frequency maps of RR interval and blood pressure of patients with severe brainstem injury secondary to cerebro-vascular disease reveal decreased heart rate modulation over 0.01–0.5 Hz. Low frequency blood pressure variability (0.01–0.12 Hz) was present although attenuated in these patients. These results suggest that heart rate and blood pressure variability are modulated by different sites within the neuraxis. An intact brainstem appears necessary for the control of heart rate modulation while low frequency blood pressure modulation may be mediated at the spinal level.[77] (See Figures 1 and 2)

Other brainstem disorders associated with autonomic dysfunction include tumours, syringobulbia, Arnold-Chiari malformation type 1, multiple sclerosis and poliomyelitis. (See Table 2). Autonomic features of brainstem tumours include intractable vomiting, orthostatic hypotension, paroxysmal hypertension and heart rate changes. [78,79] Syringobulbia may be associated with Horner's syndrome, lability of arterial pressure, and central hypoventilation. Autonomic testing may reveal abnormalities in cardiac vagal control and orthostatic hypotension[80–82] Syncope, sleep apnea, and cardio-respiratory arrest have been reported in association with the Arnold-Chiari malformation type 1.[83] A rare autonomic manifestation of brainstem dysfunction in multiple sclerosis is fulminant neurogenic pulmonary oedema due to demyelination of the area surrounding the nucleus of the tractus solitarius.[84]

A

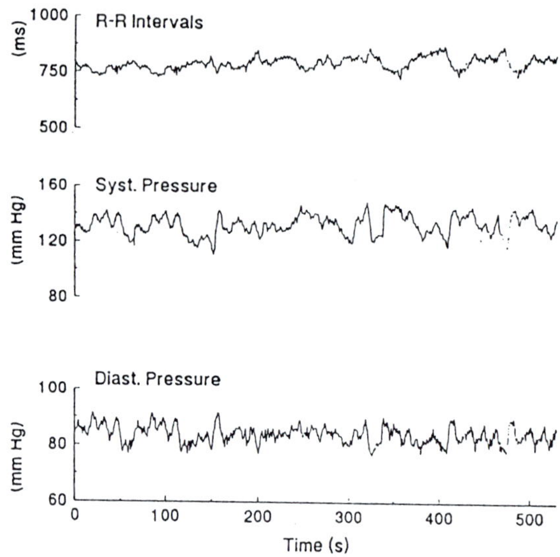

Figure 1. Signal of R-R interval, systolic and diastolic pressure (A) and corresponding time-frequency maps (B) from an elderly healthy woman at rest. (From Novak V, et al. The effect of severe brainstem injury on heart rate and blood pressure oscillations. *Clin Autonom Res* 1995; 5:24–30.)

Spinal Cord

Traumatic spinal cord injury, particularly above the T5 level, is associated with severe and disabling cardiovascular, gastrointestinal, bladder, and sexual dysfunction. These patients have both supine and orthostatic hypotension. The vasopressin and renin-angiotensin-aldosterone system play an important role in maintenance of orthostatic arterial pressure. Autonomic dysreflexia may be triggered by stimuli that arise from the bladder, bowel, trachea and skin. The manifestations of autonomic dysreflexia include vasodilatation above the level of the lesion that produces flushing of the face, chest, and upper arms and congestion of the nasopharyngeal mucosa. Severe hypertension may occur that can lead to hypertensive encephalopathy or intracranial, subarachnoid, or retinal haemorrhages. The

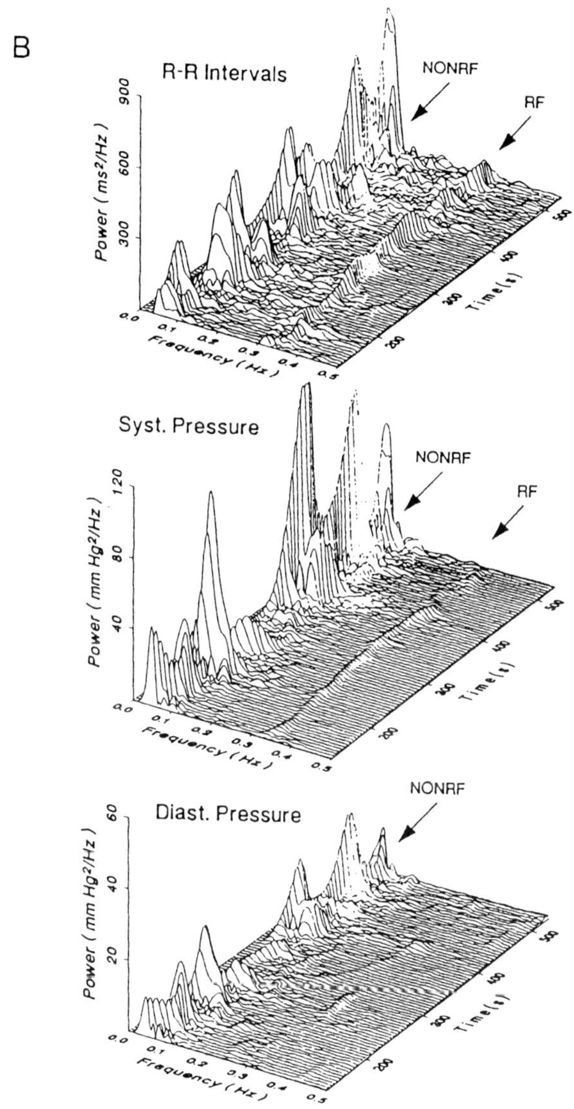

FIGURE 1 (Continued). See the legend opposite.

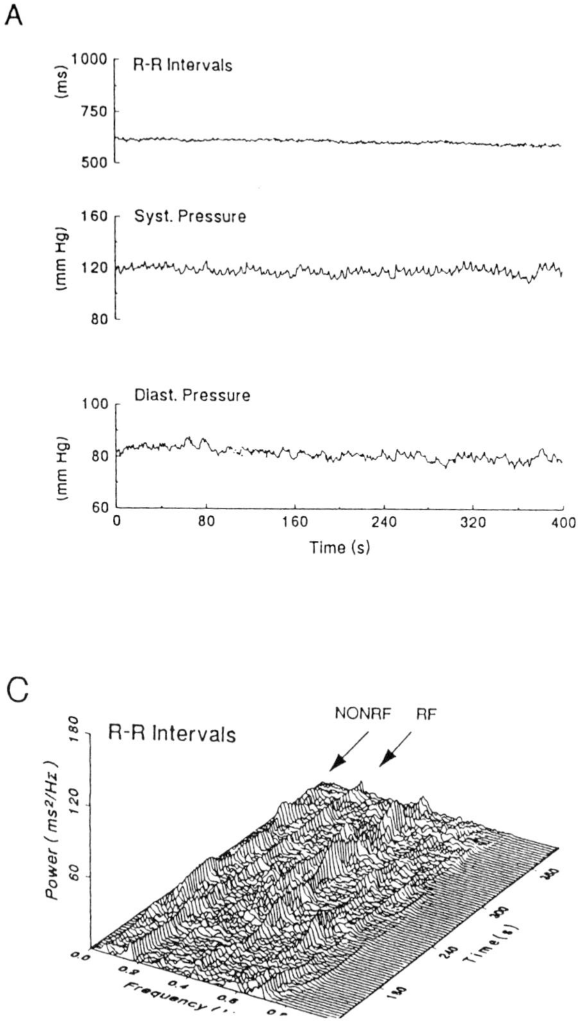

FIGURE 2. Signal of R-R intervals and systolic pressure (A) from an elderly patient with pontine haemorrhage and 'brainstem death.' Time-frequency maps of RRI, systolic and pressure (B – page opposite). Time-frequency map of R-R intervals displayed over a larger frequency range (0.01–1 HZ) (C). (From Novak V, et al. The effect of severe brainstem injury on heart rate and blood pressure oscillations. *Clin Autonom Res* 1995; **5**:24–30.)

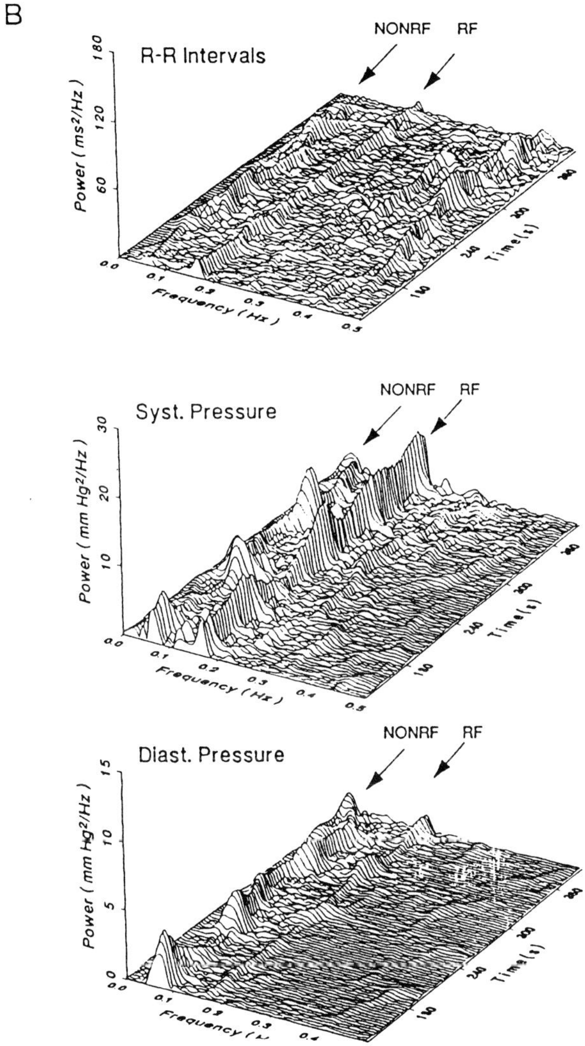

FIGURE 2 (Continued). See the legend opposite.

combined parasympathetic and sympathetic stimulation may cause potentially dangerous supraventricular and ventricular brady- and tachyarrhythmias.[85]

Studies of blood pressure and heart rate variability using autoregressive power spectral analysis has shown that low frequency heart rate and systolic blood pressure variability is reduced or absent in quadriplegic subjects while high frequency variability is relatively spared. These reports suggest that the cervical spinal sympathetic pathways play a role in the generation of heart rate and blood pressure low frequency variability. They also provide evidence that interruption of the spinal pathways linking supraspinal cardiovascular centres with the peripheral sympathetic outflow attenuates low frequency heart rate and blood pressure variability.[86–88]

Multiple sclerosis (MS) may affect autonomic pathways at the level of the spinal cord, brainstem, or diencephalon. Bladder, bowel and sexual dysfunction are prominent autonomic manifestations of MS. Symptomatic orthostatic hypotension may be present. Subclinical abnormalities in cardiovascular sympathetic and parasympathetic function are frequently detected.[89–93]

Syringomyelia produces partial interruption of sympathetic output pathways in the intermediolateral cell columns. Its autonomic manifestations include Horner's syndrome, sudomotor and vasomotor dysfunction, and trophic changes in the limbs, especially the hands. The autonomic test abnormalities are described above. Tetanus and the stiff-man syndrome are spinal cord disorders that may be associated with sympathetic hyperactivity due to impairment of synaptic inhibition of preganglionic autonomic neurones. This results in hyperpyrexia, sweating, tachycardia, hypertension, tachypnea and pupillary dilatation, usually associated with the muscle spasms.[94]

PERIPHERAL AUTONOMIC DISORDERS

PURE AUTONOMIC FAILURE

Pure autonomic failure (PAF) is an idiopathic peripheral autonomic nervous system degeneration that, in contrast to the Shy-Drager syndrome, has no motor manifestations.[19,95,96] There are also no signs or minimal signs of a somatic peripheral neuropathy. Because this disorder is slowly progressive and has a significantly better prognosis than the Shy-Drager syndrome, it is important to differentiate between these two disorders. The autonomic features of the Shy-Drager syndrome may, however, precede the other neurological signs by several years preventing an early definitive diagnosis.[97] Patients with PAF typically have a low resting plasma norepinephrine level due to degeneration or dysfunction of the postganglionic sympathetic neuron. The plasma norepinephrine level, which in normal subjects increases by 100–200% when moving from the supine to the upright position, does not change significantly in patients with PAF. There is, however, a wide scatter of results within the group of patient with PAF that often makes it difficult to classify individual patients.[18,98] Cardiovascular autonomic tests show impairment with a severity that is equivalent to that seen in patients with MSA. For example, in a controlled study of autonomic function 77% PAF patients had abnormal measures of heart rate variability in response to deep respiration (less than the 1st percentile) and 91% had an abnormal Valsalva ratio.[18]

AUTONOMIC PERIPHERAL NEUROPATHY

Autonomic dysfunction occurs with most peripheral neuropathies – particularly those that predominantly involve the small or unmyelinated fibres.[99] This review will cover those peripheral neuropathies in which autonomic dysfunction is a clinically significant manifestation. A complete list of peripheral neuropathies with autonomic manifestations is found in Table 2.

Diabetes Mellitus

Diabetic autonomic neuropathy is the most common autonomic neuropathy in the developed world. A constellation of autonomic features occur that affect cardiovascular, gastrointestinal, urogenital, thermoregulatory, sudomotor and pupillomotor function.[100,101] The cardiovascular autonomic manifestations include an increased resting heart rate that may be followed by a fixed heart rate that responds only minimally to physiological stimuli. The initial tachycardia is most likely due to due to a vagal cardiac neuropathy. The ensuing decrease in heart rate and ultimately fixed heart rate is predominantly due to progressive cardiac sympathetic nervous system dysfunction. Orthostatic hypotension occurs in diabetes as a consequence of efferent sympathetic vasomotor denervation, causing reduced vasoconstriction of the splanchnic and other peripheral vascular beds. Several authors have drawn attention to the association between increased mortality

TABLE 2.
Peripheral autonomic nervous system disorders

- PURE AUTONOMIC FAILURE
- PERIPHERAL NEUROPATHIES
  - DIABETES
  - AMYLOIDOSIS
  - GUILLAIN-BARRE SYNDROME
  - ACUTE AND SUBACUTE AUTONOMIC NEUROPATHIES
  - CHRONIC INFLAMMATORY POLYNEUROPATHY
  - HEREDITARY NEUROPATHIES
    - Hereditary sensory and autonomic neuropathies
    - Fabry's disease
    - Navajo Indian neuropathy
    - Herditary motor and sensory neuropathy
    - Tangier disease
  - INFECTIOUS DISEASES
    - Chagas disease
    - HIV neuropathy
    - Botulism
    - Diphtheria
    - Leprosy
  - TOXIC NEUROPATHIES
    - Vacor
    - Vincristine
    - Perhexilene maleate
    - cis-Platinum
    - Taxol
    - Organic solvents
    - Acrylamide
    - Heavy metals
  - CONNECTIVE TISSUE DISEASES
    - Sjogren's syndrome
    - Systemic lupus erythematosus
    - Rheumatoid arthritis
  - PERNICIOUS ANEMIA
  - PORPHYRIA
  - UREMIA
  - ALCOHOLIC NEUROPATHY
  - HEPATIC DISEASE
  - PARANEOPLASTIC NEUROPATHIES
  - EATON LAMBERT SYNDROME
  - DOPAMINE BETA HYDROXYLASE DEFICIENCY
  - ADIE'S SYNDROME

and cardiovascular autonomic dysfunction in diabetics. The five year cumulative mortality rate of patient with cardiovascular autonomic test abnormalities ranges form 27 to 56%.[102–106] There is an increased frequency of sudden death in patients with autonomic neuropathy. Proposed aetiologies for sudden unexplained death in diabetics have included cardio-respiratory arrest due to cardiac arrhythmias, silent cardiac ischaemia and sleep apnea.

Wheeler and Watkins drew attention to the loss of heart rate variability in diabetics and suggested that the difference between the maximum and minimum heart rate with deep respiration might serve as a clinical measure of cardiac vagal autonomic function[107]. In the ensuing years, other workers have advocated a variety of time domain statistical measures such as the 'inspiratory-respiratory ratio' or E:I ratio – the ratio between the shortest R-R interval during inspiration to the longest R-R interval during expiration;[108] the SD of R-R interval; and the mean successive difference (MSD), mean square successive difference (MSSD) and root mean square successive difference (rMSSD) of the RR interval.[109–111]. The mean circular resultant, a determination based on vector analysis, has been proposed as a method which is resistant to non-respiratory sources of variability such as ectopic heart beats and slow trends in heart rate. In this method, R-R intervals are recorded as time events plotted or wrapped on a circle with the periodicity of a single respiratory cycle. The distribution of the points on the circle is determined by vector analysis giving a result in arbitrary units. Regular distribution of the events on the circle would result in a small vector and denote reduced heart rate variability. In contrast, clustering of the events on the circle would result in a large vector and suggest normal heart rate variability.[112,113] Although this measure has not attained widespread use, it was the sole test of heart rate variability with respiration in the DCCT study.[114]

Ewing et al. drew attention to the large, irregular, episodic changes in heart rate that occur in normal subjects. Theses steps were quantified using a threshold value of a 50 milliseconds difference from the preceding RR interval (NN50). They demonstrated a reduced number of these steps in diabetic subjects with autonomic neuropathy, some diabetics without autonomic neuropathy and patients with cardiac transplants. The number of such steps or 'RR counts' in normal subjects shows an inverse relationship with age and the steps show a significant increase during sleep.[115,116] Twenty-four percent of diabetic subjects with normal cardiovascular reflexes, had 24 hour RR count results that were less than the lower 95% confidence limit for healthy controls related to age. Based on these results, the authors suggested that this method of heart rate variability analysis is more sensitive than the conventional tests of cardiovascular reflexes.[116] This measure and the related pNN50 (the proportion of differences in consecutive normal RR intervals that are greater than 50 milliseconds) correlates with the rMSSD and power in the high frequency of the heart rate power spectrum (> 0.15 Hz).[117–119]

Several other provocative manoeuvres have been used to induce heart rate responses and thereby provide a measure of autonomic function in diabetic subjects. The haemodynamic response to standing provides the physiological basis of a commonly performed measure of autonomic function. The heart rate response to active standing is bimodal. There is an abrupt increase in heart rate that peaks at approximately 3 seconds, followed by a more gradual increase that peaks approximately 12 seconds after standing.[120] The 30:15 ratio assesses this physiological response by measuring the ratio of the heart rate increase that occurs at approximately 15 seconds after standing to the relative bradycardia that occurs at approximately the 30 seconds after standing. This ratio provides a measure of cardiac vagal function.[121–124]

A related test measures the heart rate response to assuming the supine position. After lying down, there is a decrease in the R-R interval that is maximal around the third or fourth beat, which is followed by an increase in the R-R interval value (greater than the resting value) at approximately the 25th to 30th beat.[123,125]. Similarly, the heart rate response to squatting lengthens the R-R interval. A vagal ratio based on this test (the ratio between the R-R interval mean before squatting and the longest R-R interval after squatting) was outside the 99% confidence interval in 42% of diabetic patients and 1.3% of the control subjects. Whereas a sympathetic ratio (the ratio between the baseline R-R interval and the shortest R-R interval after standing was outside the 99% confidence interval in 40% of diabetic patients and 0.8% of the control subjects. [126,127]

There are several reports on the utility of power spectral analysis to assess diabetic autonomic neuropathy. [128–133] The spectral measures correlate with time domain tests of autonomic function. There is a gradual decline in spectral power in the high and low frequencies with progression of the diabetic autonomic neuropathy[128] (see Figure 3). Some authors have suggested that spectral analysis may be a more sensitive diagnostic technique for the evaluation of diabetic autonomic neuropathy.[133,134] Transfer function analysis, by relating the amount of heart rate variability to the lung volume amplitude change, allows a value to be reported that is normalised for the depth of breathing.[135] (See Figure 4)

Amyloid Neuropathy

Autonomic dysfunction frequently accompanies the polyneuropathy of both primary and familial amyloidosis. Autonomic dysfunction is not as common in myeloma associated amyloidosis. The variant transthyretin in which methionine substitutes for valine at position 30, is the point mutation that is the most common cause of familial amyloid polyneuropathy.[136] Other transthyretin mutations as well as mutations in apolipoprotein A1 and gelsolin also give rise to familial amyloid polyneuropathy. Amyloid is thought to cause autonomic and somatosensory dysfunction by pressure of amyloid deposits on peripheral nerves, dorsal nerve root ganglia or autonomic ganglia and by ischaemic damage due to amyloid infiltration of epineural and intraneural blood vessel walls.[137]

Patients with amyloid neuropathy typically present with distal sensory symptoms such as numbness, parasthesias and dysesthesias. Autonomic manifestations are rarely the presenting feature of amyloid neuropathy. On examination there are signs of a sensorimotor polyneuropathy that predominantly involve the small nerve fibres that mediate pain and temperature sensation. Characteristic autonomic signs and symptoms include postural hypotension, diarrhoea, constipation, faecal incontinence, disturbances in bladder function, pupillary abnormalities and erectile failure. These autonomic manifestations are similar to those described with diabetic autonomic neuropathy.[137] Sick sinus syndrome and A-V conduction deficits are frequently present. Tests assessing cardiac vagal function are often abnormal.[138]

Acute and Subacute Autonomic Neuropathies

Autonomic manifestations usually accompany the Guillain-Barre syndrome although they are usually overshadowed by motor features of that disorder.[139,140] The most frequently noted autonomic features of GBS include sinus tachycardia and other cardiac arrhythmias, EKG

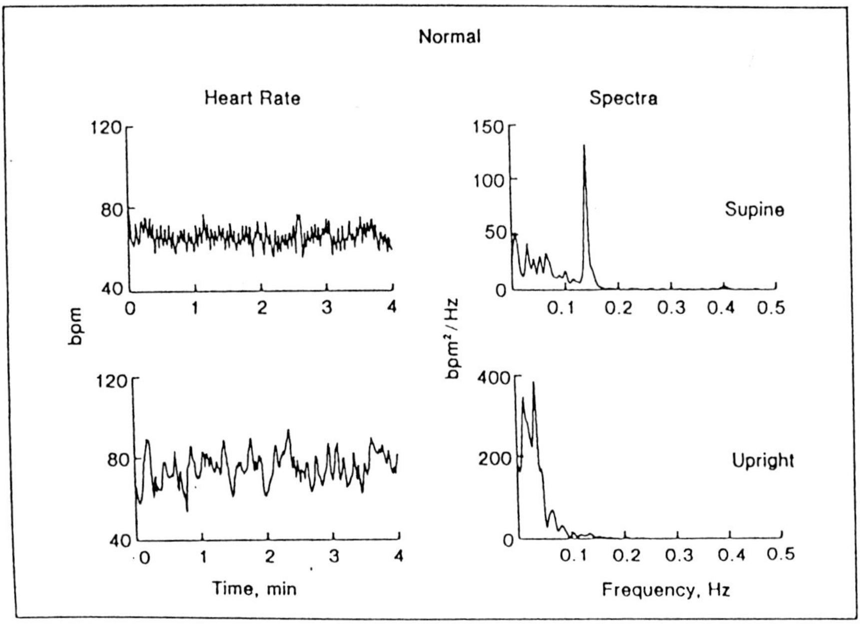

FIGURE 3. (A) The heart-rate tachogram and power spectrum of a control subject. (B – opposite) The heart-rate tachogram and power spectrum of a diabetic subject with moderate autonomic neuropathy (C – opposite) The heart-rate tachogram and power spectrum of a diabetic subject with severe autonomic neuropathy. Note the lower y-axis scale of the heart rate power spectrum (bpm indicates beats per minute). (From Freeman R, et al. Spectral analysis of heart rate in diabetic autonomic neuropathy. A comparison with standard tests of autonomic function. *Arch Neurol* 1991; **48**:185–190.)

abnormalities, sustained hypertension, blood pressure lability, bowel and bladder dysfunction, pupillomotor disturbances, sudomotor dysfunction, and vasomotor abnormalities.[139–144 145–150] Serious bradyarrhythmias and heart block may necessitate emergency interventions. Ropper et al.[139] have estimated that up to 65% of patients with typical Guillain-Barre syndrome will exhibit dysautonomic features. Tachycardia and loss of heart rate variability was present in approximately 50% of cases. Recent evidence suggests that cardiovascular autonomic test abnormalities are present in most Guillain-Barre syndrome patients. [148] Some[139,151,152] but not all[140] have suggested that dysautonomia is more frequent in patients with respiratory failure, severe motor deficits, and the axonal variant of Guillain-Barre syndrome. Autonomic test abnormalities include reduced heart rate variability with deep

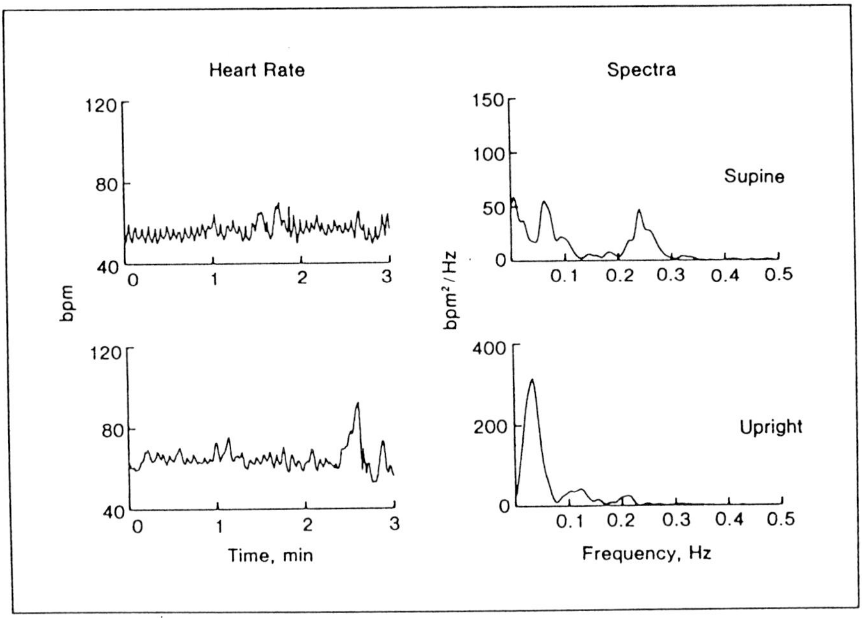

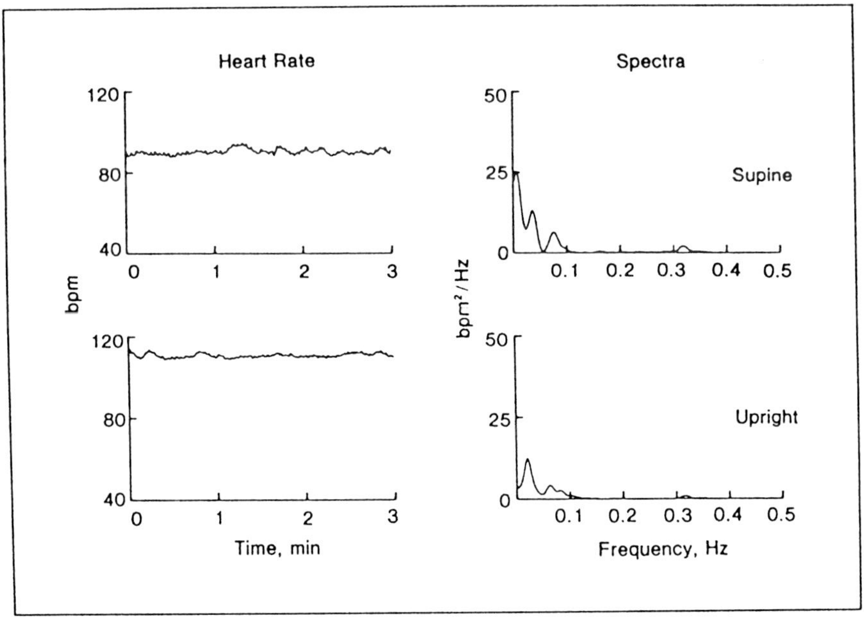

FIGURE 3 (Continued). See the legend opposite.

respiration.[148,149,153] and an abnormal Valsalva ratio[148,153,154] The abnormal autonomic test results improve in parallel with the improvement in motor function. [148,149].

Autonomic manifestations may also be the sole or predominant feature of an acute or subacute autonomic peripheral neuropathy. The hallmark of these autonomic neuropathies is the acute or subacute presentation of varying combinations of orthostatic hypotension, constipation, bladder atony, impotence, secretomotor paralysis, and blurring of vision associated with tonic pupils. Sensorimotor manifestations may accompany the autonomic manifestations but are not the predominant aspect of the presentation. The autonomic manifestations of this disorder may involve both the sympathetic and parasympathetic divisions of the autonomic nervous system (pandysautonomia) or the parasympathetic nervous system alone (cholinergic dysautonomia). [155]

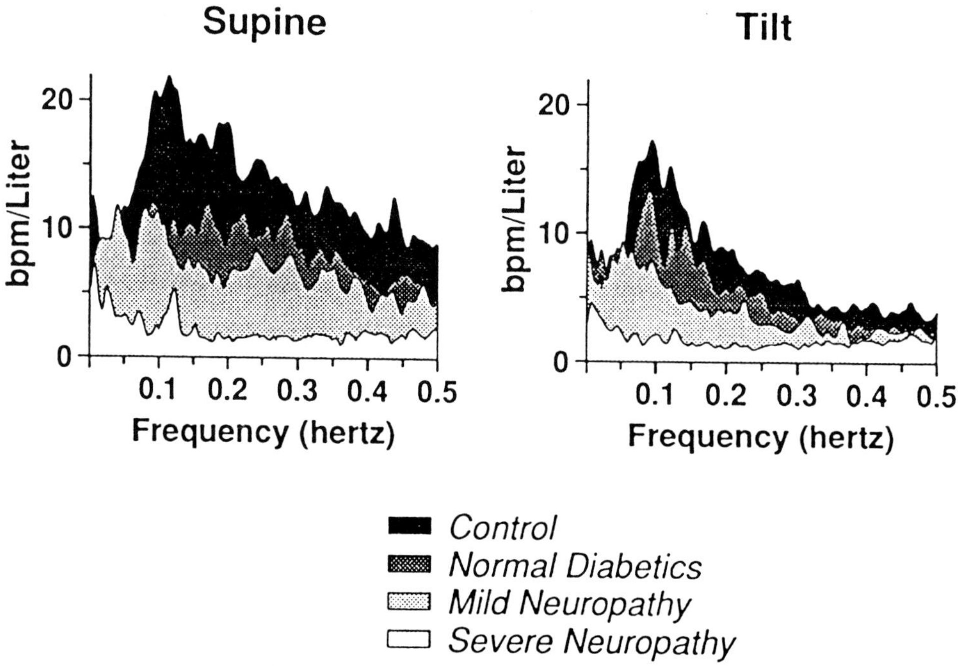

FIGURE 4. The transfer function group means for control and diabetic subjects in the supine and tilted positions. (From Freeman R, et al. Transfer function analysis of respiratory sinus arrhythmia: a measure of autonomic function in diabetic neuropathy. *Muscle & Nerve* 1995; 18:74–84.)

Hereditary Autonomic Neuropathies

The hereditary autonomic neuropathies are a heterogeneous group of disorders, some of which have significant involvement of autonomic fibers.[156] A list is provided in Table 2. The hereditary sensory and autonomic neuropathies (HSAN) are characterised by prominent sensory loss without motor involvement and by often striking dysautonomia. The autonomic manifestations are prominent in HSAN type III, or Riley-Day syndrome (familial dysautonomia). This autosomal recessive disorder is seen primarily in Ashkenazi Jewish children. The clinical features of this disease include insensitivity to pain and temperature but sparing visceral pain, absence of tears, hypoactive corneal and tendon reflexes, and absence of fungiform papillae. Poor suck and feeding, oesophageal reflux with vomiting and aspiration, and a dyscoordinated swallow may be the first clinical manifestations. Later in the course of the illness, vibratory sensory loss and impaired appendicular co-ordination manifest. Autonomic disturbances may be prominent at any point in the disease course. Autonomic manifestations include episodic hyperhidrosis, vasomotor instability with defective temperature homeostasis, postural hypotension, hypertensive crises, and supersensitivity to cholinergic and adrenergic agents. A decline in blood pressure on standing without a compensatory tachycardia is characteristic of the disorder.[156,157] Failure to increase the low frequency power of the heart rate power spectrum in response to postural change has been documented.[158]

Autonomic manifestations are modest in autosomal-dominant sensory neuropathy (HSAN type I) and autosomal-recessive sensory neuropathy (HSAN type II) with the possible exception of bladder dysfunction. These disorders are associated with severe sensory loss with acral injuries.[156,157] Anhidrotic sensory neuropathy, a disorder with autosomal recessive inheritance (HSAN type IV), manifests in the first months of life with insensitivity to pain, anhidrosis, episodes of unexplained fever, and retardation of motor development. Virtual absence of unmyelinated fibres has been noted in peripheral nerves.[157]

Infectious Diseases

The peripheral neuropathies associated with a number of infectious diseases have accompanying autonomic manifestations. Autonomic dysfunction may occur in patients with human immunodeficiency virus infection. The symptoms of dysautonomia have included orthostatic hypotension, syncope, presyncope, sweating disturbances, bladder and bowel dysfunction, and impotence.[159]

Autonomic testing of seropositive and AIDS patients has been the subject of several reports and case studies [160–167]. In a controlled study, we demonstrated significant differences in autonomic function between controls and HIV-infected patients, using multiple tests of the autonomic nervous system. A steady decline in autonomic function was noted across diagnostic groups (controls, ARC, AIDS) with the most severe autonomic dysfunction found in the patients with AIDS. The abnormalities in autonomic function correlated with signs of HIV associated nervous system disease. The three patients with the most severe autonomic dysfunction had evidence of dementia, myelopathy and distal sensory neuropathy[159] These observations may underlie the predisposition of HIV infected patients to cardiac arrhythmias such as ventricular tachycardia and torsades de pointes (often in association with medication administration) and the observed incidence of unexpected cardio-respiratory arrest[162,168–172].

The autonomic abnormalities were most prominent in tests of heart rate variability. There were significant differences between the resting heart rate in patients with AIDS and control subjects. Of the tests assessing sympathetic nervous system function, the blood pressure response to isometric exercise and the fall in mean arterial pressure on tilt-table testing revealed statistically significant difference in findings between the study groups. These differences were largely a result of significantly more abnormal test results in the AIDS subgroup. These test results have been replicated in a number of studies of similar design[167,173–175].

Chagas' disease, which is due to a parasitic infection by Trypanosoma cruzi, is associated in the late stages of illness with severe cardiovascular and gastrointestinal dysautonomia. The pathogenesis of the autonomic dysfunction is unresolved, and may be due to direct neural injury during the acute illness or a persisting immune mediated response. Reduced bowel motility, sialorrhea, megaesophagus, and megacolon are the most frequent gastrointestinal manifestations of this disease. Cardiovascular manifestations include impairment in the blood pressure response to standing, resting bradycardia, anhidrosis, conduction abnormalities, arrhythmias, cardiac failure, and cardiomegaly.[176,177] Power spectral analysis of subjects with Chagas' disease reveals decreased ability to increase low frequency power in response to manoeuvres that increase sympathetic efferent activity.

CONCLUSION

Autonomic dysfunction is common feature of central and peripheral nervous system disorders. Cardiovascular autonomic function testing plays an important role in documenting, characterising and quantifying the extent of autonomic dysfunction in these disorders. The profile of autonomic test abnormalities, however, with some exceptions, is not unique or specific to individual neurological diseases. Nevertheless, in conjunction with the clinical features of the neurological disorder and other investigations such as neuroimaging, biochemical measures and neurophysiological tests, cardiovascular autonomic function test are helpful in arriving at a specific neurological diagnosis.

REFERENCES

1. Freeman R. Autonomic nervous system. In: Samuels MA, Feske S (Eds) The ofice practice of neurology. Churchill Livingstone, New York 1995, 129–136.
2. Benarroch EE, Freeman R, Kaufmann H. Autonomic nervous system dysfunction. In: Goetz CG, Pappert EJ (Eds) Textbook of clinical neurology. Saunders, New York 1998.
3. Cechetto DF. Central representation of visceral function. *Fed Proc* 1987; **46:**17–23.
4. Cechetto DF, Saper CB. Role of the cerebral cortex in autonomic function. In: Loewy AD, Spyer KM (Eds) Central regulation of autonomic functions. Oxford University Press, New York 1990, 208–223.
5. Benarroch EE. The central autonomic network: functional organization, dysfunction, and perspective. *Mayo Clin Proc* 1993; **68:**988–1001.
6. Yasui Y, Breder CD, Saper CB, Cechetto DF. Autonomic responses and efferent pathways from the insular cortex in the rat. *J Comp Neurology* 1991; **303:**355–374.
7. Hurley KM, Herbert H, Moga MM, Saper CB. Efferent projections of the infralimbic cortex of the rat. *J Comp Neurology* 1991; **308:**249–276.
8. Pool JL, Ransohoff H. Autonomic effects on stimulating the rostral portion of the cingulate gyri in man. *J Neurophysiol* 1949; **12:**385–392.
9. Ledoux JE. Emotion and the amygdala. In: Aggleton JP (Ed) The Amygdala. Neurobiological Aspects of Emotion, Memory, and Mental Dysfunction. Wiley-Liss, New York 1992, 339-351.

10. Andresen MC, Kunze DL. Nucleus tractus solitarius – gateway to neural circulatory control. *Annu Rev Physiol* 1994; **56:**93–116.
11. Dampney RA. Functional organization of central pathways regulating the cardiovascular system. *Physiol Rev* 1994; **74:**323–364.
12. Feldman JL, Ellenberger HH. Central coordination of respiratory and cardiovascular control in mammals. *Annu Rev Physiol* 1988; **50:**593–606.
13. Chur C. Autonomic Neuroeffector Mechanism. Harvard, New York 1992.
14. Burnstock G. The changing face of autonomic neurotransmission. *Acta Physiol Scand* 1986; **126:**67–91.
15. Levy MN. Sympathetic-parasympathetic interactions in the heart. *Circ Res* 1971; **29:**437–445.
16. Warner HR, Russell RO. Effects of combined sympathetic and vagal stimulation on heart rate in the dog. *Circ Res* 1969; **24:**567–573.
17. Bannister R, Oppenheimer DR. Degenerative diseases of the nervous system associated with autonomic failure. *Brain* 1972; **95:**457–474.
18. Cohen J, Low P, Fealey R, Sheps S, Jiang NS. Somatic and autonomic function in progressive autonomic failure and multiple system atrophy. *Ann Neurol* 1987; **22:**692–699.
19. Position paper. Orthostatic hypotension, multiple system atrophy (the Shy Drager syndrome) and pure autonomic failure. *J Auton Nerv Syst* 1996; **58:**123–124.
20. Papp MI, Lantos PL. The distribution of oligodendroglial inclusions in multiple system atrophy and its relevance to clinical symptomatology. *Brain* 1994; **117:**235–243.
21. Rajput AH, Rozdilsky B. Dysautonomia in Parkinsonism: a clinicopathological study. *J Neurol Neurosurg Psychiatry* 1976; **39:**1092–1100.
22. Gross M, Bannister R, Godwin-Austen R. Orthostatic hypotension in Parkinson's disease. *Lancet* 1972; **1:**174–176.
23. Aminoff MJ, Wilcox CS, Slater JD. Autonomic defects in paralysis agitans and in the Shy-Drager syndrome. *Acta Neurologica Scandinavica – Supplementum* 1972; **51:**105–107.
24. Aminoff MJ, Wilcox CS. Assessment of autonomic function in patients with a Parkinsonian syndrome. *Br Med J* 1971; **4:**80–84.
25. Wilcox CS, Aminoff MJ. Blood pressure responses to noradrenaline and dopamine infusions in Parkinson's disease and the Shy-Drager syndrome. *Br J Clin Pharmacol* 1976; **3:**207–214.
26. Kuroiwa Y, Wada T, Tohgi H. Measurement of blood pressure and heart-rate variation while resting supine and standing for the evaluation of autonomic dysfunction. *J Neurol* 1987; **235:**65–68.
27. Khurana RK, Nelson E, Azzarelli B, Garcia JH. Shy-Drager syndrome: diagnosis and treatment of cholinergic dysfunction. *Neurology* 1980; **30:**805–809.
28. Sandroni P, Ahlskog JE, Fealey RD, Low PA. Autonomic involvement in extrapyramidal and cerebellar disorders. *Clin Auton Res* 1991; **1:**147–155.
29. Goetz CG, Lutge W, Tanner CM. Autonomic dysfunction in Parkinson's disease. *Neurology* 1986; **36:**73–75.

30. Meco G, Pratesi L, Bonifati V. Cardiovascular reflexes and autonomic dysfunction in Parkinson's disease. *J Neurol* 1991; **238**:195–199.
31. Appenzeller O, Goss JE. Autonomic deficits in parkinson's syndrome. *Arch Neurol* 1971; **24**:50–57.
32. Low PA. Composite autonomic scoring scale for laboratory quantification of generalized autonomic failure. *Mayo Clin Proc* 1993; **68**:748–752.
33. Burke WJ, Coronado PG, Schmitt CA, Gillespie KM, Chung HD. Blood pressure regulation in Alzheimer's disease. *J Auton Nerv Syst* 1994; **48**:65–71.
34. Camerlingo M, Bottacchi E, Gambaro P, D'Alessandro G, Carenini L, Mamoli A. Parasympathetic function in Huntington's disease. *Funct Neurology* 1987; **2**:227–230.
35. Novak P, Novak V, Li Z, Remillard G. Time-frequency analysis of slow cortical activity and cardiovascular fluctuations in a case of Alzheimer's disease. *Clin Auton Res* 1994; **4**:141–148.
36. Sachs C, Conradi S, Kaijser L. Autonomic function in amyotrophic lateral sclerosis: a study of cardiovascular responses. *Acta Neurol Scand* 1985; **71**:373–378.
37. Gutrecht JA. Autonomic cardiovascular reflexes in progressive supranuclear palsy. *J Auton Nerv Syst* 1992; **39**:29–35.
38. Lown B, DeSilva RA, Lenson R. Neural activity and ventricular fibrillation. *N Engl J Med* 1976; **294**:1165–1170.
39. Myers GA, Martin GJ, Magid NM, Barnett PS, Schaad JW, Weiss JS, Lesch M, Singer DH. Power spectral analysis of heart rate variability in sudden cardiac death: comparison to other methods. *IEEE Trans Biomed Eng* 1986; **33**:1149–1156.
40. Kleiger RE, Miller JP, Bigger JT, Jr., Moss AJ. Decreased heart rate variability and its association with increased mortality after acute myocardial infarction. *Am J Cardiol* 1987; **59**:256–262.
41. Van Buren J. Some autonomic concomitants of ictal automatism. A study of temporal lobe attacks. *Brain* 1958; **LXXXI**:505–528.
42. Van Buren J, Ajmone-Marsan C. A correlation of autonomic and EEG components in temporal lobe epilepsy. *Arch Neurol* 1960; **91**:683–703.
43. Devinsky O, Price BH, Cohen SI. Cardiac manifestations of complex partial seizures. *Am J Med* 1986; **80**:195–202.
44. Mulder D, Daly D, Bailey A. Visceral epilepsy. *Arch Intern Med* 1954; **93**:481–493.
45. Constantin L, Martins JB, Fincham RW, Dagli RD. Bradycardia and syncope as manifestations of partial epilepsy. *J Am Coll Cardiol* 1990; **15**:900–905.
46. Phizackerley PJR, Poole EW, Whitty CWM. Sinoauricular heart block as an epileptic manifestation. *Epilepsia* 1954; **3**:89–91.
47. Fincham RW, Shivapour ET, Leis AA, Martins JB. Ictal bradycardia with syncope: a case report. *Neurology* 1992; **42**:2222–2223.
48. Jacome DE, Seroppian ER. Ictal bradycardia. *Am J Med Sci* 1988; **295**:469–471.
49. Smaje JC, Davidson C, Teasdale GM. Sino-atrial arrest due to temporal lobe epilepsy. *J Neurol Neurosurg Psychiatry* 1987; **50**:112–113.

50. Walsh GO, Masland W, Goldensohn E. Relationship between paroxysmal atrial tachycardia and paroxysmal cerebral discharges. *Bull L A Neurol Assoc* 1972; **37**:28–35.
51. Mathew NT, Taoti GM, Mathai KV, Chandy J. Atrial fibrillation associated with seizure in a case of frontal meningioma. *Neurology* 1970; **20**:725–728.
52. Blumhardt LD, Smith PE, Owen L. Electrocardiographic accompaniments of temporal lobe epileptic seizures. *Lancet* 1986; **1**:1051–1056.
53. Keilson MJ, Hauser WA, Magrill JP. Electrocardiographic changes during electrographic seizures. *Arch Neurol* 1989; **46**:1169–1170.
54. Keilson MJ, Hauser WA, Magrill JP, Goldman M. ECG abnormalities in patients with epilepsy. *Neurology* 1987; **37**:1624–1626.
55. Frysinger RC, Engel J, Harper RM. Interictal heart rate patterns in partial seizure disorders. *Neurology* 1993; **43**:2136–2139.
56. Devinsky O, Perrine K, Theodore WH. Interictal autonomic nervous system function in patients with epilepsy. *Epilepsia* 1994; **35**:199–204.
57. Oppenheimer SM, Hachinski VC. The cardiac consequences of stroke. *Neurol Clin* 1992; **10**:167–176.
58. Korpelainen JT, Sotaniemi KA, Myllyla VV. Asymmetric sweating in stroke: a prospective quantitative study of patients with hemispheral brain infarction. *Neurology* 1993; **43**:1211–1214.
59. Barron SA, Rogovski Z, Hemli J. Autonomic consequences of cerebral hemisphere infarction. *Stroke* 1994; **25**:113–116.
60. Gilman S, Koeppe RA, Junck L, Kluin KJ, Lohman M, St Laurent RT. Benzodiazepine receptor binding in cerebellar degenerations studied with positron emission tomography. *Ann Neurol* 1995; **38**:176–185.
61. Oppenheimer SM, Gelb A, Girvin JP, Hachinski VC. Cardiovascular effects of human insular cortex stimulation. *Neurology* 1992; **42**:1727–1732.
62. Hines EA, Bannick EG. Intermittent hypothermia with disabling hyperhidrosis: report of a case with successful treatment. *Mayo Clin Proc* 1934; **9**:705–708.
63. LeWitt PA, Newman RP, Greenberg HS, Rocher LL, Calne DB, Ehrenkranz JRL. Episodic hyperhidrosis, hypothermia, and agenesis of corpus callosum. *Neurology* 1983; **33**:1122–1129.
64. Shapiro WR, Williams GH, Plum F. Spontaneous recurrent hypothermia accompanying agenesis of the corpus callosum. *Brain* 1969; **92**:423–436.
65. Carr-Locke D, Millac P. Diencephalic epilepsy in a patient with agenesis of the corpus callosum confirmed by computerised axial tomography. *J Neurol Neurosurg Psychiatry* 1977; **40**:808–814.
66. Bullard DE. Diencephalic seizures: responsiveness to bromocriptine and morphine. *Ann Neurol* 1987; **21**:609–611.
67. Solomon GE. Diencephalic autonomic epilepsy caused by a neoplasm. *J Pediatr* 1973; **83**:277–280.
68. Engel GL, Aring CD. Hypothalamic attacks with thalamic lesion. I. Physiologic and psychologic considerations. *Arch Neurol Psychiatr* 1945; **54**:37–43.

69. Fox RH, Wilkins DC, Bell JA, Bradley RD, Browse NL, Cranston WI, Foley TH, Gilby ED, Hebden A, Jenkins BS, Rawlins MD. Spontaneous periodic hypothermia: diencephalic epilepsy. *Br Med J* 1973; **2**:693–695.
70. Gambetti P, Parchi P, Petersen RB, Chen SG, Lugaresi E. Fatal familial insomnia and familial Creutzfeldt-Jakob disease: clinical, pathological and molecular features. *Brain Pathol* 1995; **5**:43–51.
71. Medori R, Montagna P, Tritschler HJ, LeBlanc A, Cortelli P, Tinuper, P, Lugaresi E, Gambetti P. Fatal familial insomnia: a second kindred with mutation of prion protein gene at codon 178. *Neurology* 1992; **42**:669–670.
72. Manetto V, Medori R, Cortelli P, Montagna P, Tinuper P, Baruzzi A, Rancurel G, Hauw JJ, Vanderhaeghen JJ, Mailleux P, et al. Fatal familial insomnia: clinical and pathologic study of five new cases. *Neurology* 1992; **42**:312–319.
73. Cortelli P, Parchi P, Contin M, Pierangeli G, Avoni P, Tinuper P, Montagna P, Baruzzi A, Gambetti PL, Lugaresi E. Cardiovascular dysautonomia in fatal familial insomnia. *Clin Auton Res* 1991; **1**:15–21.
74. Lugaresi E, Medori R, Montagna P, Baruzzi A, Cortelli P, Lugaresi A, Tinuper P, Zucconi M, Gambetti P. Fatal familial insomnia and dysautonomia with selective degeneration of thalamic nuclei. *N Engl J Med* 1986; **315**:997–1003.
75. Ito A, Omae T, Katsuki S. Acute changes in blood pressure following vascular diseases in the brain stem. *Stroke* 1973; **4**:80–84.
76. Norrving B, Cronqvist S. Lateral medullary infarction: prognosis in an unselected series. *Neurology* 1991; **41**:244–248.
77. Novak V, Novak P, deMarchie M, Schondorf R. The effect of severe brainstem injury on heart rate and blood pressure oscillations. *Clin Auton Res* 1995; **5**:24–30.
78. Hsu CY, Hogan EL, Wingfield W, Jr., Webb JG, Perot PL, Jr., Privitera PJ, Talbert OR, Balentine JD. Orthostatic hypotension with brainstem tumors. *Neurology* 1984; **34**:1137–1143.
79. Cameron SJ, Doig A. Cerebellar tumours presenting with clinical features of phaeochromocytoma. *Lancet* 1970; **1**:492–494.
80. Nogues MA, Gene R, Encabo H. Risk of sudden death during sleep in syringomyelia and syringobulbia. *J Neurol Neurosurg Psychiatry* 1992; **55**:585–589.
81. Hampton F, Williams B, Loizou LA. Syncope as a presenting feature of hindbrain herniation with syringomyelia. *J Neurol Neurosurg Psychiatry* 1982; **45**:919–922.
82. Nogues MA, Newman PK, Male VJ, Foster JB. Cardiovascular reflexes in syringomyelia. *Brain* 1982; **105**:835–849.
83. Corbett JJ, Butler AB, Kaufman B. 'Sneeze syncope', basilar invagination and Arnold-Chiari type I malformation. *J Neurol Neurosurg Psychiatry* 1976; **39**:381–384.
84. Simon RP, Gean-Marton AD, Sander JE. Medullary lesion inducing pulmonary edema: a magentic resonannce imaging study. *Ann Neurol* 1991; **30**:727–730.

85. Mathias CJ, Frankel HL. Autonomic disturbance in spinal cord lesions. In: Bannister R, Mathias CJ (Eds) Autonomic failure. Oxford University Press, Oxford 1992, 839–881.
86. Inoue K, Miyake S, Kumashiro M, Ogata H, Yoshimura O. Power spectral analysis of heart rate variability in traumatic quadriplegic humans. *Am J Physiol* 1990; **258**:H1722–H1726.
87. Inoue K, Miyake S, Kumashiro M, Ogata H, Ueta T, Akatsu T. Power spectral analysis of blood pressure variability in traumatic quadriplegic humans. *Am J Physiol* 1991; **260**:H842–H847.
88. Inoue K, Ogata H, Hayano J, Miyake S, Kamada T, Kuno M, Kumashiro M. Assessment of autonomic function in traumatic quadriplegic and paraplegic patients by spectral analysis of heart rate variability. *J Auton Nerv Syst* 1995; **54**:225–234.
89. Anema JR, Heijenbrok MW, Faes TJ, Heimans JJ, Lanting P, Polman CH. Cardiovascular autonomic function in multiple sclerosis. *J Neurol Sci* 1991; **104**:129–134.
90. Neubauer B, Gundersen HJ. Analysis of heart rate variations in patients with multiple sclerosis. A simple measure of autonomic nervous disturbances using an ordinary ECG. *J Neurol Neurosurg Psychiatry* 1978; **41**:417–419.
91. Nordenbo AM, Boesen F, Andersen EB. Cardiovascular autonomic function in multiple sclerosis. *J Auton Nerv Syst* 1989; **26**:77–84.
92. Pentland B, Ewing DJ. Cardiovascular reflexes in multiple sclerosis. *Eur Neurology* 1987; **26**:46–50.
93. Thomaides TN, Zoukos Y, Chaudhuri KR, Mathias CJ. Physiological assessment of aspects of autonomic function in patients with secondary progressive multiple sclerosis. *J Neurol* 1993; **240**:139–143.
94. van Lieshout JJ, Wieling W, Settels JJ, Karemaker JM. Cardiovascular instability and baroreflex activity in a patient with tetanus. *Clin Auton Res* 1991; **1**:5–8.
95. Freeman R. Pure autonomic failure. In: Robertson D, Biaggioni I (Eds) Disorders of the autonomic nervous system. Horwood, Luxembourg 1995, 83–106.
96. Bannister R. Multiple system atrophy and pure autonomic failure. In Low PA (Ed). Clinical autonomic disorders. Little Brown & Co, Boston 1993, 517–526.
97. Bannister R, Mathias CJ. Clinical features and investigations of the primary autonomic failure syndromes. In: Bannister R, Mathias CJ (Eds) Autonomic failure. Oxford University Press, Oxford 1992, 531–547.
98. Goldstein DS, Polinsky RJ, Garty M, Robertson D, Brown RT, Biaggioni I, Stull R, Kopin IJ. Patterns of plasma levels of catechols in neurogenic orthostatic hypotension. *Ann Neurol* 1989; **26**:558–563.
99. Miyawaki E, Freeman R. Peripheral autonomic neuropathy. In: Korczyn A (Ed) Handbook of autonomic nervous sytem dysfunction. Marcel Dekker, New York 1994, 253–282.
100. Hilsted J, Low PA. Diabetic autonomic neuropathy. In Low PA (Ed) Clinical autonomic disorders. Little, Brown & Co, Boston 1993, 423–444.

101. Tarsy D, Freeman R. The nervous system and diabetes. In: Weir G, Kahn R (Eds) Joslin's diabetes mellitus. Lea and Febiger, Philadelphia 1994, 794–816.
102. Ewing DJ, Campbell IW, Clarke BF. The natural history of diabetic autonomic neuropathy. *Q J Med* 1980; **49**:95–108.
103. O'Brien IA, McFadden JP, Corrall RJ. The influence of autonomic neuropathy on mortality in insulin-dependent diabetes. *Q J Med* 1991; **79**:495–502.
104. Navarro X, Kennedy WR, Loewenson RB, Sutherland DE. Influence of pancreas transplantation on cardiorespiratory reflexes, nerve conduction, and mortality in diabetes mellitus. *Diabetes* 1990; **39**:802–806.
105. Sampson MJ, Wilson S, Karagiannis P, Edmonds M, Watkins PJ. Progression of diabetic autonomic neuropathy over a decade in insulin-dependent diabetics. *Q J Med* 1990; **75**:635–646.
106. Rathmann W, Ziegler D, Jahnke M, Haastert B, Gries FA. Mortality in diabetic patients with cardiovascular autonomic neuropathy. *Diabet Med* 1993; **10**:820–824.
107. Wheeler T, Watkins PJ. Cardiac denervation in diabetes. *Br Med J* 1973; **4**:584–586.
108. Sundkvist G, Almer L, Lilja B. Respiratory influence on heart rate in diabetes mellitus. *Br Med J* 1979; **1**:924–925.
109. Murray A, Ewing DJ, Campbell IW, Neilson JM, Clarke BF. RR interval variations in young male diabetics. *Br Heart J* 1975; **37**:882–885.
110. Gundersen HJG, Neubauer B. A long-term diabetic autonomic nervous abnormality. *Diabetologia* 1977; **13**:137–140.
111. Ewing DJ, Borsey DQ, Bellavere F, Clarke BF. Cardiac autonomic neuropathy in diabetes: comparison of measures of R-R interval variation. *Diabetologia* 1981; **21**:18–24.
112. Weinberg CR, Pfeifer MA. An improved method for measuring heart-rate variability: assessment of cardiac autonomic function. *Biometrics* 1984; **40**:855–861.
113. Genovely H, Pfeifer MA. RR-variation: the autonomic test of choice in diabetes. *Diab/Met Rev* 1988; **4**:255–271.
114. Anonymous. The effect of intensive diabetes therapy on the development and progression of neuropathy. The Diabetes Control and Complications Trial Research Group. *Ann Intern Med* 1995, **122**:561–568.
115. Ewing DJ, Neilson JM, Travis P. New method for assessing cardiac parasympathetic activity using 24 hour electrocardiograms. *Br Heart J* 1984; **52**:396–402.
116. Ewing DJ, Neilson JM, Shapiro CM, Stewart JA, Reid W. Twenty four hour heart rate variability: effects of posture, sleep, and time of day in healthy controls and comparison with bedside tests of autonomic function in diabetic patients. *Br Heart J* 1991; **65**:239–244.
117. Bigger JT, Jr., Albrecht P, Steinman RC, Rolnitzky LM, Fleiss JL, Cohen RJ. Comparison of time- and frequency domain-based measures of cardiac parasympathetic activity in Holter recordings after myocardial infarction. *Am J Cardiol* 1989; **64**:536–538.

118. Kleiger RE, Bigger JT, Bosner MS, Chung MK, Cook JR, Rolnitzky LM, Steinman R, Fleiss JL. Stability over time of variables measuring heart rate variability in normal subjects. *Am J Cardiol* 1991; **68**:626–630.
119. Bigger JT, Jr., Fleiss JL, Steinman RC, Rolnitzky LM, Kleiger RE, Rottman JN. Correlations among time and frequency domain measures of heart period variability two weeks after acute myocardial infarction. *Am J Cardiol* 1992; **69**:891–898.
120. Borst C, Weiling W, van Brederode JFM, Hond A, DeRijk LG, Dunning AJ, Wieling W, Van Brederode JF, de Rijk LG. Mechanisms of initial heart rate response to postural change. *Am J Physiol* 1982; **243**:H676–H681.
121. Ewing DJ, Campbell IW, Murray A, Neilson JM, Clarke BF. Immediate heart-rate response to standing: simple test for autonomic neuropathy in diabetes. *Br Med J* 1978; **1**:145–147.
122. Ewing DJ, Hume L, Campbell IW, Murray A, Neilson JM, Clarke BF. Autonomic mechanisms in the initial heart rate response to standing. *J Appl Physiol,: Respir Environ Exercise Physiol* 1980; **49**:809–814.
123. Bellavere F, Cardone C, Ferri M, Guarini L, Piccoli A, Fedele D. Standing to lying heart rate variation. A new simple test in the diagnosis of diabetic autonomic neuropathy. *Diab Med* 1987; **4**:41–43.
124. Mitchell EA, Wealthall SR, Elliott RB. Diabetic autonomic neuropathy in children: immediate heart-rate response to standing. *Austr Paed J* 1983; **19**:175–177.
125. Bellavere F, Ewing DJ. Autonomic control of the immediate heart rate response to lying down. *Clin Sci* 1982; **62**:57–64.
126. Marfella R, Giugliano D, di Maro G, Acampora R, Giunta R, D'Onofrio F. The squatting test. A useful tool to assess both parasympathetic and sympathetic involvement of the cardiovascular autonomic neuropathy in diabetes. *Diabetes* 1994; **43**:607–612.
127. Marfella R, Salvatore T, Giugliano D, di Maro G, Giunta R, Torella R, Juchmes J, Scheen A, Lefebvre PJ. Detection of early sympathetic cardiovascular neuropathy by squatting test in NIDDM. *Diabetes Care* 1994; **17**:149–151.
128. Freeman R, Saul JP, Roberts MS, Berger RD, Broadbridge C, Cohen RJ. Spectral analysis of heart rate in diabetic autonomic neuropathy. A comparison with standard tests of autonomic function. *Arch Neurol* 1991; **48**:185–190.
129. Weise F, Heydenreich F, Runge U. Heart rate fluctuations in diabetic patients with cardiac vagal dysfunction: a spectral analysis. *Diab Med* 1988; **5**:324–327.
130. Lanting P, Faes TJ, Heimans JJ, ten Voorde BJ, Nauta JJ, Rompelman O. Spectral analysis of spontaneous heart rate variation in diabetic patients. *Diab Med* 1990; **7**:705–710.
131. Thomaseth K, Cobelli C, Bellavere F, Balzani I, De Masi G, Bax G, Carenza P. Heart rate spectral analysis for assessing autonomic regulation in diabetic patients. *J Auton Nerv Syst* 1990; **30**:S169–S171.

132. Pagani M, Malfatto G, Pierini S, Casati R, Masu AM, Poli M, Guzzetti S, Lombardi F, Cerutti S, Malliani A. Spectral analysis of heart rate variability in the assessment of autonomic diabetic neuropathy. *J Auton Nerv Syst* 1988; **23**:143–153.
133. Comi G, Sora MG, Bianchi A, Bontempi B, Gianoglio P, Cerutti S, Micossi P, Canal N. Spectral analysis of short-term heart rate variability in diabetic patients. *J Auton Nerv Syst* 1990; **30**:S45–S49.
134. Lishner M, Akselrod S, Avi VM, Oz O, Divon M, Ravid M. Spectral analysis of heart rate fluctuations. A non-invasive, sensitive method for the early diagnosis of autonomic neuropathy in diabetes mellitus. *J Auton Nerv Syst* 1987; **19**:119–125.
135. Freeman R, Cohen RJ, Saul JP. Transfer fuction analysis of respiratory sinus arrythmia – a measure of autonomic function in diabetic neuropathy. *Muscle Nerve* 1995; **18**:74–84.
136. Saraiva MJM, Costa PP, Goodman DS. Biochemical marker in familial amyloidotic polyneuropathy, Portuguese type: family studies of transthyretin (prealbumin) – methionine-30 variant. *J Clin Invest* 1985; **76**:2171–2177.
137. Kyle RA, Dyck PJ. Amyloidosis and neuropathy. In: Dyck PJ, Thomas PK, Griffin JW, Low PA, Podulso JF (Eds) Peripheral Neuropathy. Saunders, Philadelphia 1993, 1294–1309.
138. Niklasson U, Olofsson BO, Bjerle P. Autonomic neuropathy in familial amyloidotic polyneuropathy. A clinical study based on heart rate variability. *Acta Neurologica Scandinavica* 1989; **79**:182–187.
139. Ropper AH, EFM Wijdicks, BT Truax. Guillain Barre Syndrome. F. A. Davis, Philadelphia 1991.
140. Tuck RR, McLeod JG. Autonomic dysfunction in Guillain-Barre syndrome. *J Neurol Neurosurg Psychiatry* 1981; **44**:983–990.
141. Appenzeller O, Marshall J. Vasomotor disturbance in Landry-Guillain-Barre syndrome. *Arch Neurol* 1963; **9**:368–372.
142. Birchfield RI, Shaw CM. Postural hypotension in the Guillain-Barre syndrome *Arch Neurol* 1964; **40**:149–157.
143. Lichtenfeld P. Autonomic dysfunction in the Guillian-Barre syndrome. *Am J Med* 1971; **50**:772–780.
144. Keane JR. Tonic pupils with acute ophthalmoplegic polyneuritis. *Ann Neurol* 1977; **2**:393–396.
145. Cortelli P, Contin M, Lugaresi A, Baruzzi A, Montagna P. Severe dysautonomic onset of Guillain-Barre syndrome with good recovery. A clinical and autonomic follow-up study. *Ital J Neurol Sci* 1990; **11**:159–162.
146. Davies AG, Dingle HR. Observations on cardiovascular and neuroendocrine disturbance in the Guillain-Barre syndrome. *J Neurol Neurosurg Psychiatry* 1972; **35**:176–179.
147. Edmonds ME, Sturrock RD. Autonomic neuropathy in the Guillain-Barre syndrome. *Br Med J* 1979; **2**:668–668.
148. Flachenecker P, Wermuth P, Hartung HP, Reiners K. Quantitative assessment of cardiovascular autonomic function in Guillain-Barre syndrome. *Ann Neurol* 1997; **42**:171–179.

149. Persson A, Solders G. R-R variations in Guillain-Barre syndrome: a test of autonomic dysfunction. *Acta Neurol Scand* 1983; **67**:294–300.
150. Zochodne DW. Autonomic involvement in Guillain-Barre syndrome: a review. *Muscle Nerve* 1994; **17**:1145–1155.
151. Winer JB, Hughes RAC. Identification of patients at risk of arrhythmia in the Guillain-Barre syndrome. *Q J Med* 1988; **257**:735–739.
152. Feasby TE, Gilbert JJ, Brown WF, Bolton CF, Hahn AF, Koopman WF, Zochodne DW. An acute axonal form of Guillain-Barre polyneuropathy. *Brain* 1986; **109**:1115–1126.
153. Singh NK, Jaiswal AK, Misra S, Srivastava PK. Assessment of autonomic dysfunction in Guillain-Barre syndrome and its prognostic implications. *Acta Neurol Scand* 1987; **75**:101–105.
154. Tuck RR, McLeod JG. Autonomic dysfunction in the Landry-Guillain-Barre syndrome. *Clinical & Experimental Neurology* 1978; **15**:197–203.
155. Hart RG, Kanter MC. Acute autonomic neuropathy. Two cases and a clinical review. *Arch Intern Med* 1990; **150**:2373–2376.
156. Thomas PK. Autonomic involvement in inherited neuropathies. *Clin Auton Res* 1992; **2**:51–56.
157. Dyck PJ. Neuronal atrophy and degeneration predominantly affecting peripheral sensory and autonomic neurons. In: Dyck PJ, Thomas PK, Griffin JW, Low PA, Podulso JF (Eds) Peripheral Neuropathy Volume II. Saunders, Philadelphia 1993, 1065–1093.
158. Maayan C, Axelrod FB, Akselrod S, Carley DW, Shannon DC, Shannon CD. Evaluation of autonomic dysfunction in familial dysautonomia by power spectral analysis. *J Auton Nerv Syst* 1987; **21**:51–58.
159. Freeman R, Roberts MS, Friedman LS, Broadbridge C. Autonomic function and human immunodeficiency virus infection. *Neurology* 1990; **40**:575–580.
160. Cohen JA, Laudenslager M. Autonomic nervous system involvement in patients with human immunodeficiency virus infection. *Neurology* 1989; **39**:1111–1112.
161. Cohen JA, Miller L, Polish L. Orthostatic hypotension in human immunodeficiency virus infection may be the result of generalized autonomic nervous system dysfunction. *J Acquir Immune Defic Syndr* 1991; **4**:31–33.
162. Craddock C, Pasvol G, Bull R, Protheroe A, Hopkin J. Cardiorespiratory arrest and autonomic neuropathy in AIDS. *Lancet* 1987; **2**:16–18.
163. Evenhouse M, Haas E, Snell E, Visser J, Paul L, Gonzalez R. Hypotension infection with the human immunodeficiency virus (letter). *Ann Intern Med* 1987; **107**:598–599.
164. Lin-Greenberger A, Taneja-Uppal N. Dysautonomia and infection with the human immunodeficiency virus (letter). *Ann Intern Med* 1987; **106**:167.
165. Mulhall B, Jennens I. Testing for neurological involvement in HIV infection (letter). *Lancet* 1987; **II**:1531.
166. Villa A, Foresti V, Confalonieri F. Autonomic neuropathy and HIV infection (letter). *Lancet* 1987; **1**:915.
167. Villa A, Cruccu V, Foresti V, Guareschi G, Tronchi M, Confalonieri F. HIV related functional involvement of autonomic nervous system. *Acta Neurologica* 1990; **12**:14–18.

168. Stein KM, Haronian H, Mensah GA, Acosta A, Jacobs J, Kligfield P. Ventricular tachycardia and torsades de pointes complicating pentamidine therapy of Pneumocystis carinii pneumonia in the acquired immunodeficiency syndrome. *Am J Cardiol* 1990; **66**:888–889.
169. Wharton JM, Demopulos PA, Goldschlager N. Torsade de pointes during administration of pentamidine isethionate. *Am J Med* 1987; **83**:571–576.
170. Cohen AJ, Weiser B, Afzal Q, Fuhrer J. Ventricular tachycardia in two patients with AIDS receiving ganciclovir (DHPG). *Aids* 1990; **4**:807–809.
171. Luginbuhl LM, Orav EJ, McIntosh K, Lipshultz SE. Cardiac morbidity and related mortality in children with HIV infection. *JAMA* 1993; **269**:2869–2875.
172. Eisenhauer MD, Eliasson AH, Taylor AJ, Coyne PE, Jr., Wortham DC. Incidence of cardiac arrhythmias during intravenous pentamidine therapy in HIV-infected patients. *Chest* 1994; **105**:389–395.
173. Villa A, Foresti V, Confalonieri F. Autonomic nervous system dysfunction associated with HIV infection in intravenous heroin users. *Aids* 1992; **6**:85–89.
174. Rüttimann S, Hilti P, Spinas GA, Dubach UC. High frequency of human immunodeficiency virus-associated autonomic neuropathy and more severe involvement in advanced stages of human immunodeficiency virus disease. *Arch Intern Med* 1991; **151**:2441–2443.
175. Welby SB, Rogerson SJ, Beeching NJ. Autonomic neuropathy is common in human immunodeficiency virus infection. *J Infect* 1991; **23**:123–128.
176. Iosa D, Dequattro V, De-Ping Lee D, Elkayam U, Caeiro T, Palmero H. Pathogenesis of cardiac neuro-myopathy in Chagas' disease and the role of the autonomic nervous system. *J Auton Nerv Syst* 1990; **30**:S83–S88.
177. Iosa D, Dequattro V, De-Ping Lee D, Elkayam U, Palmero H. Plasma norepinephrine in Chagas' cardioneuromyopathy: A marker of progressive dysautonomia. *Am Heart J* 1989; **117**:882–887.

Chapter Fifteen

CLINICAL PERSPECTIVE

Yee Guan Yap and A John Camm

Sudden cardiac death, largely resulting from ventricular arrhythmias, is a major cause of death in the first year after acute myocardial infarction (AMI).[1,2] The mortality rate of AMI in the first year ranges from between 5% to 15%.[3,4] While the definition of sudden death is still debated[5] it is recognised that most of these events (84%) are due to rapid ventricular tachyarrhythmias.[6,7]

The identification of patients at risk of developing life-threatening ventricular arrhythmias following AMI remains a challenge. Considerable effort has focused on the use of non-invasive risk parameters in identifying these patients who are at high risk from ventricular arrhythmias. All the known risk parameters such as left ventricular ejection fraction, signal-averaged ECG and ventricular premature complexes are limited in their usefulness due to low sensitivity and a high number of false positive results. Recently, it has been shown that myocardial infarction can interact with the function of the autonomic nervous system which may play a significant role in the pathogenesis of ventricular arrhythmias. In the animal model, myocardial ischaemia stimulates an increase in sympathetic activity and increases the likelihood of ventricular fibrillation, especially when the parasympathetic activity is reduced.[8,9] The increased sympathetic activity may facilitate the 3 major mechanisms involved in the genesis of malignant arrhythmias, i.e. enhanced automaticity, triggered automaticity and reentry.[10,11] Enhanced vagal tone, on the other

Malik M (Ed) Clinical guide to cardiac autonomic tests, 393–414.

hand, will produce an increase in the ventricular fibrillation threshold, and thus provides an antifibrillatory effect.[10]

The markers of cardiac autonomic status such as heart rate variability (HRV) and baroreflex sensitivity (BRS) may provide a more accurate means of risk stratification in patients at risk of ventricular arrhythmias, especially in patients after myocardial infarction. Evidence has also indicated that HRV may have a wider role in risk assessment in other high risk groups, such as patients with chronic heart failure, cardiac transplantation or diabetes mellitus.

PHYSIOLOGY

In a normal heart, the heart rate is not regular but is determined by the balance between vagal and sympathetic activity. HRV is a measure of such spontaneous variability of heart rate and periodic change of the heart rate rhythm. HRV should probably be defined as heart period variability. Several factors such as respiratory sinus arrhythmia and blood pressure regulation can affect heart rate.[12] Others including thermoregulation, the renin-angiotensin system and circadian rhythm can also affect the heart rate.[12] The analysis of HRV is an established non-invasive method of assessing autonomic influence on the heart rate at the level of the sinus node.[13] As a general rule, an increased sympathetic or decreased parasympathetic tone are reflected in decreased indices of HRV, while decreased sympathetic or increased parasympathetic activity are reflected in increased indices of HRV.[12] When stationary data are analysed, in the frequency domain of HRV (see below), the so-called high frequency (HF) component of physiologic HRV is almost exclusively mediated by vagal activity, whereas the low frequency (LF) component of HRV is under the influence of both sympathetic and vagal activity.[14] Both HF and LF oscillations interact to modulate target functions and determine their rhythmic properties. In other words, HRV represents a measure of sympathovagal modulation.

While HRV is a measure of sympathovagal modulation of the autonomic nervous system, BRS assesses the ability of the autonomic nervous system to respond to a stimulus with an increase in acetycholine release (i.e. vagal reflex activation).

METHODS FOR MEASURING CARDIAC AUTONOMIC ACTIVITY

While there are several methods available for measuring autonomic activity of the heart; e.g. plasma catecholamine levels, pharmacological challenges, HRV and BRS analysis, it is clear that some measures are superior to others.[13] Plasma catecholamine levels have been widely used previously but unfortunately, the changes in catecholamine levels do not reflect autonomic activity at the cardiac level.[13] This is due to the high specificity of the autonomic nervous system and localised changes in cardiac sympathetic activity may not significantly alter the overall plasma catecholamine levels. The most commonly use methods are HRV and BRS analysis. The measurement of HRV is usually performed by recording the electrocardiogram (ECG) using the 24-hour Holter ambulatory ECG. The variability of the RR data can be quantified using either time or frequency domain measurements. Time domain analysis involves simple statistical calculations or geometric reconstruction on a set of interbeat intervals and provides a measure of the degree of variability of the heart rate.[12] In contrast, frequency domain analysis provides an idea of the underlying rhythm and information on the amount of the overall variance in heart rate resulting from periodic oscillations of heart rate at various frequencies. Frequency domain analysis requires specialised spectral analysis such as Fourier transformation or autoregression.[12] Fourier analysis reduces the variance of the input RR data into the variance attributable to each specific frequency. Several algorithms can be employed for frequency measures to assess the standard deviation, centre frequency, and amplitude of the oscillatory components within the signal variability. Frequency domain measures are better than time domains in resolving parasympathetic and sympathetic influence of the heart rate.

BRS can be assessed by studying either the reflex heart rate response to a physiological activation or deactivation of the baroreceptors by a variety of mechanical or pharmacological manipulations, or by analysing the spontaneous fluctuations of the arterial pressure in a steady-state condition.[15] The most well accepted standardised technique involves a bolus injection of phenylephrine as described by Smyth et al[16] and is of proven value in assessing BRS. The slope of the regression line of the beat-to-beat change in blood pressure (ΔBP) plotted against the subsequent change in the RR interval (ΔRR) following phenylephrine infusion is taken as the BRS. A large increase in RR interval for a given blood pressure increase reflects strong vagal reflexes. On the other hand, a small increase in the RR interval for same blood pressure increase reflects weak vagal reflexes.

EPIDEMIOLOGY

The evidence that disturbance of autonomic nervous system activity is of prognostic significance in a community-based population that is free of clinically apparent heart disease comes from 2 large population-based studies. The Framingham Heart Study Group showed that in a population that was apparently free of coronary heart disease or congestive heart failure, depressed HRV was associated with subsequent cardiac events (angina, myocardial infarction, coronary heart disease death or congestive heart failure).[17] A one-standard deviation decrement in the standard deviation of total normal RR intervals (natural log transformed) was associated with a hazard ratio of 1.47 for these new cardiac events.[17] The mechanism for such an association is as yet unknown although the study suggested that autonomic imbalance as manifested by reduced HRV predisposed to subclinical cardiac disease. Interestingly, in this study, a short-term recording of 2 hours of ambulatory ECG was used instead of a 24-hour ECG. Malik et al found that arbitrary short-term recordings were not sufficient compared with a 24-hour ECG because the specificity for predicting arrhythmia risk was small.[18] The predictive power of short term HRV is still not absolutely clear. Nevertheless, in a separate epidemiological study, the Framingham Heart Study Group demonstrated that reduced HRV also predicted an increased risk for all-cause mortality in the elderly population (mean age, 72 years).[19]

In another epidemiological study, Algra followed-up a cohort of 6693 consecutive patients for up to 2 years and examined the specific role of heart rate variability in relation to sudden death.[20] The patients underwent 24-hour ECGs to investigate their symptoms of palpitations, dizziness, syncope, angina (65%); to assess the effect of antiarrhythmic therapy (8%); to assess the risk after MI (10%); or to investigate the cause of transient ischaemic attacks or strokes (7%). 245 of these patients died suddenly. Of these, 193 patients that were in sinus rhythm had their HRV parameters derived from the 24-hour ECGs and compared with that of a random sample 230 patients from the study cohort. The results showed that patients with low short-term RR interval variability (mean during 24 hours of per-minute standard deviation of RR intervals < 25 ms, corresponding to SDNN) had a 4.1-fold higher risk for sudden death than patients with higher short-term variability (≥ 40ms). After adjustment for age, cardiac dysfunction and history of myocardial infarction, the relative risk was 2.6. Similarly, patients with reduced long-term RR interval variability (standard deviation during 24 hours of per-minute means of RR intervals < 8 ms) had a 4.4-fold increased risk (2.2 after adjustment for other risk factors) of sudden death than patients with higher long-term variability (≥ 12 ms). The authors concluded that the reduction in the short-term RR interval variability, which corresponds to low parasympathetic activity, independently increased the risk of sudden death.

This epidemiological evidence suggested that abnormal autonomic activity, as detected by abnormal HRV is predictive of subsequent cardiac events and sudden death. Thus, HRV may provide prognostic information beyond that provided by the evaluation of traditional cardiovascular disease risk factors in a seemingly disease free community population.

Post-Myocardial Infarction

HRV and Risk Stratification After Myocardial Infarction

HRV is depressed following an AMI, reflecting an increased tonic sympathetic activity saturating the ability to modulate heart rate. Increased heart rate leaves little room for heart rate variability. The predominant sympathetic activity enhances cardiac electrical instability. The full mechanism by which HRV is transiently reduced after AMI is not clear. However, one hypothesis suggests that the changes in the geometry of a beating heart due to necrotic and non-contracting segments may abnormally increase the firing of sympathetic afferent fibres by mechanical distortion of the sensory endings, this in turn attenuates the activity of vagal fibres directed to the sinus node. Another explanation is that there is a reduced responsiveness of sinus node to neural modulation. Casolo and colleagues found that HRV assessed during the acute phase of myocardial infarction was significantly related to clinical and haemodynamic indices of severity, such as peak creatine kinase, LVEF, and Killip class.[21]

There is compelling evidence that HRV is a powerful risk predictor of survival post-MI (Figure 1). Kleiger et al demonstrated that a depressed HRV (SDNN < 50 ms) was associated with a 5.3 times increase in all-cause mortality in the first 4 years after AMI than

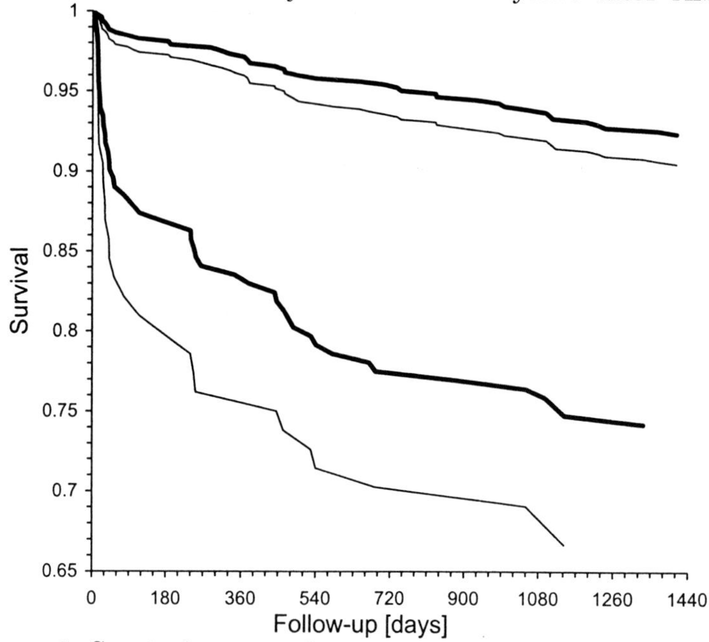

FIGURE 1. Survival curves of post-MI patients according to 24-hr heart rate variability of HRV index values with dichotomy points of 15 (fine lines) and 20 (bold lines) units (St. George's Postinfarction Research Survey Programme).

SDNN > 100 ms, irrespective of demographics, New York Heart Association functional class, left ventricular ejection fraction (LVEF) and ventricular ectopic activity.[22] Cripps et al showed that HRV was not only capable of predicting the total cardiac mortality after infarction but also identified patients at risk of sustained symptomatic ventricular tachycardia.[23] Indeed, Odemuyiwa et al demonstrated that depressed HRV expressed as HRV Triangular Index (HRVi), was a better predictor of sudden death and symptomatic ventricular tachycardia than LVEF, whereas both HRV and LVEF performed equally well in predicting all-cause mortality in the postinfarction patients.[24] This suggests that HRV is a stronger predictor of arrhythmic events (sudden cardiac death and ventricular tachycardia) than non-arrhythmic mortality.

TABLE 1
Univariate predictors of cardiac arrhythmic and nonarrhythmic mortality after myocardial infarction

Risk factor	Mortality		
	Cardiac	Arrhythmic	Nonarrhythmic
HRV	8.8 (4.5–16.9)*	7.5 (3.3–16.9)*	8.5 (2.9–24.1)*
SAECG	2.4 (1.0–5.6)†	2.6 (11.1–7.8)†	2.0 (0.6–6.9)
VE	4.9 (2.4–8.5)*	2.5 (1.1–5.7)†	8.2 (3.1–21.8)*
VT	2.6 (1.3–5.3)‡	2.9 (1.2–6.9)†	1.9 (0.6–5.9)
LVEF	3.0 (1.6–5.5)*	1.8 (0.8–3.9)	6.1 (2.1–17.3)*

The table shows risk ratios (95% confidence intervals). HRV – heart rate variability index < 20 units, SAECG – positive signal-averaged ECG, VE – ventricular ectopic beats > 10 per hour, VT – nonsustained ventricular tachycardia, LVEF – left ventricular ejection fraction < 40%. * p< 0.001, †p< 0.01, ‡p< 0.05 (adapted from Hartikainen et al[25]).

Hartikainen and co-workers provided further insight into the use of established risk parameters such as HRV, signal-averaged electrocardiogram (SAECG), ventricular arrhythmias and LVEF as predictors of arrhythmic and non-arrhythmic death.[25] They demonstrated that arrhythmic death was associated predominantly with depressed HRV and runs of ventricular tachycardia, whereas low ejection fraction, ventricular ectopic beats, and depressed HRV predicts non-arrhythmic death (Table 1). These findings can be explained by our current understanding of the pathophysiology of depressed HRV following a myocardial infarction.[26] The reduced LVEF reflects a liability to cardiac failure and is likely to identify patients at risk of non-sudden death. On the other hand, reduced HRV implies a lack of an autonomic protection against ventricular arrhythmias and would therefore be more powerful in identifying those patients who are at high risk of sudden death or arrhythmic complication.[26] Among all the time domain measures of HRV, Bigger et al showed that standard deviation of 5-minute RR intervals (SDANN) was the strongest predictor for arrhythmic deaths, even after adjustment for other risk factors such as age, New York Heart Association functional class, pulmonary rales in the coronary care unit, left ventricular ejection fraction and frequency of ventricular arrhythmias on Holter recordings[27].

HRV is unaffected by thrombolytic therapy. It continues to retain its independent prognostic significance in post-infarction patients of all ages treated with thrombolysis.[28] On the other hand, beta-blocker increases the RR variance and the normalised power of the high-frequency component (vagal tone) and reduces the low frequency component (sympathetic activity). This may partly explain the beneficial effect of these drugs after myocardial infarction.[29]

Predictive accuracy of HRV in postinfarction patients

For postinfarction risk stratification, the predictive value of HRV is modest, although it appears to be better than other recognised risk factors. Differences in the timing of investigations, method of analysis and specific endpoints make comparisons of predictive accuracy between studies difficult. For instance, Kleiger et al showed that a standard deviation of all normal RR intervals (SDNN) < 50ms had a sensitivity of 33.8% and a specificity of 87.9% for all-cause mortality.[22] On the other hand, Odemuyiwa demonstrated that for a HRVi ≤ 30U, the sensitivity was 75% with a specificity of 76% for the prediction of arrhythmic events.[24] Farrell et al showed that using triangular interpolation of the RR histogram (that is, HRVi), a baseline width < 20 units had a sensitivity of 92% for arrhythmic events (sudden death or life-threatening ventricular arrhythmias) but a specificity of 77% which resulted in a large number of false-positive results and a low predictive accuracy of 17%.[30]

The length of recording increases the predictive power of depressed HRV. Thus, 24-hour recordings appear to be better than 5-minute recording.[31] However, a 5-minute recording may be useful for screening purposes. Among all the HRV measures, time domain HRV provides better prognostic information than the rest, particularly SDNN and HRV triangular index (total number of all RR intervals divided by the height of the histogram of all RR intervals).[31] A high risk group may be selected by the dichotomy limits of SDNN< 50ms or HRVi< 15U. It is recommended that HRV should be measured approximately 1 week after the index infarct.[31]

To improve the predictive value of HRV, HRV should be combined with other risk factors because of their independent risk prediction for mortality and arrhythmic complication.[31] Multivariate analysis showed that the most sensitive combination for the prediction of arrhythmic events was a reduced HRV and a positive SAECG.[30] Such combination had a sensitivity of 58%, a positive predictive accuracy of 33% and a relative risk of 18 (95% CI, 7–34). This was superior to

TABLE 2
Predictive accuracy of HRV combined with other parameters

	Sen	Spc	PPA	NPA	RR
HRV & LP	58	93	33	93	18.5 (6.7–34)
HRV & VE_{10}	50	94	34	96	12.6 (5.7–28.1)
HRV & REP	50	96	43	77	15.0 (6.7–33.9)
HRV & LVEF	42	90	22	91	6.3 (2.8–14.1)
LVEF & LP	25	94	19	94	4.7 (1.9–11.9)
LVEF & VE_{10}	33	93	19	94	5.9 (2.5–13.7)
LVEF & REP	25	91	15	95	5.2 (2.2–12.1)
LVEF & VE_{10} & LP	20	97	28	97	6.9 (2.6–18.5)
HRV & VE_{10} & LP	38	97	43	96	13.9 (6.1–31.9)
HRV & REP & LP	29	99	58	95	23.5 (9.6–57.3)

Sen – sensitivity, Spc – specificity, PPA – positive predictive accuracy, NPA – negative predictive accuracy, RR – risk ratio. The values of sensitivity, specificity and predictive accuracies are in per cent. For risk ratios, 95% confidence intervals are given in brackets. HRV – heart rate variability index < 20 units; LP– positive late potentials; LVEF – left ventricular ejection fraction < 40%; REP – repetitive ventricular forms; VE10 – >10 ventricular premature beats per hour (Adapted from Farrell et al[30]).

other combinations including those incorporating LVEF, exercise ECG, ventricular ectopic beats frequency and repetitive ventricular forms (Table 2).

BRS and Risk Stratification in Patients with Coronary Artery Disease

Earlier studies showed that BRS was more reduced in postinfarction patients with ventricular tachycardia on Holter monitoring or during an exercise stress test than in those without arrhythmias.[32] BRS is also markedly reduced in patients with inducible sustained ventricular tachycardia.[33] Indeed, BRS appears to be an independent risk factor for predicting cardiac death and the risk of dying was 17 times greater for those patients who had a depressed BRS 1 month after a myocardial infarction.[32]

Furthermore, a reduction in BRS appears to correlate with the severity of coronary artery disease. BRS is significantly more reduced in postinfarction patients with 3 vessels disease than in those with single vessel coronary artery disease.[34] In patients with stable coronary disease, the decrease in BRS is correlated to the extent and severity of coronary narrowing.[34] The latest evidence from the recently published ATRAMI study on BRS by Motara et al confirmed that the presence of an open infarct related artery was associated with a higher BRS and lower incidence of markedly depressed BRS (<3ms/mm Hg), therefore reducing the risk of post-MI mortality.[35] The association between infarct related artery patency and BRS was more evident in anterior than in inferior MI. In the same study, multivariate regression analysis showed that the age of the patient and patency of the infarct related artery were the major independent determinants of BRS, while left ventricular ejection fraction was only weakly related to BRS. These data offer new insights into the mechanisms by which coronary artery patency may affect cardiac electrical stability and survival.

Odemuyiwa et al showed that the mean BRS at 6 weeks postinfarction was higher in patients treated with thrombolysis compared with non-treated patients although this difference disappeared at 6 weeks and 3 months after the event.[36] This finding has now been substantiated by the preliminary results from a separate ATRAMI report which showed that there were fewer patients with a markedly depressed BRS in patients who received thrombolytic treatment compared with those who did not.[37]

Predictive Accuracy of BRS

Farrell et al studied the prognostic significance of BRS in 122 postinfarction patients and reported a sensitivity of 89%, a specificity of

91% and a positive predictive accuracy of 44% for predicting arrhythmic events (including sudden death).[38] In the ATRAMI study, preliminary results on 1085 patients confirmed that depressed BRS was significantly related to mortality with a relative risk of 5.5 and 2.7 on univariate and multivariate analyses respectively. This demonstrated that analysis of BRS contributes significantly to the early identification of post-infarction patients at higher risk of arrhythmic death.[39]

The ATRAMI Study

The evidence so far suggests that HRV is an important prognostic tool particularly in post-infarction patients. A prospective multicentre study to determine the prognostic implications of HRV is needed. In 1991, a large, prospective multicentre trial, ATRAMI (Autonomic Tone and Reflexes in Acute Myocardial Infarction), investigating the additional and independent prognostic values of HRV and BRS for cardiac mortality in patients after myocardial infarction (<28 days). The trial, conducted in 25 centres in Europe, North America and Japan and enrolled 1284 patients is now completed. The full results published recently,[40] showed that during 21 months of followed-up, low value of HRV (SDNN <70 ms) and BRS (< 3.0 ms/mm Hg) carried a significant multivariate risk of cardiac mortality of 3.2 and 2.8 respectively. The association of low SDNN and BRS further increased the risk; the 2-year mortality was 17% when both were below the cutoffs and 2% when both were well preserved (SDNN > 105 ms, BRS > 6.1 ms/mm Hg). Furthermore, the association of low SDNN or BRS with left ventricular ejection fraction (LVEF) below 35% carried a relative risk of 6.7 or 8.7 respectively, compared with LVEF above 35% and less compromised SDNN (≥ 70 ms) and BRS (≥ 3 ms). Thus, the ATRAMI results confirm that after MI, the analysis of autonomic markers has significant prognostic value independently of established clinical predictors such as LVEF and ventricular arrhythmias. The combination of low values of autonomic markers and reduced LVEF may help to identify a group of post-MI patients at high risk of cardiac death.

CHRONIC HEART FAILURE

Earlier studies showed that patients with chronic heart failure (CHF) also demonstrated autonomic dysfunction similar to that seen in AMI,[41,42] with a withdrawal of parasympathetic tone together with chronic activation of the adrenergic system and impaired arterial baroreflex sensitivity.

HRV and Functional Severity in Chronic Heart Failure

Casolo et al examined the behaviour of HRV with advancing severity of heart failure in 80 patients (20 patients for each NYHA functional class) with CHF secondary to coronary artery disease.[43] They showed that CHF patients with worsening NYHA functional class had a significant progressive decrease in HRV in both time and frequency domains. Among the spectral components, high frequency (HF) power decreased by almost 5 times and the low frequency (LF) power decreased by 3 times from NYHA class I to class II, with less impressive reduction in higher functional class. Similarly, SDNN, SDANN and HRVi measures also decreased with worsening functional class but in a stepwise manner. Furthermore, they also found a strong correlation between all the measures of HRV and LVEF. They suggested that HRV might potentially be used to detect patients with CHF, particularly those with mild to moderate failure and to evaluate their response to therapeutic interventions.

Other workers also demonstrated a link between HRV and LVEF.[29,44,45] Szabo and colleagues found that time domain parameters and low frequency power had significant correlation with NYHA functional class while high frequency power correlates with LVEF.[44] They also demonstrated an association between peak oxygen consumption and HRV.

However, despite the above evidence, the relationship between indices of HRV and clinical severity of CHF remains controversial. Other investigators had failed to detect such a relationship.[46,47] Kienzle et al examined 33 patients with NYHA class III to IV CHF and found that HRV only correlated weakly with cardiac output, but all the other parameters of HRV (both time and frequency domains) did not correlate with LVEF or functional class.[47] Rich and colleagues also failed to find any correlation between HRV and LVEF.[46]

Thus, in light of this contradictory evidence, the role of HRV in the clinical assessment of patients with CHF is not yet clear.

HRV and Risk of Ventricular Arrhythmias in CHF

While HRV has been shown to be a predictor of arrhythmic risk in post-MI patients, such prediction is less clear in patients with CHF. Fei et al analysed the HRV in 40 patients with CHF (15 had non-sustained ventricular tachycardia/VTns), 15 patient with VTns but clinically normal heart and 15 controls.[48] They found that in patients with VTns but a normal heart, the frequency of ventricular extrasystoles was significantly related to the high frequency component of HRV On the contrary, there was no difference in HRV between the controls, normal heart with non-sustained VT group and the CHF group, particularly in the CHF group; there was no difference in HRV between those with and without VTns. At 1 year follow-up, there was no significant difference between survivors (n=34) and those that died suddenly (n=6) in the CHF group. In a prospective study, Hoffmann et al assessed the relationship between HRV and subsequent major arrhythmic events (sustained ventricular tachycardia, ventricular fibrillation and sudden death) in 71 patients with idiopathic dilated cardiomyopathy.[49] During a mean follow-up of 15 ± 5 months, there was no difference in both time- and frequency-domain indices of HRV between patients with (10 patients) and patients without major arrhythmic events. These results suggested that in patients with CHF, there was a diverse mechanisms in the pathogenesis of ventricular arrhythmias and impaired autonomic tone was not a predominant mechanisms. A larger study is needed to clarify the role of HRV in predicting CHF patients at high risk of ventricular arrhythmias.

HRV and Prognosis in CHF

In a recent study, Yi and colleagues analysed the time domain measures of HRV in 64 patients with dilated cardiomyopathy, 19 relatives of DCM patients with left ventricular enlargement (i.e. possible early DCM) and 33 healthy controls.[50] They showed that HRV (i.e. SDNN) was reduced in DCM patients compared to controls or relatives. Furthermore, among the DCM patients, HRV was much lower in patients who developed progressive heart failure than in those who remained clinically stable (Figure 2). At 12 months follow-up, DCM patients with SDNN < 50 ms had a significantly lower survival rate free of progressive heart failure than those with SDNN > 50 ms, independent of other clinical variables (Figure 3). This suggests that HRV might be a useful non-invasive marker of DCM deterioration. At a dichotomy limit of SDNN < 50 ms, the sensitivity, specificity, positive predictive accuracy and negative predictive accuracy of HRV were 46%, 97%, 93% and 70% respectively. However, HRV was not able to differentiate between relatives of DCM patients with left ventricular enlargement, who might have an early pre-clinical form of DCM and normal control. This makes HRV a less useful tool in family screening of DCM.

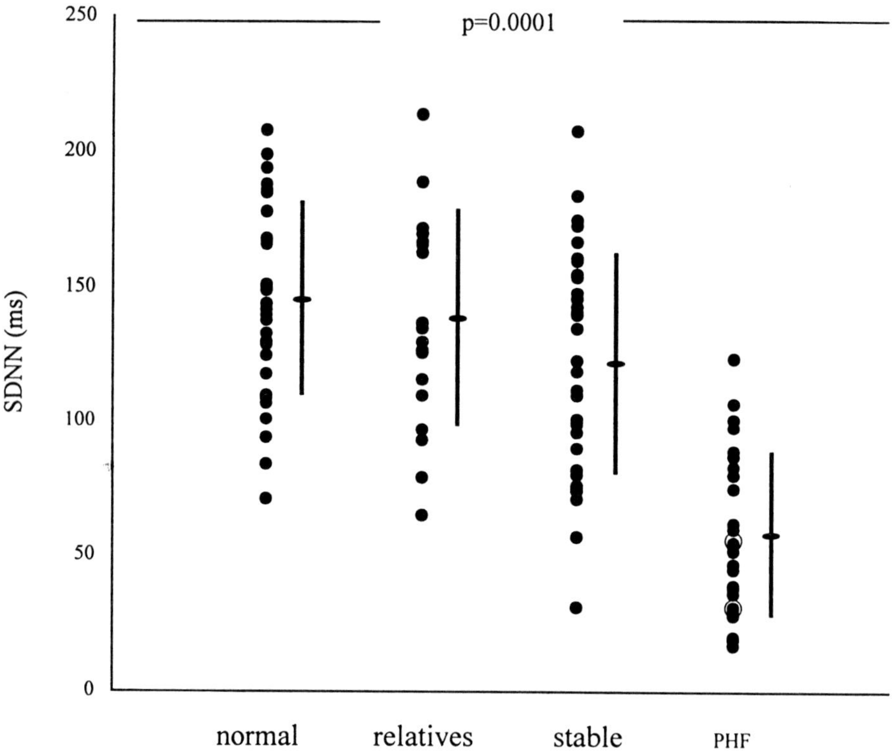

FIGURE 2. Heart rate variability (SDNN) in normal control, relatives with left ventricular enlargement, clinical stable DCM patients, DCM patients with progressive heart failure (From Yi et al[50], with permission).

For the prediction of survival, several studies have shown that HRV has an independent value.[51,52,53] Szabo et al examined 159 patients with ischaemic or idiopathic dilated cardiomyopathy and LVEF < 0.40.[51] During a follow-up of 23 months (range 1 to 67), they found that decreased HRV (SDNN < 108 ms and pNN50 (% of subsequent RR differences > 50 ms) < 2%) had an independent predictive value for all-cause cardiac mortality. Furthermore, both pNN50 < 2% and low frequency power >14 ms^2 had a strong correlation with an increased risk of death due to progressive pump failure. In a smaller study, Jiang et al followed-up 26 patients for an average of 17 months, and found that SDNN ≤ 53.4 and SDANN ≤ 41.3 were significantly associated with a greater risk of events (sudden death, death due to pump failure or ventricular tachycardia).[52] This observation by Jiang et al was supported by a study by Polikowski and colleagues.[53] In this study, Polikowski and colleagues prospectively followed up 102 patients with moderate to severe CHF

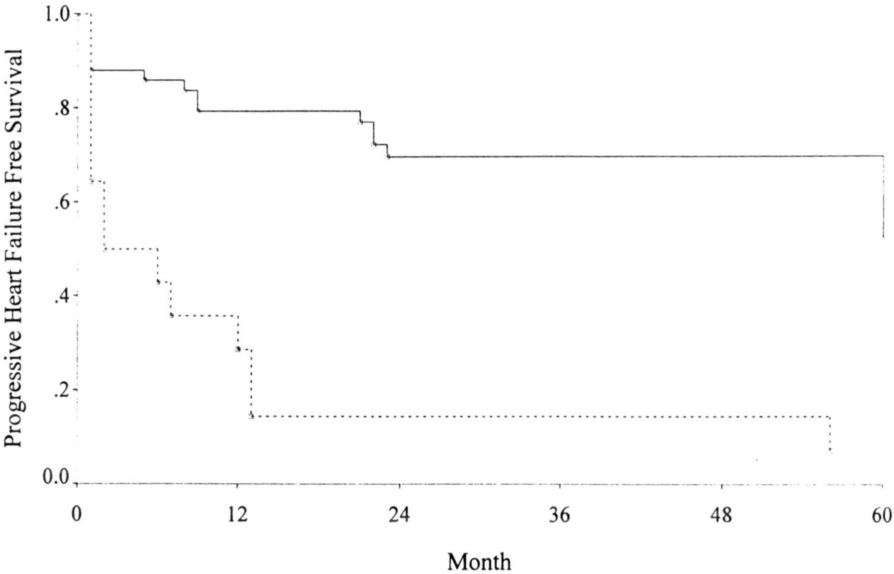

FIGURE 3. Survival curve for development of progressive heart failure in patients with DCM. Solid line – patients with SDNN> 50ms; dashed line – patients with SDNN < 50 ms. The difference between the survival is highly significance (p= 0.0001) (Adapted from Yi et al[50], with permission).

(NYHA class II to IV) of mixed aetiology. In a follow-up of 584 ± 405 days, HRV parameters (SDNN, SDANN, LF) were found to predict survival independently of NYHA functional class, LVEF, peak oxygen consumption, and ventricular tachycardia on Holter monitoring. The 1-year survival in patients with SDNN < 100 ms was 78% when compared with 95% in those with SDNN > 100 ms. Indeed, by combining SDNN < 100 ms and a peak oxygen consumption < 14 ml/kg/min allowed identification of a high risk group of patients with particularly poor prognosis, a 1-year mortality of 37 %. HRV has an important prognostic implication in patients with moderate to severe CHF. Together with other variables such as LVEF and peak oxygen consumption, HRV may have a role in identifying CHF patients with adverse prognosis for consideration of cardiac transplantation.

Cardiac Transplantation

Denervation of the heart after heart transplantation abolishes the discrete spectral peaks seen in untransplanted subject.[54] HRV increases with time after transplant. Kaye et al studied 15 cardiac transplantation recipients and 25 healthy control of similar age. They showed that HRV measured by total power was significantly reduced in transplant recipients compared with control, with a trend towards increasing spectral power up to 8 years after transplantation.[55] Halpert et al demonstrated that compared with patients in the early post-transplant period, patients > 36 months after transplant had an increased mean of 5-minutes standard deviation of RR intervals and higher ratio of low- to high-frequency power (LF/HF), indicating that the reinnervation is mainly sympathetic in origin.[56] Preliminary results showed that the development of allograft rejection might significantly increase the log total power of HRV.[54] The result however needs confirmation in a larger study.

Other Clinical Conditions

Diabetes Mellitus

In patients with diabetes mellitus, a reduction in time-domain measures of HRV seems to precede clinical expression of autonomic neuropathy and indicates an adverse prognosis.[57] Therefore, early subclinical detection of autonomic dysfunction is important for stratifying high risk patients for subsequent management. There are three basic methods from which to choose to investigate diabetic patients for autonomic neuropathy: a) simple bedside RR interval methods, b) long-term time-domain measures, which are more sensitive and reproducible than the short-term tests, and c) frequency-domain analysis performed under short-term steady state conditions which is useful in separating sympathetic from parasympathetic abnormalities.[31]

Alcoholic Neuropathy

Alcohol has acute and chronic autonomic neurotoxicity. In healthy subjects, acute intake of a moderate dose of alcohol (1g/kg body weight) was shown to decrease HF spectral power of HRV and BRS index in the HF component[58] although it did not change the systolic blood pressure or blood pressure variability. In drug-free patients with alcohol dependency, there was a significant increase in heart rate and a reduction in the high frequency spectral power at 3 weeks after admission to a closed ward for treatment.[59] These results suggest that acute intake of alcohol temporarily diminishes vagal modulation of the heart rate in healthy subjects and in alcohol-dependent patients even after 3 weeks of drug-free period. However, whether such observations have any implications on the prognosis, psychotherapeutic or psychopharmacological treatment strategy remains to be seen.

Chagas' Disease

Chagas' disease is a chronic parasitic infection caused by Trypanosoma Cruzi. It can cause acute or chronic myocarditis, or cardiomyopathy after an interval of 10–30 years. In can also cause neuropathies on cardiovascular, digestive and autonomic nervous systems. Guzzetti et al studied HRV in 19 patients with chronic Chagas' disease without heart failure (7 with positive serology only and 12 with positive serology and bundle branch block and/or arrhythmias on ECG).[60] They found that compared with the controls, all chronic Chagas' patients did not show any increase in the LF spectral power of HRV during sympathetic stimulation such as standing and handgrip exercise. They suggested that in Chagas' disease, there is a reduced capability to sympathetic responses, even in the absence of heart failure. However, while it is generally agreed of the role of the severely reduced cardiac function in the late stage of the disease, it is still uncertain how significant this autonomic neuropathy contributes to the prevalence of arrhythmias and sudden cardiac death in this condition. Hence, the diagnostic and prognostic value of HRV in the evaluation of autonomic neuropathy in patients with Chagas' disease will require further studies.

CONCLUSION

HRV and BRS have enhanced our understanding of the role of the autonomic nervous system in the disease mechanism and mode of death of patients following AMI, in particular death from arrhythmias. The ATRAMI study has demonstrated the relative and combined predictive values of HRV and BRS in patients after myocardial infarction. There is increasing evidence that HRV may have a wider role of risk assessment in other high risk groups such as patients with chronic heart failure or diabetes mellitus.

It is hoped that with better means of stratifying patients at high risk of ventricular arrhythmias and sudden death irrespective of the aetiology, appropriate preventive therapeutic interventions and risk factor modification can be instituted.

REFERENCES

1. Rosenthal ME, Oseran DS, Gang E, Peter T. Sudden cardiac death following acute myocardial infarction. *Am Heart J* 1985; **109:**865–875.
2. Bigger JT Jr, Heller CA, Wenger TL, Weld FM. Risk stratification after acute myocardial infarction. *Am J Cardiol* 1978; **42:**202–210.
3. Frishman WH, Furberg CD, Friedewald WT. Beta-adrenergic blockade for survivors of acute myocardial infarction. *N Engl J Med* 1984; **310:**830–837.
4. The Multicenter Postinfarction Research Group. Risk stratification and survival after myocardial infarction. *N Engl J Med* 1983; **309:**331–336.
5. Hinkle LE, Thaler HT. Clinical classification of cardiac deaths. *Circulation* 1982; **65:**457–464.
6. Bayes de Luna A, Coumel P, Leclercq JF. Ambulatory sudden cardiac death: mechanisms of production of fatal arrhythmia on the basis of data from 157 cases. *Am Heart J* 1989; **117**: 151–159.
7. Rapaport E. Sudden cardiac death. *Am J Cardiol* 1988; **62**: 31–61.
8. Schwartz PJ, Billman GE, Stone HL. Autonomic mechanisms in ventricular fibrillation induced by myocardial ischaemia during exercise in dogs with healed myocardial infarction. *Circulation* 1984; **69:**790–800.
9. Schwartz PJ, Vanoli E, Stramba-Badiale M, De Ferrari GM, Billman GE, Foreman RD. Autonomic mechanisms and sudden death: new insight from analysis of baroreceptor reflexes in conscious dogs with or without a myocardial infarction. *Circulation* 1988; **78:**969–979.
10. Vanoli E, De Ferrari GM, Stramba-Badiale M, Hull Jr SS, Foreman RD, Schwartz PJ. Vagal stimulation and prevention of sudden death in conscious dogs with a healed myocardial infarction. *Cir Res* 1991; **68**: 1471–1481.
11. Fetsch T, Reinhardt L, Makijarvi M, Bocker D, Block M, Borggrefe M, Breithardt G. Heart rate variability in time domain after myocardial infarction. *Clin Science* 1996; **91:**136–140.
12. Stein PK, Bosner MS, Kleiger RE, Conger BM. Heart rate variability: a measure of cardiac autonomic tone. *Am Heart J* 1994; **127:**1376–1381.
13. Schwartz PJ, Randall WC, Anderson EA, Engel BT, Freidman M, Hartley LH, Pickering TG, Thoresen CE. Sudden cardiac death: Nonpharmacologic interventions. *Circulation* 1987; **76:**1215–1219.

14. Malik M, Camm AJ. Components of heart rate variability-what they really mean and what we really measure. *Am J Cardiol* 1993; **72**:821–822.
15. La Rovere MT, Montara A, Schwartz PJ. Baroreflex sensitivity. *J Cardiovas Electrophysiol* 1995; **6**:761–774.
16. Smyth HS, Sleight P, Pickering GW. Reflex regulation of arterial pressure during sleep in man. A quantitative method for assessing baroreflex sensitivity. *Cir Res* 1969; **24**:109–121.
17. Tsuji H, Larson MG, Venditti FJ, Manders ES, Evans JC, Feldman CL, Levy D. Impact of reduced heart rate variability on risk for cardiac events: the Framingham Heart Study. *Circulation* 1996; **94**:2850–2855.
18. Malik M, Farrell T, Camm AJ. Circadian rhythm of heart rate variability after acute myocardial infarction and its influence on the prognostic value of heart rate variability. *Am J Cardiol* 1990; **66**:1049–1054.
19. Tsuji H, Venditti FJ, Manders ES, Evans JC, Larson MG, Feldman CL, Levy D. Reduced heart rate variability and mortality risk in an elderly cohort: the Framingham Heart Study. *Circulation* 1994; **90**:878–883.
20. Algra A, Tijssen JGP, Roelandt JRTC, Pool J, Lubsen J. Heart rate variability from 24-hour electrocardiography and the 2-year risk for sudden death. *Circulation* 1993; **88**:180–185.
21. Casolo GC, Stroder P, Signorini C, Calzolari F, Zucchini M, Balli E, Sulla A, Lazzerini S. Heart rate variability during the acute phase of myocardial infarction. *Circulation* 1992; **85**:2073–2079.
22. Kleiger RE, Miller JP, Bigger JT, Moss AJ, and the Multicenter Post-Infarction Research Group. Decreased heart rate variability and its association with increased mortality after acute myocardial infarction. *Am J Cardiol* 1987; **59**:256–262.
23. Cripps TR, Malik M, Farrell TG, Camm AJ. Prognostic value of reduced heart rate variability after myocardial infarction: clinical evaluation of a new analysis method. *Br Heart J* 1991; **65**:14–19.
24. Odemuyiwa O, Malik M, Farrell T, Bashir Y, Poloniecki J, Camm AJ. Comparison of the predictive characteristics of heart rate variability index and left ventricular ejection fraction for all-cause mortality, arrhythmic events and sudden death after acute myocardial infarction. *Am J Cardiol* 1991; **68**:434–439.
25. Hartikainen JEK, Malik M, Staunton A, Poloniecki J, Camm AJ. Distinction between arrhythmic and nonarrhythmic death after acute myocardial infarction based on heart rate variability, signal-averaged electrocardiogram, ventricular arrhythmias and left ventricular ejection fraction. *J Am Coll Cardiol* 1996; **28**:296–304.
26. Malik M, Camm AJ. Heart rate variability and clinical cardiology. *Br Heart J* 1994; **71**:3–6.
27. Bigger T Jr. Fleiss JL, Steinman RC, Rolnitzky LM, Kleiger RE, Rottman JN. Correlation among time and frequency domain measures of heart rate variability two weeks after acute myocardial infarction. *Am J Cardiol* 1992; **69**:891–898.

28. Zuanetti G, Neilson JMM, Latini R, Santoro E, Maggioni AP, Ewing DJ; on behalf of GISSI-2 investigators. Prognostic significance of heart rate variability in post-myocardial infarction patients in the fibrinolytic era. *Circulation* 1996; **94:**432–436.
29. Sandrone G, Mortara A, Torzillo D, La Rovere MT, Malliani A, Lombardi F. effects of beta blockers (atenolol or metoprolol) on heart rate variability after acute myocardial infarction. *Am J Cardiol* 1994; **74:**340–345.
30. Farrell TG, Bashir Y, Cripps T, Malik M, Poloniecki J, Bennett D, Ward DE, Camm AJ. Risk stratification for arrhythmic events in postinfarction patients based on heart rate variability, ambulatory electrocardiographic variables and the signal-averaged electrocardiogram. *J Am Coll Cardiol* 1991; 18:687–697.
31. Task Force of the European Society of Cardiology and the North American Society of Pacing and Electrophysiology. Heart rate variability: standards of measurement, physiological interpretation, and clinical use. *Circulation* 1996; **93:**1043–1065.
32. La Rovere MT, Specchia G, Mortara A, Schwartz PJ. Baroreflex sensitivity, clinical correlates and cardiovascular mortality among patients with a first myocardial infarction. A prospective study. *Circulation* 1988; **78:**816–824.
33. Farrell TG, Paul V, Cripps TR, Malik M, Bennett ED, Ward D, Camm AJ. Baroreflex sensitivity and electrophysiological correlates in patients after myocardial infarction. *Circulation* 1991; **83:**945–952.
34. Katsube Y, Saro H, Naka M, Kim BH, Kinoshita N, Koretsure Y, Hori M. Baroreflex sensitivity in patients with stable coronary artery disease is correlated with the severity of arterial narrowing. *Am J Cardiol* 1996; **78:**1007–1010.
35. Mortara A. Specchia G. La Rovere MT, Bigger JT Jr, Marcus FI, Camm AJ, Hohnloser SH, Nohara R, Schwartz PJ. Patency of infarct-related artery: effect of restoration of anterograde flow on vagal reflexes. ATRAMI (Autonomic Tone and Reflexes After Myocardial Infarction) Investigators. *Circulation* 1996; **93:**1114–1122.
36. Odemuyiwa O, Farrell T, Staunton A, Sneddon J, Poloniecki J, Bennett D, Malik M, Camm AJ. Influence of thrombolytic therapy on the evolution of baroreflex sensitivity after myocardial infarction. *Am Heart J* 1993; **125:**285–291.
37. La Rovere MT, Mortara A, Bigger JT Jr, Marcus FI, Hohnloser SH, Nohara, Camm AJ, Schwartz PJ on behalf of the ATRAMI investigators. Effect of thrombolytic therapy on baroraceptors reflexes. *Eur Heart J* 1994; **15:**446.
38. Farrell T, Odemuyiwa O, Bashir Y, Cripps TR, Malik M, Ward D, Camm AJ. Prognostic value of baroreflex sensitivity testing after acute myocardial infarction. *Br Heart J* 1992; **67:**129–132.
39. La Rovere MT, Bigger JT Jr, Marcus FI, Mortara A, Camm AJ, Hohnloser SH, Nohara R, Schwartz PJ on behalf of the ATRAMI Investigators. Prognostic value of depressed baroreflex sensitivity. The ATRAMI study. *Circulation* 1995; **91:**2676.
40. Eckberg DL, Drabinsky M, Braunwald E, Defective cardiac parasympathetic control in patients with heart disease. *N Engl J Med* 1971; **285**:877–883.

41. La Rovere MT, Bigger JT, Marcus FI, Mortara A, Schwartz PJ, for the ATRAMI Investigators. Baroreflex sensitivity and heart-rate variability in prediction of total cardiac mortality after myocardial infarction. *Lancet* 1998; **351**:478–484.
42. Hasking GJ, Esler MD, Jennings GL, Burton D, Johns JA, Korner PI. Norepinephrine spillover to plasma in congestive heart failure: Evidence of increased cardiorenal and total sympathetic nerve activity. *Circulation* 1986; **73**:615–621.
43. Casolo GC, Stroder P, Sulla A, Chelucci A, Freni A, Zerauscheck M. Heart rate variability and functional severity of congestive heart failure secondary to coronary artery disase. *Eur Heart J* 1995; **16**:360–367.
44. Szabo BM, van Veldhuisen DJ, Brouwer J, Haaksma J, Lie KL. Relation between severity of disease and impairment of heart rate variability parameters in patients with chronic congestive heart failure secondary to coronary artery disease. *Am J Cardiol* 1995; **76**:713–716.
45. Nolan J, Flapan AD, Capewell S, MacDonald TM, Neilson JMM, Ewing DJ. Decreased cardiac parasympathetic activity in chronic congestive heart failure and its relation to left ventricular function. *Br Heart J* 1992; **67**:482–485.
46. Rich MW, Saini JS, Kleiger RE, Carney RM, teVelde A, Freeland KE. Correlation of heart rate variability with clinical and angiographic variables and late mortality after coronary angiography. *Am J Cardiol* 1988; **62**:714–717.
47. Kienzle M, Ferguson DW, Birkett CL, Myers GA, Berg WJ, Marino DJ. Clinical, haemodynamic and sympathetic neural correlates of heart rate variability in congestive heart failure. *Am J Cardiol* 1992; **69**:761–767.
48. Fei L, Keeling PJ, Gill JS, Bashir Y, Statters D, Poloniecki J, McKenna WJ, Camm AJ. Heart rate variability and its relation to ventricualr arhythmias in congestive heart failure. *Br Heart J* 1994; **71**:322–328.
49. Hoffmann J, Grimm W, Menz V, Knop U, Maisch B. Heart rate variability and major arrhythmic events in patients with idiopathic dilated cardiomyopathy. *PACE* 1996; **19**:1841–1844.
50. Yi G, Goldman JH, Keeling PJ, Reardon M, McKenna WJ, Malik M. Heart rate variability in idiopathic dilated cardiomyopathy: relation to disease severity and prognosis. *Heart* 1997; **77**:108–114.
51. Szabo BM, van Veldhuisen DJ, van der Veer N, Brouwer J, De Graeff PA, Crijns HGM. Prognostic value of heart rate variability in chronic congestive heart failure secondary to idiopathic or ischaemic dilated cardiomyopathy. *Am J Cardiol* 1997; **79**:978–980.
52. Jiang W, Hathaway WR, McNulty S, Larsen RL, Hansley K, Zhang Y, O'Connor CM. Ability of heart rate variability to predict prognosis in patients with advanced congestive heart failure. *Am J Cardiol* 1997; **80**:808–811.

53. Ponikowski P, Anker SD, Chua TP, Szelemej R, Piepoli M, Adamopoulos S, Webb-Peploe K, Harrington D, Banasiak W, Wrabec K, Coates AJS. Depressed heart rate variability as an independent predictor of death in chronic congestive heart failure secondary to ischaemic or idiopathic dilated cardiomyopathy. *Am J Cardiol* 1997; **79:**1645–1650.
54. Sands KEF, Appel ML, Lilly LS, Schoen FJ, Mudge GH, Cohen RJ. Power spectral analysis of heart rate variability in human cardiac transplant recipients. *Circulation* 1989; **79:**76–82.
55. Kaye DM, Esler M, Kingwell B, McPherson G, Esmore D, Jennings G. Functional and neurochemical evidence for partial cardiac sympathetic reinnervation after cardiac transplantation in humans. *Circulation* 1993; **88:**1110–1118.
56. Halpert I, Goldberg AD, Levine AB, Levine TB, Kornberg R, Kelly C, Lesch M. Reinnervation of the transplanted human heart as evidenced from heart rate variability studies. *Am J Cardiol* 1996; **77:**180–183.
57. Pagani M, Malfatto G, Pierini S, Casati R, Masu AM, Poli M, Guzzetti S, Lombardi F, Cerutti S, Malliani A. Spectral analysis of heart rate variability in the assessment of autonomic diabetic neuropathy. *J Auton Nerv Syst* 1988; **23:**143–153.
58. Koskinen P, Virolainen J, Kupari M. Acute alcohol intake decreases short-term heart rate variability in healthy subjects. *Clin Science* 1994; **87:**225–230.
59. Rechlin T, Orbes I, Weis I, Kaschka WP. Autonomic cardiac abnormalities in alcohol-dependent patients admitted to a psychiatric department. *Clin Auton Res* 1996; **6:**119–122.
60. Guzzetti S, Iosa D, Pecis M, Bonura L, Prosdocimi M, Malliani A. Impaired heart rate variability in patients with chronic Chagas' disease. *Am Heart J* 1991; **121:**1727–1734.

INDEX

D

E

Q

R

S

W

X